Evaluation of Persistence and Long-Range Transport of Organic Chemicals in the Environment

Other titles from the Society of Environmental Toxicology and Chemistry (SETAC):

Environmental Contaminants and Terrestrial Vertebrates: Effects on Populations, Communities, and Ecosystems
Albers, Heinz, Ohlendorf, editors
2000

Ecotoxicology of Amphibians and Reptiles
Sparling, Linder, Bishop, editors
2000

Natural Remediation of Environmental Contaminants: Its Role in Ecological Risk Assessment and Risk Management
Swindoll, Stahl, Ells, editors
2000

Development of Methods for Effect-Driven Cumulative Effects Assessment Using Fish Population: Moose River Project
Munkittrick, McMaster, Van Der Kraak, Portt, Gibbons, Farwell, Gray, editor s
2000

Multiple Stressors in Ecological Risk and Impact Assessment: Approaches to Risk Estimation
Ferenc and Foran, editors
2000

Evaluating and Communicating Subsistence Seafood Safety in a Cross-Cultural Context: Lessons Learned from the Exxon Valdez Oil Spill
Field, Fall, Nightwander, Peacock, Varanasi, editors
2000

Reproductive and Developmental Effects of Contaminants in Oviparous Vertebrates
Di Giulio and Tillitt, editors
1999

Linkage of Effects to Tissue Residues: Development of a Comprehensive Database for Aquatic Organisms Exposed to Inorganic and Organic Chemicals
Jarvinen and Ankley, editors
1999

Atmospheric Deposition of Contaminants to the Great Lakes and Coastal Waters
Baker, editor
1997

For information about any SETAC publication, including SETAC's international journal, *Environmental Toxicology and Chemistry*, contact the SETAC Office nearest you.

1010 N. 12th Avenue
Pensacola, Florida, USA 32501-3367
T 850 469 1500 F 850 469 9778
E setac@setac.org

Avenue de la Tosion d'Or 67
B-1060 Brussels, Belgium
T 32 2 772 72 81 F 32 2 770 53 83
E setac@setaceu.org

http://www.setac.org

Environmental Quality Through Science®

Evaluation of Persistence and Long-Range Transport of Organic Chemicals in the Environment

Edited by:

Gary Klečka

Robert Boethling

James Franklin

Les Grady

Diana Graham

Philip H. Howard

Kurunthachalam Kannan

Robert J. Larson

Don Mackay

Derek Muir

Dik van de Meent

SETAC Special Publications Series

Current Coordinating Editor of SETAC Books

C.G. Ingersoll

U.S. Geological Survey

Publication sponsored by the Society of Environmental Toxicology and Chemistry (SETAC)

SETAC PRESS

Cover by Michael Kenny Graphic Design and Advertising
Indexing by Celia McCoy

Library of Congress Cataloging-in-Publication Data

Evaluation of persistence and long-range transport of organic chemicals in the environment/ edited by Gary Klečka..[et al.].

p. cm. --

Includes bibliographical references and index.

ISBN 1-880611-22-8 (alk. paper)

1. Environmental chemistry. 2. Environmental modeling. 3. Chemicals--environmental aspects I. Klečka, Gary Michael. II. Pellston Workshop on Criteria for Persistence and Long-range Transport of Organic Chemicals in the Environment (1998 : Fairmont Hot Springs, B.C.)

TD193 .P43 2000
628.5'2--dc21 99-461976

International Standard Book Number 1-880611-22--8
Printed in the United States of America
06 05 04 03 02 01 00 99 10 9 8 7 6 5 4 3 2 1

∞
The paper used in this publication meets the minimum requirements of the American National Standard for Information Sciences—Permanence of Paper for Printed Library Materials, ANSI Z39.48-1984

Reference Listing: Klečka G,Boethling B, Franklin J, Grady L, Graham D, Howard PH, Kannan K, Larson B, Mackay D, Muir D, van de Meent D, editors. 2000. Evaluation of persistence and long-range transport of organic chemicals in the environment. Published by the Society of Environmental Toxicology and Chemistry (SETAC), Pensacola, Florida, USA. 400 p.

You must teach your children that the ground beneath their feet is the ashes of our ancestors. So that they will respect the land, teach your children that the earth is our mother. Whatever befalls the earth befalls the children of the earth. If we destroy the land, destroy the water, and foul the air, we destroy ourselves.

— Chief Seattle, 1858

The SETAC Special Publications Series

The SETAC Special Publications Series was established by the Society of Environmental Toxicology and Chemistry (SETAC) to provide in-depth reviews and critical appraisals on scientific subjects relevant to understanding the impact of chemicals and technology on the environment. The series consists of books on topics reviewed and recommended by the Publications Advisory Counsil and approved by the SETAC Board of Directors for their importance, timeliness, and contribution to multidisciplinary approaches to solving environmental problems. The diversity and breadth of subjects covered in the series reflect the wide range of disciplines encompassed by environmental toxicology, environmental chemistry, and hazard and risk assessment. These volumes attempt to present the reader with authoritative coverage of the literature, as well as paradigms, methodologies, and controversies; research needs; and new developments specific to the featured topics. All books in the series are peer reviewed for SETAC by acknowledged experts.

SETAC Special Publications are useful to environmental scientists in research, research management, chemical manufacturing, chemical regulation, risk assessment and education, as well as to students considering careers in these areas. The Series provides information for keeping abreast of recent developments in familiar subject areas and for rapid introduction to principles and approaches in new subject areas.

SETAC would like to recognize the past SETAC Special Publications Series editors:

T.W. La Point, Institute of Applied Sciences
University of North Texas, Denton, TX

B.T. Walton, U.S. Environmental Protection Agency
Research Triangle Park, NC

C.H. Ward, Department of Environmental Sciences and Engineering
Rice University, Houston, TX

Contents

List of Figures

List of Tables

Preface

"Evaluation of Persistence and Long-Range Transport of Organic Chemicals in the Environment" documents the outcome of a workshop and consensus-building process. The workshop was conceived by a small group of people in the public and private sectors who met informally as a steering committee for more than a year to discuss the need for a scientific foundation for the criteria and process used for the evaluation of the environmental behavior of persistent organic pollutants.

The urgency and impact of the workshop are well established. For example, 1998 saw the initiation of a process by which chemicals would be controlled internationally for the first time, as nations of the world entered into treaty negotiations on persistent organic pollutants under the auspices of the United Nations Environment Program. Furthermore, this is the global counterpart to many other national and regional initiatives. While the Steering Committee recognized that in addition to the environmental attributes, toxicity and bioaccumulation potential are equally important in these discussions, the workshop specifically focused on the scientific fundamentals, criteria, and processes used for the evaluation of persistence and long-range transport.

With this background the workshop brought together experts from academia, industry, government, and non-governmental organizations to reach consensus on where we are today and what we can accomplish with the current scientific understanding, as well as what should be done in the future as we address this issue. If this workshop helps bring together those who devote their energies to the science, regulation, and management of chemicals to work together more effectively toward a common goal of deciding how we must manage chemicals on our planet, it will have been a success. Perhaps future historians will note that as the millennium drew to a close, the worldwide scientific community took a first, exploratory, but determined step towards this challenging task. As such, the workshop and its proceedings are welcome evidence that the broad environmental community is striving purposefully towards this goal.

Acknowledgments

The Society of Environmental Toxicology and Chemistry (SETAC), the organizing committee of the Pellston Workshop on Criteria for Persistence and Long-Range Transport of Organic Chemicals in the Environment, and the editors of this book wish to thank the workshop participants for their input and hard work. The chapters were written by workgroups, whose members are acknowledged on page xviii.

The Pellston Workshop on Criteria for Persistence and Long-Range Transport of Organic Chemicals in the Environment was made possible through the financial support of the following organizations:

American Crop Protection Association

Canadian Chemical Producers Association

Chlorine Chemistry Council of the Chemical Manufacturers Association

Department of Indian Affairs and Northern Development

Environment Canada

EuroChlor Federation

Exxon Corporation, USA

The Procter & Gamble Company

U.S. Environmental Protection Agency.

The workshop organizers wish to express gratitude to Dan Woltering for his advice and assistance with the workshop proposal. We also graciously thank the following individuals for their thorough and thoughtful reviews of this volume:

James Devillers, CTIS;

Rebecca Dickhut, Virginia Institute of Marine Sciences;

Ron Hites, Indiana University;

Gerrit Schüürmann, UFZ Centre for Environmental Research;

Jim Spain, United States Air Force; and

Geoffrey Tyndall, National Center for Atmospheric Research.

Finally, we thank the SETAC staff for its support, in particular, Greg Schiefer, Linda Longsworth, Melissa Winter, Stacey Hagman, Chris Englert, Kristine Tabb, and Rod Parrish.

About the Editors

Gary Klečka is an environmental scientist with The Dow Chemical Company in Midland, MI. His research interests include the microbial degradation of organic chemicals, environmental fate and transport processes, wastewater treatment, and soil and groundwater remediation. Of particular interest is the role microorganisms play in the natural attenuation and recycling of chlorinated organic chemicals in the environment. He received his B.A. (1975) and Ph.D. (1980) in Microbiology from the University of Texas at Austin.

Bob Boethling is an environmental scientist in the Office of Pollution Prevention and Toxics, U.S. Environmental Protection Agency, in Washington, DC. His main interests are in environmental fate assessment and the development of estimation methods for transformation processes such as biodegradation as well as the physical properties of chemicals that control transport and partitioning. His professional *raison d'être* is to ensure the appropriate application of environmental science in the public interest. He is a native of Minneapolis, Minnesota, and received his B.A. (1971) and Ph.D. (1976) degrees in Microbiology from the University of California at Los Angeles.

James Franklin, senior scientist, has been working since 1970 at the Research and Development Center of Solvay S.A., Brussels, Belgium. His research activities during this time have focused mainly on developing production processes and applications for organic chlorine and fluorine compounds and on studying the environmental impact of these products. The following environmental issues are of particular interest to him: ozone depletion, global warming, and atmospheric degradation mechanisms for organic halogen compounds. He received his B.A. (1964) from Cambridge University, UK, and his Ph.D. (1969) from the Université Libre de Bruxelles, Belgium. His Ph.D. thesis and a post-doctoral fellowship at the University of Alberta, Edmonton, Canada, were in the field of gas-phase kinetics.

Les Grady is R. A. Bowen Professor of Environmental Engineering at Clemson University in Clemson, SC. His research has focused on the microbial processes used for the treatment of wastewaters and he is coauthor of the textbook *Biological Wastewater Treatment*. Biodegradation of synthetic organic compounds has been his major research interest for the past two decades, with emphasis on the quantification of kinetic parameters. He is also interested in microbial ecology and the interactions between microbes in natural and engineered environments. He received B.A. (1960), B.S. (1961), and M.S. (1963) degrees from Rice University and a Ph.D. (1968) from Oklahoma State University.

Diana Graham is an environmental scientist with Zeneca Ag Products in Richmond, CA. Her research interests include environmental fate and transport processes, especially fate in air, and exposure of workers to pesticides during application and harvesting. Of particular interest is the fate and transport of persistent pesticides. She received her Ph.D. (1979) in Analytical Chemistry from the University of Hawaii.

Philip H. Howard was trained as an organic chemist (Ph.D., 1970, Syracuse University). At Syracuse Research Corporation (SRC), he has served as project director on hundreds of contracts and grants with the U. S. Environmental Protection Agency (USEPA), other government agencies, and industry for developing environmental fate databases and determining correlations between structure and environmental fate properties. Dr. Howard is presently Director of the Environmental Science Center at SRC and project director for contracts for 1) ATSDR for toxicology profile development; 2) the National Library of Medicine, to enhance the Environmental Fate/Exposure Sections of the Hazardous Substances Data Bank; 3) the Office of Pesticide Programs, USEPA, to review industrial submissions on fate for pesticide registration; and 4) the Office of Pollution, Prevention and Toxics, to provide support for chemical selection for PBTs and Pollution Prevention.

Kurunthachalam Kannan is a Visiting Assistant Professor at the National Food Safety and Toxicology Center, Michigan State University, East Lansing, MI. His research focuses on the distribution, behavior, and fate of organic and organometallic contaminants in the environment, and human and wildlife exposure pathways and effects. He received his B.Sc. (1986) in Agricultural Sciences from Tamil Nadu Agricultural University, Coimbatore, India, and Ph.D. (1994) in Environmental Chemistry and Ecotoxicology from Ehime University, Matsuyama, Japan.

Robert J. Larson is an environmental scientist/microbial ecologist with The Procter & Gamble Company in Cincinnati, OH. His research interests include the characterization of microbial diversity in natural habitats, microbial degradation kinetics in natural and engineered systems, environmental fate and transport processes, and the use of knowledge-based computer systems to predict biodegradation pathways for novel compounds. He received his B.S. degree in Biology from the University of Puget Sound (1969), and, following two years of military service, his M.S. (1974) and Ph.D. (1975) in Bacteriology from the University of Wisconsin at Madison. He joined Procter & Gamble in 1976 after Post-Doctoral training in the Department of Molecular and Cell Biology at Cornell University.

Don Mackay is the Director of the Canadian Environmental Modeling Center at Trent University, Peterborough, Ontario, Canada. He received his Ph.D. in Chemical Engineering (1961) from what is now the University of Strathclyde. His primary interest is the development, application, validation, and dissemination of mass-balance models describing the fate of chemicals in the environment. These models include descriptions of bioaccumulation in a variety of organisms, water-quality models of contaminant fate in lakes, rivers, sewage-treatment plants, soils, and vegetation. He has developed a series of multimedia mass-balance models employing the fugacity concept that are widely used for assessment of chemical fate in national regions and in the global environment.

Derek Muir is a Research Scientist with Environment Canada's National Water Research Institute in Burlington, Ontario, Canada. He leads an environmental chemistry research group at NWRI involved in measurements of persistent organic pollutants and mercury in the aquatic and terrestrial environment from arctic and sub-arctic regions of Canada and the circumpolar Arctic, as well as in the Great Lakes. His research interests also include the bioaccumulation and bioavailability of halogenated organic pollutants in the aquatic environment under field and laboratory conditions with special emphasis on food chain transfer and exposure assessment. He received his B.Sc. (1970) and Ph.D. (1977) in Agricultural Chemistry from McGill University in Montreal.

Dik van de Meent is an environmental chemist with the National Institute of Public Health and the Environment (RIVM) in Bilthoven, The Netherlands. His main interests are fate and effects of chemicals in the environment. He has contributed to this field by developing the multimedia fate model SimpleBox. His mission at RIVM is to quantify effects of toxic chemicals on biodiversity. Van de Meent obtained an Engineering degree in Organic Geochemistry (1976) and an Ph.D. in Environmental Science (1982), both from Delft University of Technology.

List of Workshop Participants*

Workshop Chairman

Gary Klečka[1]
The Dow Chemical Company, USA

Persistence in Air Workgroup

Roger Atkinson
University of California at Riverside, USA

James Franklin[1]
Solvay Research and Technology, Belgium

Phil Howard[1]
Syracuse Research Corporation, USA

John Orlando
National Center for Atmospheric Research, USA

Christian Seigneur
Atmospheric and Environmental Research, USA

Timothy Wallington[2]
Ford Motor Company, USA

Cornelius Zetzsch[2]
Fraunhofer-Institut für Toxikologie und Aerosolforschung, Germany

Persistence in Soils, Water, and Sediments Workgroup

Larry Forney
Biologisch Centrum, The Netherlands

Les Grady, Jr.[1,2]
Clemson University, USA

Robert Larson[1, 2]
The Procter & Gamble Company, USA

Shigeki Masunaga
Yokohama National University, Japan

Willie Peijnenburg
RIVM, The Netherlands

Lee Wolfe
U.S. Environmental Protection Agency, USA

Partitioning, Inter-media Transfer and Long-range Transport Workgroup

Anders Andren
University of Wisconsin-Madison, USA

Joseph DePinto
State University of New York at Buffalo, USA

Katherine Fox
Unilever Research, United Kingdom

Stefan Haderlein
EAWAG/ETH Zurich, Switzerland

Donald Mackay[1, 2]
Trent University, Canada

Michael McLachlan
University of Bayreuth, Germany

Louis Thibodeaux[2]
Louisiana State University, USA

Multimedia Modeling Workgroup

Michael Matthies
University of Osnabrueck, Germany

Tom McKone[2]
University of California, USA

Tom Parkerton[2]
Exxon Biomedical Sciences, Inc, USA

Richard Purdy
3M Company, USA

Martin Scheringer
ETH Zentrum, Switzerland

* Affiliations were current at the time of the workshop.

[1] Steering Committee members

[2] Session chairpersons

Dik van de Meent[1]
RIVM, The Netherlands

Frank Wania
WECC, Canada

Ambient Monitoring Workgroup

Davide Calamari[2]
University of Insubria, Italy

Kevin Jones[2]
Lancaster University, United Kingdom

Kurunthachalam Kannan[1]
Michigan State University, USA

André Lecloux
EuroChlor Federation, Belgium

Mats Olsson
Swedish Museum of Natural History, Sweden

Michael Thurman
U.S. Geological Survey, USA

Paolo Zannetti
Exponent Failure Analysis Associates, USA

Case Studies Workgroup

Robert Bailey
Bailey Associates, USA

Terry Bidleman
Atmospheric Environment Service, Canada

Jordi Dachs
Rutgers University, USA

Steve Eisenreich[2]
Rutgers University, USA

Robie Macdonald[2]
Fisheries and Oceans, Canada

Derek Muir[1]
Environment Canada, Canada

Framework and Criteria Workgroup

Robert Boethling[1]
U.S. Environmental Protection Agency, USA

John Buccini
Environment Canada, Canada

Christina Cowan-Ellsberry[2]
The Procter & Gamble Company, USA

Diana Graham[1]
Zeneca Ag Products, USA

Bjorn Hansen[2]
European Chemicals Bureau, Italy

Michael Murray
National Wildlife Federation, USA

Larry Rampy
Chemical Manufacturers Association, USA

Bruce Rodan
U.S. Environmental Protection Agency, USA

Bo Wahlström
UNEP Chemicals, Switzerland

Executive Summary

Persistent organic pollutants (POPs) are bioaccumulative and toxic organic compounds of natural or anthropogenic origin that resist photolytic, chemical, and biological degradation. They are characterized by low water solubility and high lipid solubility, resulting in bioaccumulation in fatty tissues of living organisms. POPs may be transported in the environment at low concentrations by movement of fresh and marine waters. Recent evidence has shown that because of their semi-volatility, POPs are also transported long distances in the atmosphere, resulting in widespread distribution around the earth, including in regions where they never have been produced or used. Thus, both humans and ecosystems may be exposed, in many cases for extended periods of time.

In recent years, several national and international accords have been initiated or completed to identify a variety of persistent, bioaccumulative, and toxic substances (PBTs) and to require action to control the use and/or release of these substances, including both organic compounds (POPs) and those containing metals. Examples of regional or global initiatives that are directed at identifying POPs and/or PBTs and developing risk management measures to control the exposure of humans and ecosystems include the Oslo-Paris (OSPAR) Convention for the Protection of the North-East Atlantic, the Canada-USA Great Lakes Water Quality Agreement and Binational Toxics Strategy, the United Nations Economic Commission for Europe (UNECE) Convention on Long-range Transboundary Air Pollution, and the United Nations Environment Program (UNEP) Global Convention on POPs. In addition to these intergovernmental accords, several national governments and non-governmental organizations have developed and implemented regulatory and non-regulatory initiatives to address the use and/or release of POPs and PBTs. While the list is not exhaustive, it illustrates that PBTs have been and will continue to be the subject of considerable attention for scientists and policymakers.

To foster the development of a sound scientific foundation for identifying and evaluating POPs and PBTs, the international "Pellston Workshop on Criteria for Persistence and Long-range Transport of Organic Chemicals in the Environment," sponsored by the Society of Environmental Toxicology and Chemistry (SETAC), was held 13–19 July 1998, to discuss scientific issues related to persistence and long-range transport of organic chemicals. The workshop had broad involvement from academia, government, industry, and other interested organizations. While it is widely recognized that substances that exhibit a combination of properties (i.e., persistence, bioaccumulation, and toxic potency) should be the focus of societal concern and regulatory management efforts, this workshop specifically restricted its attention to the need for a scientific foundation for the criteria and processes used to evaluate persistence and long-range transport of POPs and PBTs.

The specific objectives of the workshop were to discuss the issues, reach consensus, and make recommendations based on the current state of the science concerning the evaluation of persistence and long-range transport of organic chemicals in the environment. The workshop followed the format of previous SETAC Pellston Workshops, and it was limited to about 50 participants with broad international perspective and expertise in atmospheric and environmental chemistry, partitioning and transport, multimedia modeling, and environmental assessment. To address the objectives, the workshop was organized into 8 individual sessions. A general plenary was held on the first day in which invited, authoritative discussion papers were presented to stimulate dialogue. Based on expertise, each participant was assigned to a small working group, although interaction among the working groups was encouraged. A major goal of each working group was to develop a summary of its deliberations, including discussions of the state of the science pertinent to that working group, a critical analysis of available approaches, and methods to integrate these approaches into a framework for assessing persistence and potential for long-range transport. Key points from the discussions are summarized below.

Quantitative Determination of Persistence in Air

The persistence of organic compounds in the atmosphere is governed by the rates at which the compounds are removed by chemical and/or physical processes. Chemical degradation is dominated by reaction with the hydroxyl radical (OH) in most cases, but for some compounds reaction with ozone (O_3) or the nitrate radical (NO_3) or destruction by direct photolysis may also be significant; therefore, these processes should be assessed. Physical removal from the atmosphere can occur by wet or dry deposition of gaseous or particle-borne species.

Satisfactory methods exist for experimental determination of the rate constants for reaction with OH, O_3, and NO_3 (with an accuracy of about ± 30 %), as well experimental determination of UV-visible absorption cross-sections (at least for the more volatile compounds). For the less-volatile compounds (which include many POP-type substances), new procedures are beginning to be explored and validated. Determining photolysis quantum yields is a time-consuming process for which no standard procedure exists.

Estimation of rate constants for reaction with OH by using Atkinson's empirical method allows prediction within a factor of 2 for 90% of the approximately 600 compounds used in its development. For compound classes for which no experimental data are available, however, the uncertainties in rate constants estimated by this method could well be a factor of 5 or greater. The accuracy of estimation methods for reactions of O_3 and NO_3 is generally lower than that with OH, and order-of-magnitude uncertainties must be expected. No reliable methods exist for quantitative prediction of absorption cross-sections or quantum yields.

Many POP-type compounds present in the atmosphere are to a large extent adsorbed on particles. There are severe limitations, however, to our understanding of the impact of adsorption on chemical degradation. In particular, reaction with OH may not cease when the substance is in the adsorbed state, as has been assumed in the literature. Furthermore, some inorganic constituents of atmospheric particulates may photocatalyze degradation of adsorbed organics.

The main factors affecting the removal of chemicals from the atmosphere by wet and dry deposition are known. They include water solubility, particle size (for species adsorbed on particulates), nature of the receiving surface (for dry deposition), and intensity and duration of precipitation (for wet deposition). Deposition processes therefore can be modeled, with varying degrees of sophistication, and their rates estimated. In addition, revolatilization must be considered.

The overall atmospheric half-life, taking into account chemical destruction and physical removal, can be calculated at the screening level by using a simple box model of the atmosphere and the earth's surface and by assuming global, regional, or local average values of input parameters such as OH concentrations, temperature, solar radiation intensity, aerosol densities, deposition velocities, etc. For more detailed evaluations, the overall atmospheric half-life can be calculated by using increasingly sophisticated models of the atmosphere coupled to other media. These models should take into account spatial and temporal variations in the input parameters and provide a more realistic description of atmospheric transport and inter-media partitioning.

While persistence is probably easier to quantify for the atmosphere than for other environmental compartments, there are still severe limitations to both our predictive capacity for purely gas-phase processes and to our basic understanding of processes involving substances adsorbed on particles. Furthermore, few experimental data exist for POP-type compounds. Finally, it is important to note that in the winter months at mid-latitudes, the very low atmospheric OH concentrations will affect the half-lives of practically all organic compounds.

Quantitative Determination of Persistence in Soils, Surface Waters, and Sediments

The persistence of organic compounds in soils, surface waters, and sediments is principally governed by the rates at which they are removed by physical, chemical, and biological processes. Biotic degradation processes include phytodegradation and microbial biodegradation. Abiotic degradation processes include hydrolysis, direct and indirect photolysis, and oxidation/reduction reactions.

A variety of experimental methods are available for measuring the biotic and abiotic degradation of organic compounds. These methods are generally organized in tiers, from screening level to more complex, and are typically implemented in a specific sequence by using standardized test procedures. Standard protocols are well-developed for some biotic process like microbial biodegradation and for some abiotic degradation processes like hydrolysis and photolysis. However, in all cases, kinetic constants or half-lives generated by standardized test protocols are relevant only to the experimental conditions used and must be normalized for physical, chemical, or biological factors to allow extrapolation to different environmental compartments. When experimental data are not available, empirical approaches or quantitative structure-activity relationships (QSARs) can be used to estimate biotic or abiotic kinetic constants. The use of QSAR predictions, however, can be supported only for materials that have similar chemical structures and are biotically or abiotically degraded by the same mechanism.

A wide variety of physical, chemical, biological, and environmental factors are known to influence biotic and abiotic degradation rates. In many cases, these factors can be identified by expert judgment and are quantified by using relatively simple mathematical expressions. For example, the dominant factor controlling the rate of biodegradation is the number of competent microorganisms. This number can vary by several orders of magnitude for different tests and different environmental compartments. However, in many cases, the number of competent microorganisms can be estimated by using standard assumptions. Similar approaches can be used to identify and quantify other key variables affecting biotic and abiotic degradation processes.

The quantitative determination of persistence in soil, water, and sediment environments is largely controlled by 2 factors: 1) the accurate determination of biotic and abiotic rate coefficients and 2) the extrapolation of these rate coefficients (in the form of transformation half-lives) to relevant environmental compartments. Because a number of factors influence abiotic and biotic processes, the determination and extrapolation of transformation rate coefficients are complex. One of the goals of the workgroup was to reduce this complexity to manageable levels and to develop a technically sound approach for estimating persistence in soil, water, and sediment compartments that was broadly applicable to a wide range of organic chemicals. The recommended approach utilizes multiple data sources and databases (QSARs, experimental determination, monitoring, etc.) and encompasses both screening and detailed evaluations. The key outputs are first-order rate constants and the corresponding half-lives for both biotic and abiotic transformations in soil, water, and sediment compartments. Moreover, methods are suggested for calculation of the fundamental second-order rate expressions by normalizing the first-order rate constants for the known or estimated concentration of the biological or chemical reactant, that is,

$$(dS/dt)_b = -k_b\,[B]\,[S] \text{ and } (dS/dt)_a = -k_a\,[A]\,[S],$$

where

k_b and k_a are the true second-order biotic and abiotic transformation coefficients, respectively;

B is the biomass concentration for biotic reactions;

A is the reactant concentration for abiotic reactions; and

S is the aqueous phase substrate concentration in the environmental compartment under consideration.

Thus, k_b and k_a are not "lumped" pseudo–first-order constants, i.e., expressions that contain the concentration of some reactant as a hidden term. Lumped first-order values are highly specific to individual systems and cannot be extrapolated to different systems without considerable (i.e., order-of-magnitude) uncertainty.

The expression of biological and chemical reactions by using second-order kinetics was proposed to minimize the effects of environmental heterogeneity and to reduce uncertainty in the extrapolation of persistence data. By converting system-specific, pseudo-first-order rate constants to chemical-specific, second-order rate constants, the latter become, in effect, intrinsic chemical properties. This approach allows persistence data to be developed and extrapolated to a variety of different environmental compartments, as long as B or A are known or can be reasonably estimated from experimental data.

Partitioning, Inter-media Transfer, and Long-Range Transport

The inter-media partitioning and transport workgroup initially addressed the environmental media or compartments that play a key role in controlling the fate of chemicals and hence their persistence. At the minimum level of detail is a 3-compartment system of air, water, and soil, with soil possibly including sediment. More realistic is a system consisting of air, water, soil, and sediment compartments in which there is also treatment of subcompartments such as aerosols in air and suspended solids in water. Such a system probably captures the principal phases of accumulation and transport, at least for most chemicals and environmental conditions. The most difficult issue is that of including vegetation as a sub-compartment of terrestrial soils. Related issues were reviewed by this workgroup, including the separate treatment of fresh and marine water, the vertical segmentation of the atmosphere, the possible inclusion of ground water, and the effects of snow and ice.

This workgroup also reviewed the phase-partitioning properties of chemicals and suggested approaches for estimating the appropriate partition coefficients and their temperature dependence. This aspect of environmental chemistry is one of the more advanced branches of environmental science. For most chemicals, there is a fairly good predictive capability, although there is a need to improve predictive correlations for certain classes of polar chemicals and certain media such as vegetation. Quantifying the partitioning of speciating chemicals such as phenolic substances remains problematic.

Inter-media transport processes were reviewed, including exchanges between air and soil, air and water, air and vegetation, soil and water, and water and bottom sediments. Robust transport equations exist in most cases, but there remains doubt about parameter values. This is an area in which models are essential because chemical fluxes cannot be measured directly and must be inferred from measurements of related transport processes such as sedimentation rates. There is a need to document more extensively the range of transport parameters such as mass transfer coefficients or deposition velocities that are encountered in the environment in order that evaluative models use appropriate values or ranges of values.

In general, the workgroup determined that for most chemicals (especially the non-polar substances) the key media that control persistence have been identified and the partitioning and transport processes that are essential components of any mass-balance model can be expressed by using fairly robust, reliable expressions. The magnitudes of uncertainties associated with predictions of transport and partitioning are believed to be known with acceptable accuracy. There are concerns, however, about the real possibility that predictive equations and models may be wrongly applied, especially to less-frequently modeled substances. Thus, there is no substitute for insights and interpretations provided by the experienced environmental scientist or engineer.

Modeling Persistence in a Multimedia Environment

Based on discussions within the multimedia modeling workgroup, a set of working recipes was proposed for calculating the persistence and transport potentials of chemicals in the environment. The workgroup operationally defined "persistence in the environment" as the result of model-calculated overall residence time at steady state in a multimedia environment. A simple Level III multimedia, mass-balance model was proposed for this calculation. Only irreversible losses were considered, which consist primarily of chemical transformation of parent compound but also include transport to deep ocean layers, burial in deep sediments, and unidirectional transport to deep ground water.

Persistence in the environment and potential for long-range transport are related in the sense that time is required for transport, i.e., the residence time in the mobile media (air or water) must be sufficiently long for a chemical to be transported over long distances. A strong correlation between persistence and transport distance among chemicals with different chemical properties has been demonstrated, but this correlation is not strong enough to predict one from the other. A separate definition of transport potential was proposed by using the same multimedia modeling approach for defining persistence. While a global-scale, closed model was proposed for calculating persistence, a regional, open model was proposed for calculating transport potential. Advective transport fluxes of a chemical out of the system are controlled by wind speed and ocean currents, and they depend on the linear scale of the system. Loss by transformation, however, is proportional to the area scale of the system.

The task of assessing the persistence and transport potential of both new and existing chemicals would be a major undertaking, requiring the collection of chemical properties for use in multimedia model simulations for each chemical or the worldwide collection of data in multiple environmental media at each sample site. As an alternative, a classification system is proposed that considers sets of chemical properties that would result in the classification of a chemical as "persistent" and as "having the potential for long-range transport." One method for conducting such an analysis involves the use of a set of binary classification trees, e.g., the Classification and Regression Tree (CART) approach. An introduction to

CART is presented along with a simple illustration of how the approach can be applied with multimedia models to assess the link between an output classification such as persistence and the range of input values that give rise to this classification.

Recommendations of the workgroup are as follows:

- useful measures of persistence and transport potential can be derived from half-lives in specific media and inter-media partition coefficients by using Level III multimedia models;
- two-dimensional plots of persistence versus transport potential are useful for determining agreed-on concern and no-concern chemicals;
- given appropriate protection criteria, e.g., geographic extent, overall global residence time, etc., the parameter domains that give rise to undesired persistence and transport potential can be derived by using CART analysis; and
- although there is no scientific rationale for selection of persistence and transport criteria, science can help in the process of developing criteria that are coherent.

Role of Ambient Monitoring

Monitoring was recognized as an essential tool in the evaluation of persistence and long-range transport. Monitoring is defined as the long-term and standardized measurement, observation, evaluation, and reporting of the environment in order to define status, trends, and mass-flows. Monitoring studies have been essential in the discovery, description, and understanding of the environmental behavior of POPs. Systematic air monitoring studies have shown that a range of chemicals are subject to long-range transport, while monitoring of biological samples has demonstrated a range of compounds that are persistent and prone to bioaccumulation. In addition to identifying chemicals of concern, multimedia monitoring programs also have demonstrated the effectiveness of various emission reductions, phase-outs, and elimination measures in reducing the environmental concentrations of pollutants. Although time-trend data are being assembled for different ecosystems in different regions of the globe, only a few detailed case studies exist.

In the future, multimedia monitoring will be essential for assessing equilibrium conditions, mass balances, fluxes, persistence, and long-range transport. Collaborations between environmental scientists and modelers will be essential for defining values for the range of input parameters required by multimedia models. Such collaborations should be iterative and should foster both linked and parallel efforts between monitoring and modeling. In addition, more attention should be devoted to defining emission sources and environmental loadings. Ultimately, the goal of environmental monitoring should be the elucidation of the biogeochemical cycles of persistent organic compounds through mass balance calculations.

The workgroup also recommended establishment of a global network consisting of a limited number of key sampling stations that would sample a variety of media in remote areas in different continents and climatic zones, in both the Northern and Southern hemispheres. A continuous annual sampling program is essential for defining temporal trends in the concentrations of chemicals in the environment. Global specimen banks are also needed to verify previous analytical results, to intercalibrate new analytical techniques, to develop correction factors that allow the use of historical data, and to keep a record for future investigations of new substances. Pathways of degradation and metabolites should be studied also, and low-cost analytical and real-time monitoring techniques should be developed.

Case Studies on Persistence and Long-range Transport of POPs

Traditionally, persistence has been viewed as a property of chemicals with slow environmental degradation rates that resulted in the distribution of a chemical in the environment. In contrast, a more specific measure of persistence is given by the response of environmental chemical concentrations to source reduction. This workgroup examined a series of case studies that illustrate and emphasize the role played by the receiving environment in determining the persistence of a chemical. The case studies were drawn from arctic, temperate, and tropical locations and include hexachlorocyclohexane (HCH), polychlorinated biphenyls (PCBs), and chlorobornanes (CHBs) in the Arctic Ocean; PCBs and CHBs in Lake Superior; atrazine in Lake Michigan; polycyclic aromatic hydrocarbons (PAHs) and PCBs in the western Mediterranean Sea; PCBs in the Baltic Sea; HCH and PCBs in Lake Malawi (Africa) and the Vellar Estuary (India); and PCBs in Great Britain's soils. A common finding was that biogeochemical processes in the receiving environment are as crucial to the manifestation of persistence as are the chemical properties themselves. The lessons learned from each of these studies lead to insights of how environmental reservoirs become loaded and what controls their eventual clearance when sources are reduced or removed.

Regional- or global-scale atmospheric transport is an important pathway for contaminants in most of these systems. For semi-volatile compounds like PCBs and the lighter PAHs, the atmosphere plays a crucial role during the loading and clearing phases of aquatic reservoirs. Appreciable partitioning into the air from water, plants, or soil is important for the most rapid transport in the atmosphere. Polychlorinated biphenyls are presently in the clearing phase for many temperate reservoirs but, as part of that process, other more distant regions like the Arctic Ocean continue to be loaded. As a consequence, these distant regions ultimately set global boundaries on persistence. Although partition coefficients suggest that particle fluxes (especially of organic carbon) should provide an important route of clearance for PCBs in aquatic environments, rapid mineralization of particulate organic carbon can circumvent that process and lead to recycling in lakes or to transport to deep oceans rather than permanent burial in sediments. Sea–air–land interactions are important in the fate of POPs, especially in systems heavily influenced by riverine inputs.

The budget for PAHs in the Mediterranean Sea illustrates that the pathway of the chemicals into an aquatic reservoir can dominate their later behavior; for example, light PAHs mimic PCBs, and the air–water exchange is important, whereas heavier PAHs arrive as soot and behave like particles in both the air and the water. For HCH and CHB, strong partitioning from air into water is a master control, and the relatively low Henry's Law constant for cold water explains the global phenomenon of high concentrations in the Arctic Ocean when compared to other regions, as well as the regional phenomenon of high CHB concentrations in Lake Superior. For CHBs in Lake Superior, proximity to large soil reservoirs that continue to supply the atmosphere will further delay clearance, a factor that may also be relevant for parts of the Arctic Ocean. The poor partitioning of HCH and CHB onto particles in aquatic reservoirs, combined with limited degradation, delays clearance.

Atrazine provides a surprising illustration of how relatively biodegradable chemicals can accumulate in environments in which their degradation rate is low. High use, coupled with a long lifetime under certain conditions, has resulted in substantial buildup in local reservoirs, such as Lake Michigan, that are affected by local runoff and regional atmospheric transport and deposition. Because of its long half-life in oligotrophic waters, atrazine concentrations may ultimately increase in the oceans.

In the tropics, volatilization and degradation assume much greater importance in determining the behavior of POPs than in temperate and arctic regions. Lakes in the tropical regions of the Southern Hemisphere probably provide a sink for some chemicals (e.g., HCH) because of degradation and low prevailing concentrations in water when compared to the atmosphere. Other compounds may be rapidly recycled (e.g., PCB). The behavior of persistent chemicals in tropical reservoirs is poorly studied, and further work is clearly required.

Soils and plants provide important reservoirs that resupply the atmosphere via volatilization and delay clearance after sources have been removed. Upon re-volatilization, POPs may enter aquatic or terrestrial food chains.

Framework and Criteria for Evaluation

This workgroup addressed the development of a framework for evaluation of the environmental behavior of chemicals with respect to their persistence and long-range–transport potential. It is envisioned that the evaluation process will be part of the broader framework for identifying persistent, bioaccumulative, and toxic substances. Furthermore, although the primary focus was on current regional and global regulatory initiatives, the framework for substance evaluation may prove useful also at the national level as part of a procedure for early identification of potentially problematic substances and in industry for evaluating alternative substances before marketing. The principles underlying this framework are that the process 1) is based on sound scientific principles, 2) allows for expert judgment and flexibility, and 3) incorporates the logic and spirit of the precautionary principle. Within this context, application of the precautionary principle means that if the evidence (i.e., available data and the quality and relevance of such data) for making the decision is not conclusive or is contradictory, the substance would proceed to the next tier of evaluation.

The recommended framework for substance evaluation is based on a tiered approach, proceeding from screening to more detailed evaluations. In the screening phase (Tier 1), the focus is on persistence alone, recognizing the fact that a substance must be sufficiently persistent in the environment in order to be transported long distances. The primary objective of the screening evaluation is to eliminate substances that are clearly nonpersistent. Half-lives for persistence in the various environmental compartments (air, water, soil, sediment) are compared to previously established criteria, and substances whose half-lives in the environmental media are lower than the criteria values will be judged to be nonpersistent and/or unlikely to undergo long-range transport. The accuracy of and confidence in the results of the screening evaluation can be greatly improved by including information on the likely distribution of the substance among the environmental compartments. This distribution will indicate which half-lives are most relevant for the screening evaluation of a particular substance and which are not. Further, because half-lives can vary greatly with environmental (and laboratory test) conditions, it is important to understand 1) the reference conditions assumed for the criteria, 2) the conditions under which measured half-lives for evaluated substances were determined, 3) the quality of the measurements, and 4) the distribution (range) of measured half-lives. Because of the inherent uncertainties in the measured and/or estimated half-lives, it is important to recognize that the criteria should be applied as guidelines and not as immutable quantities or "bright lines."

Substances that are identified in Tier 1 for further evaluation are examined in more detail in Tier 2 in order to 1) confirm, refute, and/or clarify the results of the Tier 1 evaluation; 2) determine whether the substance that meets the persistence criteria has the potential for long-range transport; and 3) identify additional factors that warrant consideration, such as the influence of climatic and other regional characteristics. By using multimedia environmental models, substances whose overall persistence and transport potentials are determined to be less than the criteria values will be judged to be non-persistent and/or unlikely to undergo long-range transport. These models can represent local, regional, or global scales, depending on the scope of the evaluation. Monitoring data also plays an important role in determining whether a substance is persistent and subject to long-range transport. When such data indicate that a substance resides in remote regions at environmentally significant concentrations, these substances should proceed directly into the detailed evaluation.

Because it is recognized that substances must exhibit a combination of properties to be of concern, additional tiers of evaluation are required to designate a substance as a POP or PBT. Toxic potency and bioaccumulation potential are considered in Tier 3. The final tiers of risk assessment and risk management allow for consideration of all relevant attributes of a chemical, including exposure concentrations, availability in the environment, and any new information that may become available. These latter elements of the framework were beyond the scope of this workshop.

Because persistence and transport criteria must be selected before any substances can be evaluated, the workgroup also discussed 4 approaches for criteria selection that may be used individually or in combination. Because the scientific understanding of the environmental behavior of chemicals is continually advancing, the criteria and processes used for evaluation should be based on up-to-date and sound science. One approach to establishing criteria is simply to rely on precedent, i.e., use criteria previously set for other similar initiatives. However, there are ways in which science can better inform the process. The empirical approach uses available data for one or more chemical properties for a wide range of substances, specifically including substances already widely acknowledged as PBTs or POPs, to derive criteria for these properties. The modeling approach takes advantage of the fact that most substances of potential concern are likely to partition into more than one environmental compartment (i.e., they are multimedia substances). Properties such as overall persistence, calculated by using multimedia fate models, can be graphed versus chemical properties to assist in establishing internally consistent criteria or to identify constraints in the criteria values. Finally, a theoretical approach can be envisioned in which multimedia models and computerized data classification algorithms are used to help derive relevant criteria values. This approach takes advantage of tools such as Monte Carlo and CART analysis. However, further work is needed to understand the practical application of this approach.

Additional considerations in applying the framework were also discussed. Transparency and openness must be characteristics of both the substance evaluation and criteria selection processes. Transparency is the ability of educated laypersons and policymakers to understand the process, whereas openness is the opportunity for all interested parties to comment and provide input on an ongoing basis. Other considerations relate to the availability and quality of data. The effectiveness of any process for substance evaluation will be affected by the availability, relevance, accuracy, and confidence in the required input data. Guiding principles for data evaluation are that properties should pertain to environmentally relevant conditions and that measured parameter values are always preferred over estimated values if the data are of sufficient quality and are representative of the conditions of interest. In addition, any data that are estimated must be clearly identified as such, and all parties should be permitted to bring any and all relevant data into the process. Finally, because of the inherent uncertainties in environmental properties, expert judgment is essential for both the substance evaluation process and the development of criteria, especially when deciding which data will be used, how the data are interpreted, and which approaches will be used to analyze the data. Because of the limitations in resources, expert involvement may need to be limited; therefore, it is important at each step of the process to identify where expert judgment can be used most effectively and efficiently.

Summary and Conclusions

The Workshop was successful in bringing together an international group of experts, each of whom brought individual insights into aspects of the formidable task of evaluating persistence and long-range transport of organic chemicals. The current state of the art in this subject area was discussed in detail, and consensus was reached and has been captured in these proceedings. It is clear from this review that as a result of research over recent decades, we now know a great deal about how chemicals behave in the environment. In some specific areas, such as atmospheric oxidation and inter-media partitioning, reliable predictive capabilities exist for many, but not all substances. In other areas, such as inter-media transport and the transformation kinetics in soils, waters, and sediments, a fairly clear picture exists of the key process mechanisms and some predictive capability exists. However, there is often considerable uncertainty about the extrapolation of values for key parameters, including transformation rate constants, half-lives, and transport coefficients, that have been determined for specific chemicals under specific environmental conditions. Continuing efforts on measurement and interpretation are needed to enhance our ability to forecast chemical fate, especially for less-studied chemicals. Because the actual persistence and long-range transport of a chemical depends on these transport and transformation processes, it is self-evident that society must exploit the available and emerging science when reaching decisions about chemical management. To ignore the science would be unthinkable and would inevitably lead to distorted decision-making.

A major challenge exists to marshall the scientific information in a format suitable for informing the regulatory process. It is likely that most progress will be made by building, testing, and modifying a variety of mass-balance models and decision-support systems that can predict persistence and long-range transport characteristics. But the models must be supported or validated by monitoring data and by demonstrated successful application to specific case studies. A 3-pronged, scientific approach of general monitoring, documentation of case studies, and successful modeling of these situations can create a climate of confidence that we understand the science that dictates the persistence and long-range transport attributes of substances of present concern and can, with some measure of reliability, predict and prioritize these attributes for new and emerging chemicals.

The key features of the overall evaluation framework for identifying PBTs have been identified. The most effective framework is likely to be a tiered process in which a logical sequence of decisions is made based on sound scientific principles, guided by expert judgment. Early assessment of persistence and long-range transport is desirable, preferably prior to consideration of bioaccumulation and toxicity. Regardless of the system adopted, it must be transparent so that the public can understand how decisions are being made, and it must be open so that all interested parties can provide input.

An important message from the workshop, one that perhaps is not directly articulated in the report, is that there is a widespread, earnest desire among the environmental science community to contribute to improved decision-making at local, national, and international levels in the expectation that this will lead to more effective management of those substances that are persistent and may be transported long distances in the environment. This desire is shared by the industrial, governmental, and academic sectors that comprise SETAC. Where gaps in knowledge and in evaluation techniques exist, there is enthusiasm to advance environmental science in these areas. What is needed is to bring together the needs and those who can satisfy those needs in a continuing collaborative effort. The Pellston Workshop on Criteria for Persistence and Long-Range Transport of Organic Chemicals in the Environment has initiated this process. The participants and sponsors hope it will continue.

It will be a challenging task to evaluate the large number of chemical substances in commerce for persistence and long-range transport. The workshop has not only identified how existing environmental science can contribute to this process, but also it has demonstrated the compelling need to exploit this science and has identified where the key advances must be made to enhance this contribution.

It is the hope of the participants and sponsors that this volume will provide timely input into national and international deliberations and will be valuable in shaping policies and in directing future scientific efforts.

Chapter 1

Introduction

Gary Klečka and John Buccini

Persistent organic pollutants (POPs) are bioaccumulative and toxic organic compounds of natural or anthropogenic origin that resist photolytic, chemical, or biological degradation. They are characterized by low water solubility and high lipid solubility, resulting in bioaccumulation in fatty tissues of living organisms. Persistent organic pollutants may be transported in the environment at low concentrations by movement of fresh and marine waters. Recent evidence also has shown that because of their semi-volatility, POPs are transported long distances in the atmosphere, resulting in widespread distribution around the earth, including regions in which they have never been produced or used. Thus, both humans and ecosystems may be exposed, in many cases for extended periods of time.

In recent years, several national and international accords have been initiated or completed to identify a variety of persistent, bioaccumulative, and toxic substances (PBTs) and to require action to control the use and/or release of these substances, including both organic compounds and those containing metals. The following list includes some of the major regional and global initiatives that are directed at identifying persistent toxic substances and developing risk-management measures to control the exposure of humans and ecosystems. While the list is not exhaustive, it does show that PBTs have been and will continue to be the subject of considerable attention for scientists and policymakers.

- The Oslo and Paris Commissions Convention for the Protection of the North-East Atlantic (1992) agreed that discharges and emissions of substances that are toxic, persistent, and bioaccumulative and that could reach the marine environment should, regardless of their source, be reduced to levels that are not harmful with the aim of their elimination.
- The Fourth International Conference on the Protection of the North Sea (1995) set a goal to eliminate, within one generation (25 years), releases of PBTs and another goal to reduce releases of metals to achieve environmental concentrations that approach background levels.
- The Barcelona Convention for Protection of the Mediterranean Sea against Pollution (1995) has resulted in actions being taken to reduce releases of metals and organic pollutants.

Evaluation of Persistence and Long-Range Transport of Organic Chemicals in the Environment. Gary Klečka et al., editors.

 ISBN 1-880611-22-8

- The Helsinki Convention for the Protection of the Baltic has adopted the targets of the Fourth International Conference on the Protection of the North Sea.
- The Canada-U.S. Great Lakes Water Quality Agreement (1972) and the recent Great Lakes Binational Toxics Strategy (1997) emphasize action on POPs and other persistent toxic substances.
- The North American Council for Environmental Cooperation (NACEC) Resolution 95-9 on the Sound Management of Chemicals (1995) gives immediate priority to persistent, toxic substances and has resulted in the development of continental actions plans for dichlorodiphenyltrichloroethane, chlordane, polychlorinated biphenyls, and mercury.
- The United Nations Economic Commission for Europe (UNECE) Convention on Long Range Transboundary Air Pollution (1998) addresses 3 heavy metals (lead, cadmium, mercury) and 16 POPs.
- The United Nations Environment Program (UNEP) Global Program of Action for the Protection of the Marine Environment from Land-based Activities (1995) addressed numerous sources of pollution, including sources relating to metals and POPs.
- A UNEP global convention on POPs has been proposed; this initiative began formally on 29 June 1998, with the commencement of intergovernmental negotiations to develop, by the year 2000, a global, legally binding convention with measures to reduce or eliminate the releases of 12 specified POPs. One task to be addressed is the development of scientific criteria and a procedure for selecting candidate substances that would be considered for future international action.

In addition to these intergovernmental accords, several national governments and non-governmental organizations have developed and implemented regulatory and non-regulatory initiatives to address the use and/or release of POPs and PBTs:

- In the U.S., the U.S. Environmental Protection Agency (USEPA) has recently issued a final rule (USEPA 1999a) in which several PBTs were added to the Toxics Release Inventory (TRI), and reporting thresholds were lowered for other PBTs that already were listed. In another action, the USEPA has published persistence, bioaccumulation, and toxicity criteria for the evaluation of new chemical (Premanufacture Notice) substances with PBT characteristics, along with a tiered testing scheme for suspected PBTs (USEPA 1999b).
- In Canada, the federal government issued a formal challenge, in the early 1990s under the Accelerated Reduction/Elimination of Toxics program, for industry to voluntarily reduce emissions of PBT substances that meet specific criteria. Further, the Canadian Toxic Substance Management Plan has

identified PBT criteria and has targeted substances that meet the criteria for virtual elimination.

- In the private sector, the U.S. Chemical Manufacturers Association (CMA) has published detailed guidance on PBT identification, risk characterization, and pollution prevention for member companies (CMA 1996).

To foster the development of a sound scientific foundation for identifying, evaluating, and selecting organic substances, an international workshop, sponsored by the Society of Environmental Toxicology and Chemistry, was held 13–19 July 1998, to discuss scientific issues related to persistence and long-range transport of organic chemicals. The workshop had broad involvement from academia, government, industry, and other interested organizations. While it is widely recognized that substances that exhibit a combination of properties (i.e., persistence, bioaccumulation, and toxic potency) should be the focus of societal concern and regulatory management efforts (Carey et al. 1998), this workshop specifically restricted its attention to the need for a scientific foundation for the criteria and process used to evaluate persistence and long-range transport of POPs and PBTs.

The specific objectives of the workshop were to discuss the issues, reach consensus, and make recommendations based on the current state of the science concerning the evaluation of persistence and long-range transport of organic chemicals in the environment. In meeting these objectives, the workshop addressed a number of questions, including the following:

- Is it feasible to classify chemicals according to persistence and potential for long-range transport? With what level of discrimination should chemicals be classified?
- What types of data are needed for the evaluation of chemicals? What is the proper role of estimated values versus laboratory-derived data? How does one proceed when data for critical inputs are missing?
- What level of uncertainty is acceptable with respect to each parameter, and how should uncertainty be accounted for in the process of substance evaluation?
- Is it realistic to assess persistence and long-range transport potential on the basis of specific chemical properties and transformation half-lives, or is single-medium or multimedia modeling required?
- What are the proper roles of local, regional, and global monitoring data?
- Are the schemes that have been proposed for classifying chemicals scientifically valid? How could they be improved?

The workshop followed the format of previous Pellston Workshops and was limited to about 50 participants with broad international perspective and expertise in atmospheric and environmental chemistry, partitioning and transport, multimedia modeling, and environmental assessment. To address the objectives, the workshop was organized into 8 individual sessions. A general plenary was held on the first day, during which invited, authoritative discussion papers were presented to stimulate dialogue. Based on expertise, each participant was assigned to a small working group, although interactions among the working groups were encouraged. During the workshop, summary working group presentations were made to the entire group in plenary session, and comments from the participants were incorporated into the developing document. A major goal of each working group was to develop a summary of their deliberations, including discussions of the state of the science pertinent to that working group; critical analyses of available approaches; and methods to integrate these approaches into a framework for assessing persistence and potential for long-range transport.

The chapters of this book coincide with summaries prepared by each of the working groups and provide a comprehensive account of the status of this branch of environmental science and management. It is hoped that the proceedings will be useful to the scientific community and to policymakers and that they will provide timely input into national and international discussions on the criteria and process used to evaluate the environmental behavior of organic compounds.

References

Barcelona Convention for the Protection of the Mediterranean Sea against Pollution, Barcelona Resolution. June 1995.

Carey J, Cook P, Giesy J, Hodson P, Muir D, Owens JW, Solomon K, editors. 1998. Ecotoxicological risk assessment of the chlorinated organic chemicals. Pensacola FL: Society of Environmental Toxicology and Chemistry (SETAC). 397 p.

[CMA] Chemical Manufacturers Association. 1996. PTB Policy Implementation Guidance. Product Risk Management Guidance for PTBs. Chemical Manufacturers Association, Arlington, VA. February 1996.

Environmental Leaders 2 – Accelerated Reduction/Elimination of Toxics. Progress Report; En 40-492/1-1997E. Government of Canada, Environment Canada. January 1997.

Fourth International Conference on Protection of the North Sea, Esbjerg Declaration. June 1995.

Great Lakes Binational Toxics Strategy: Canada-United States Strategy for the Virtual Elimination of Persistent Toxic Substances in the Great Lakes Basin. Great Lakes National Program Office: Chicago IL. 7 April 1997.

[NACEC] North American Council for Environmental Cooperation. Resolution 95-5 on the Sound Management of Chemicals, Mexico City, Mexico. 13 October 1995.

Oslo and Paris Conventions for the Prevention of Marine Pollution. Final Declaration of the Ministerial Meeting of the Oslo and Paris Commissions, Paris, France. 21–22 September 1992.

Toxic Substances Management Policy. Persistence and Bioaccumulation Criteria; En 40-499/2-1995E. Government of Canada, Environment Canada. June 1995.

[UNECE] United Nations Economic Commission for Europe. Protocol to the 1979 Convention on Long-Range Transboundary Air Pollution on Persistent Organic Pollutants, Aarhus, Denmark. 24 June 1998.

[UNEP] United Nations Environment Program. Global programme of action for the protection of the marine environment from land-based activities. 23 October–3 November 1995. Washington, DC.

[USEPA] U.S. Environmental Protection Agency. 1999a. Persistent bioaccumulative toxic (PBT) chemicals; Final Rule. 40 CFR Part 372. 64 FR 5866. 29 October 1999.

[USEPA] U.S. Environmental Protection Agency. 1999b. Category for persistent, bioaccumulative, and toxic new chemical substances. 64 FR 60194. 4 November 1999.

Chapter 2

Quantitative Determination of Persistence in Air

James Franklin, Roger Atkinson, Philip H. Howard, John J. Orlando, Christian Seigneur, Timothy J. Wallington, Cornelius Zetzsch

The atmosphere is a highly oxidizing environment in which degradation of pollutants proceeds in successive steps. Eventually, the pollutant either is oxidized completely, e.g., the conversion of methane into CO_2 and H_2O, or is converted into partially oxidized species that are removed via wet and/or dry deposition to the earth's surface. All of the energy for atmospheric chemistry comes from solar radiation that is absorbed by various components of the atmosphere.

Direct photolysis, gas-phase, solution-phase, and heterogeneous reactions all play important roles in atmospheric chemistry. For direct photolysis to occur in the troposphere (0 to 15 km altitude), the chemical must absorb light at wavelengths greater than 290 nm (the short wavelength cutoff for solar radiation in the troposphere) and, having absorbed radiation, must undergo chemical change (i.e., must have a nonzero quantum yield). To determine the rate of degradation by direct photolysis, the rate of light absorption needs to be calculated using tabulations of solar intensity versus wavelength and the UV absorption cross-sections for the chemical. The photolysis half-life then can be determined by using the quantum yield at each wavelength and will be dependent upon the time of the year, time of the day (for short-lived compounds), latitude, and cloud cover.

The driving force for most of the chemistry that occurs in the troposphere is the formation of hydroxyl (OH) radicals primarily via photolysis of ozone to form $O(^1D)$ atoms that react with water vapor:

$$O_3 + h\nu\ (\lambda \sim 290\text{–}320\ \text{nm}) \rightarrow O(^1D) + O_2(^1\Delta_g) \qquad \text{(Reaction 2-1)},$$

Chapter Preview

Evaluation of Persistence and Long-Range Transport of Organic Chemicals in the Environment. Gary Klečka et al., editors.
 ISBN 1-880611-22-8

$$O(^1D) + H_2O \rightarrow 2\,OH \qquad \text{(Reaction 2-2)},$$

where "hν" represents a photon of wavelength λ.

OH radicals react with almost all chemicals emitted into the atmosphere. The OH radical concentration in the atmosphere varies with location, time of day, season, and meteorological conditions. A reasonable annual global tropospheric average is $1.0 \times 10^6\ cm^{-3}$ (Krol et al. 1998; Miller et al. 1998).

Although only 10% of atmospheric ozone resides in the troposphere, it has a profound impact on tropospheric chemistry, largely through its role in the production of OH radicals (Reactions 2-1 and 2-2). Ozone concentrations in the troposphere vary from 20 to 40 ppbv (40 to $80 \times 10^{-6}\ g\ m^{-3}$) for a remote pristine site to 100 to 200 ppbv (200 to $400 \times 10^{-6}\ g\ m^{-3}$) in a highly polluted urban environment. Ozone is a reactive molecule that readily adds to carbon–carbon double bonds (Atkinson and Carter 1984; Atkinson 1994), and this reaction can be an important removal mechanism for unsaturated organic compounds. There are 2 sources of tropospheric ozone: 1) transport from the stratosphere in meteorological events known as "tropospheric folding" and 2) peroxy radical reactions that oxidize NO to NO_2. An example is the oxidation of CO initiated by OH radicals, which are regenerated via reaction of H atoms with O_2 (in the presence of a third body, M, required to remove the energy liberated by the bond formation) and the subsequent reaction of HO_2 with NO:

$$OH + CO \rightarrow H + CO_2 \qquad \text{(Reaction 2-3)},$$

$$H + O_2 + M \rightarrow HO_2 + M \qquad \text{(Reaction 2-4)},$$

$$HO_2 + NO \rightarrow OH + NO_2 \qquad \text{(Reaction 2-5)}.$$

Photolysis of NO_2 leads to the formation of O_3 by the following sequence of reactions:

$$NO_2 + h\nu\ (\lambda < 420\ nm) \rightarrow NO + O \qquad \text{(Reaction 2-6)},$$

$$O + O_2 + M \rightarrow O_3 + M \qquad \text{(Reaction 2-7)}.$$

$$\text{net} \qquad CO + 2O_2 \rightarrow CO_2 + O_3$$

Analogous reaction schemes, leading to net formation of O_3, occur for the tropospheric photooxidation of methane and other organic compounds in the presence of NO_x. Recent modeling calculations suggest that, on a global scale, transport from the stratosphere accounts for approximately 10% of tropospheric ozone, while in situ photochemical formation is the source of the remaining 90% (Warneck 1988; Crutzen 1995).

At night, the nitrate radical (NO_3) takes over from the OH radical as the dominant radical species that initiates degradation reactions. The nitrate radical is formed by the reaction of NO_2 with O_3. NO_3 radicals absorb strongly in the visible and are photolyzed rapidly during the day. Hence, NO_3 radicals are present in significant quantities only at night and can, in rare and localized cases, reach levels on the order of 10^{10} molecule cm^{-3} (Atkinson 1991a).

Organic compounds that enter the troposphere can be removed by photolysis, reaction with the OH radical (primarily during the daylight hours), reaction with the NO_3 radical (primarily during nighttime), and reaction with O_3. All organic compounds except the chlorofluorocarbons and other saturated organics that do not contain H atoms react with the OH radical. NO_3 radicals react with a large number of organic compounds, with the NO_3 radical reaction being a potentially important tropospheric loss process for compounds containing >C=C< or -C≡C- bonds, organosulfur compounds, phenolic compounds, and certain nitrogen-containing compounds (Atkinson 1991a, 1994). O_3 reacts to a significant extent only with organic compounds containing >C=C< and -C≡C- bonds (e.g., alkenes, haloalkenes, alkynes, and oxygen-containing compounds with >C=C< bonds) and with certain nitrogen-containing compounds (Atkinson and Carter 1984; Atkinson 1994).

The rate at which organic chemicals react with photooxidants is dependent upon the rate constant, which is related to the chemical structure, and the photooxidant concentration. The rate constant is somewhat temperature-dependent. The OH concentration varies with the chemical composition of the air mass (OH radical concentrations appear to be no higher than those in non-urban areas [George et al. 1999]), time of day (being highest in the afternoon), latitude and season (which affect solar flux), and ambient humidity. Therefore, the half-life of an organic reacting with hydroxyl radicals can vary considerably depending upon the location, season, and time of the day. For example, Webster et al. (1998) have calculated that tetrachlorobiphenyl has atmospheric degradation half-lives with respect to reaction with OH ranging from 1.2 d (tropical noon, 30 °C, $[OH] = 6 \times 10^6$ molecule cm^{-3}, rate constant = 1.1×10^{-12} cm^3 $molecule^{-1}$ s^{-1}) to 6.0 y (polar night, –20 °C, $[OH] = 6 \times 10^3$ molecule cm^{-3}, rate constant = 6.2×10^{-13} cm^3 $molecule^{-1}$ s^{-1}). Gas-phase organics also may be removed from the atmosphere by wet and/or dry deposition to land or water surfaces.

Many chemicals that are being considered for listing as persistent organic pollutants (POPs) are semi-volatile, like tetrachlorobiphenyl, and will be partially or totally adsorbed on particulate matter. Reactivity of adsorbed chemicals is not well understood, but some preliminary results are discussed in a later section. Also, adsorption on particulate matter affects the atmospheric residence time due to wet or dry deposition. For dry deposition, the residence time of the particle is strongly dependent upon its diameter. Although this is not degradation of the chemical, it is a major consideration for overall residence time in the atmosphere.

The semi-volatile organic compound (SOC) content of the aerosol is known to be in equilibrium between the gas and particle phases (Junge 1975; Gill et al. 1983; Seinfeld and Pandis 1997). Gas-particle partitioning is discussed in greater detail in Chapter 4. Neglecting hydrophilic or lipophilic effects, the particle-borne portion of an SOC depends on its subcooled-liquid vapor pressure and on the aerosol density, characterized by the aerosol surface area per unit air volume. Thus, the plasticizer di(2-ethylhexyl)phthalate will be particle-borne in urban areas and will vaporize as it is transported to remote areas in which the aerosol density is lower. The rather volatile pesticide lindane will vaporize rapidly from soil and remain in the gas phase even in urban areas. Figure 2-1 shows the particle-borne portion of SOCs as a function of liquid-phase vapor pressure (subcooled-liquid vapor pressure in the case of compounds that are solids at ambient temperature) for a variety of POPs and for n-alkanes with corresponding vapor pressures, taken from Bidleman (1988).

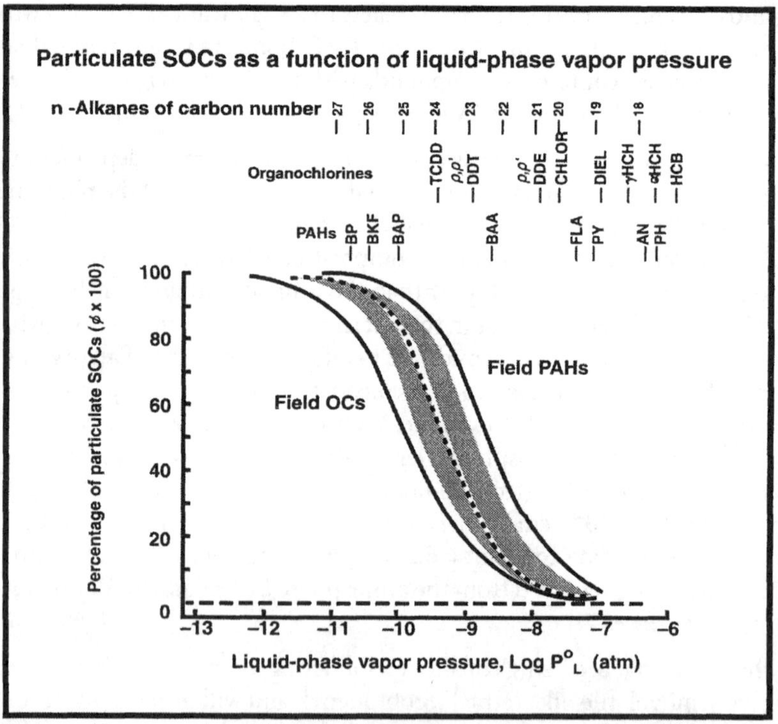

Figure 2-1 Aerosol-borne portion of SOCs (polycyclic aromatic compounds and chlorinated POPs) versus liquid-phase vapor pressure. Reprinted with permission from Bidleman (1988), ©1988 American Chemical Society. The shaded area shows the expected partitioning in the presence of aerosols with surface areas in the ranges typical of clean background to polluted urban concentrations. The solid lines show results from field observations for organochlorine compounds and polycyclic aromatic compounds.

Considerable progress has been made recently in the description of sorption of semi-volatiles on particles (absorption or adsorption). As an alternative to the use of the subcooled-liquid vapor pressure (Pankow 1987), the octanol–air partition coefficient (Finizio et al. 1997; Pankow 1998; Harner and Bidleman 1998) is found to describe the observed partitioning of SOCs in the aerosol phase of the urban atmosphere (Liang et al. 1997; Harner and Bidleman 1998) by uptake into the organic fraction. The octanol–air partition coefficient is analogous to the Henry's Law constant describing the uptake of water-soluble gases into aqueous droplets and deliquescent particles.

Many of the pesticides in use are SOCs, as can be seen in Figure 2-2, which gives a histogram of the vapor pressures of 570 pesticides listed in a recent review (Tomlin 1995). In urban areas (and even more so in the presence of soil dust during application) more than 50% of the compounds are expected to occur mainly in the aerosol state, and about half of these particle-borne SOCs will vaporize to the gas phase if the aerosol density decreases during long-range transport to remote areas (vapor pressures between 10^{-4} and 1 Pa are most abundant in pesticides and the majority of the POPs have vapor pressures > 10^{-5} Pa).

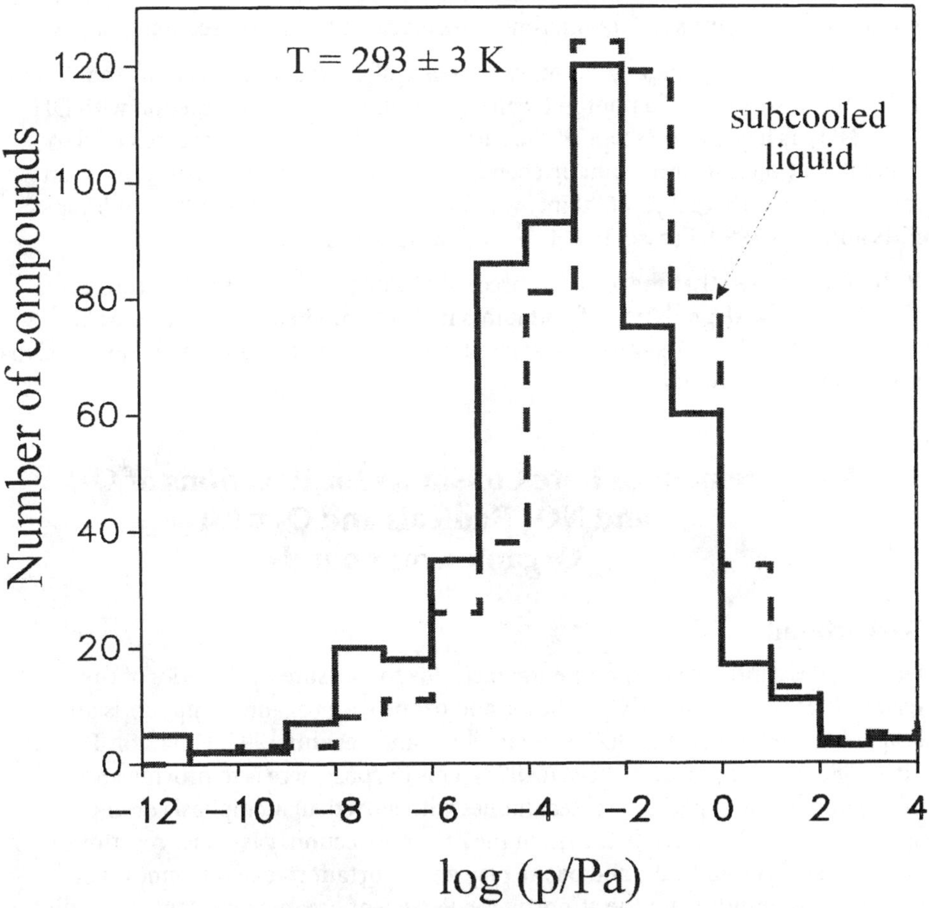

Figure 2-2 Histogram of vapor pressures (and subcooled-liquid vapor pressures, estimated by W. Palm, Fraunhofer Institute, Hanover, Germany) of pesticides currently in use (Tomlin 1995).

Hydrolysis appears to be unimportant in the vapor phase, even for compounds that are known to hydrolyze rapidly in aqueous media (Butler and Snelson 1979) and is therefore not discussed here. Hydrolysis in the condensed phase will be important only for a limited number of chemicals that have susceptible functional groups, such as esters, carbamates, epoxides, and organophosphates (Lyman et al. 1990) and carbonyl halides and acyl halides (Wallington et al. 1994). The pH and temperature control the rates of hydrolysis. A change in pH by 1 unit usually results in an order-of-magnitude change in half-life for acid- or base-catalyzed hydrolysis, and a 10 °C change in temperature results in a change in the rate constant by a factor of ~ 2.5.

The rates of the various loss processes discussed above are necessary input into the estimation of the persistence of a pollutant in the atmosphere. The "residence" lifetime corresponds to the mean time spent by the chemical in the atmosphere and is the reciprocal of the sum of the first-order rate constants for chemical reaction, photolysis, dry deposition, and wet deposition. Approaches for estimating residence lifetimes and half-lives (0.69 × lifetimes) are discussed in a later section.

A discussion of the generally complex sequence of degradation reactions that follow either the absorption of a photon by an organic molecule, or its reaction with OH, O_3 or NO_3, is beyond the scope of the current overview. The reader is referred to general textbooks on atmospheric chemistry and specialized monographs for an introduction to this subject (Atkinson 1994; Seinfeld and Pandis 1997; Finlayson-Pitts and Pitts 1999; Brasseur et al. 1999; Warneck 1988, 1999).

It should be noted that the term "indirect photolysis" is often used by ecotoxicologists to describe degradation of pollutants initiated by chemical reaction with species, such as OH, O_3, and NO_3, whose formation depends ultimately on reactions involving solar radiation.

Measurement of Rate Constants for Reactions of OH and NO_3 Radicals and O_3 with Organic Compounds

Background

Detailed descriptions of experimental methods to measure the kinetics of the reactions of OH radicals, NO_3 radicals, and ozone with organic compounds are available in the literature (OECD 1992; Pilling and Seakins 1995; Finlayson-Pitts and Pitts 1986; Wallington and Nielsen 1998). Our purpose here is to provide an overview of these experimental techniques with a particular emphasis on their applicability to POPs. As discussed in the previous section, gas-phase reactions with OH radicals, NO_3 radicals, and ozone play an important role in the removal of organic compounds from the atmosphere. Persistent organic pollutants typically react slowly with NO_3 radicals and ozone, and the gas-phase atmospheric chemistry of POPs often is dominated by reaction with OH radicals. Accordingly, we will focus on the measurement of rate constants of reactions of POPs with OH radicals. For completeness, the methods used to measure NO_3 radical and ozone kinetics are mentioned.

The rate constant of a chemical reaction is the proportionality constant relating its rate to the reactant concentrations. For example, consider the reaction of hydroxyl radicals with methane:

$$OH + CH_4 \rightarrow H_2O + CH_3 \quad \text{(Reaction 2-8).}$$

The rate of reaction (Reaction 2-8) is directly proportional to the concentration of OH radicals and CH_4. Hence, in a closed system

$$\frac{-d[CH_4]}{dt} = k_8 [OH] [CH_4] \quad \text{(Equation 2-1),}$$

where k_8 is the rate constant (or rate coefficient) for Reaction 2-8. Reaction 2-8 has been the subject of numerous experimental investigations using a wide variety of techniques and experimental conditions and it is well established that $k_8 = 2.45 \times 10^{-12} \exp(-1775/T)$ cm^3 $molecule^{-1}$ s^{-1} (NASA 1997). The atmospheric loss of methane is dominated by reaction with OH radicals. At the average tropospheric temperature of 275 K, $k_8 = 3.9 \times 10^{-15}$ cm^3 $molecule^{-1}$ s^{-1}, which combined with an estimate for the global average atmospheric OH concentration of 1×10^6 molecule cm^{-3} (Krol et al. 1998; Miller et al. 1998), yields an atmospheric half-life for methane of approximately 6 y.

All experimental systems used to measure rate constants for reactions of OH and NO_3 radicals with organic compounds have 2 elements. First, one needs to produce the reactive radical in question from some precursor molecule. Many methods are used to generate radicals: the most common are flash photolysis, discharge flow, pulse radiolysis, and steady-state photolysis. Second, one needs to monitor change in some property of the chemical system that is related directly or indirectly to the presence of the radicals. A wide variety of methods can be used to monitor the progress of reaction: the most common methods are UV/visible absorption, resonance fluorescence, laser-induced fluorescence, mass spectroscopy, Fourier transform infrared spectroscopy (FTIR), and gas chromatography. There are many possible combinations of the two elements. Flash photolysis resonance fluorescence, flash photolysis laser-induced fluorescence (FP-LIF), discharge flow mass spectroscopy, pulse radiolysis UV absorption spectroscopy, and steady-state photolysis relative rate are the most commonly used techniques.

It is useful to draw a distinction between absolute- and relative-rate techniques. In absolute techniques the rate constant in question is measured directly by following the change in concentration of one of the reactants as a function of time. For example the rate constant for Reaction 2-8 can be measured using the FP-LIF technique, wherein the loss of OH radicals in the presence of excess CH_4 is monitored as a function of time. In relative-rate studies the rate constant of interest is measured relative to a reference reaction whose rate constant has been established by absolute methods. In all absolute studies the reaction time, and concentration of organic, must be measured accurately; there are no such requirements for relative-rate studies. Absolute-rate methods are generally capable of greater accuracy than are relative methods, while relative-rate methods are generally capable of greater precision than are absolute methods. The potential for absolute methods to achieve greater accuracy is a consequence of the directness of the measurement. Relative-rate methods appear to be more useful for low-volatility compounds such as POPs.

Under ideal conditions, absolute rate methods are capable of measuring rate constants for reactions of OH radicals with organic compounds to an accuracy of approximately ± 10%, relative-rate methods can achieve an overall accuracy of ± 15% (with the majority of the uncertainty associated with the reference rate constant). Consistent data from at least two independent studies (preferably using different techniques) are needed before a rate constant can be considered to be well-established. Absolute- and relative-rate studies provide complementary pieces of information concerning the kinetics of radical reactions. When used in combination, absolute and relative rate methods can provide a wealth of information concerning the atmospheric persistence of organic compounds.

Persistent organic pollutants have vapor pressures of typically 10^{-2} to 10^{-8} Torr (1 to 10^{-6} Pa) at room temperature. Typical background decays of OH resulting from loss to reaction walls, diffusion and flow out of the viewing zone, and reaction with impurities in the diluent gases are in the range 1 to 20 s^{-1}. To provide accurate kinetic data, pseudo–first-order radical-loss rates in the presence of an organic of at least 3 s^{-1} are required (Zetzsch 1983). Large organic compounds such as POPs generally react with OH radicals with rate constants of 10^{-12} to 10^{-11} cm^3 $molecule^{-1}$ s^{-1}, and, therefore, concentrations on the order of 10^{12} to 10^{13} molecule cm^{-3} ($\approx 10^{-4}$ Torr, 10^{-2} Pa) are required. The requirement for a substantial gas-phase concentration of organic reactant severely limits the ability of absolute rate techniques to provide kinetic data for semi-volatile organic compounds at ambient temperature.

The 2 most important experimental techniques used to study the kinetics of the reactions of OH radicals with organic compounds are flash photolysis and relative rate methods. We will discuss each of these methods.

Flash photolysis studies

In flash photolysis experiments a reactant gas mixture, containing at least one photolabile component, is subjected to a short (10^{-9} to 10^{-6} s) duration high intensity flash of UV/visible light. Photolysis of the photolabile component creates radicals that are then monitored as a function of time. Hydroxyl radicals can be generated from several precursors:

$H_2O + h\nu\ (\lambda < 195\ nm) \rightarrow OH + H$ (Reaction 2-9),

$H_2O_2 + h\nu\ (\lambda = 248\ nm) \rightarrow 2\ OH$ (Reaction 2-10),

$HNO_3 + h\nu \rightarrow OH + NO_2$ (Reaction 2-11),

$N_2O + h\nu\ (\lambda = 193\ nm) \rightarrow O(^1D) + N_2$ (Reaction 2-12),

$O_3 + h\nu\ (\lambda < 320\ nm) \rightarrow O(^1D) + O_2$ (Reaction 2-1),

$O(^1D) + H_2O \rightarrow 2\ OH$ (Reaction 2-2).

The reactant gases are thoroughly mixed prior to photolysis, and, if the photolysis light is uniform, a homogeneous concentration of radicals is formed. Data are acquired immediately after the photolysis pulse. The simplicity of the FP technique imposes few limitations on the operating conditions of temperature and pressure, which are typically 250 to 500 K and 25 to 1000 Torr. Detection systems need to have a rapid response because of the real-time nature of the FP technique: common detection methods include resonance fluorescence, laser-induced fluorescence, and ultraviolet absorption.

A flash photolysis continuous-wave, laser-induced fluorescence system is described in detail by Tully et al. (1986). A conventional gas handling system is used to prepare He/N_2O/H_2O/reactant gas mixtures, which are flowed slowly through a heated reactor mounted inside a vacuum housing. An excimer laser photolysis source perpendicular to one face of the reactor is used to photolyze N_2O giving $O(^1D)$ atoms that react with H_2O to give OH radicals (Reactions 2-12 and 2-2). An intracavity-doubled, single-mode, continuous-wave ring dye laser excites OH fluorescence, which is detected by using a photomultiplier mounted at right angles to both laser beams. The decay of OH radicals is monitored in the presence of varying amounts of the organic:

$OH + organic \rightarrow products$ (Reaction 2-13).

Conditions are chosen such that the organic is in great excess ([organic] >> [OH]) and loss of OH radicals follows pseudo–first-order kinetics. Plots of ln $\{[OH]_{to}/[OH]_t\}$ versus time give pseudo–first-order rate constants (k^{1st}) for OH loss. Plots of k^{1st} versus [organic] then give a slope equal to the bimolecular rate constant k_{13}. Experiments can be performed as a function of temperature and total pressure to map out the kinetic behavior of the reaction and to gain insight into its mechanism.

Flash photolysis is a powerful and versatile method used to study the kinetics and mechanisms of reactions of OH radicals with organic compounds. However, as noted above, the low vapor pressures of POPs place a significant constraint on the usefulness of this method to study the reactivities of POPs with OH radicals.

Relative-rate method

Environmental chamber systems often are used to conduct relative-rate studies of the kinetics of radical reactions. Such systems consist of large (0.1 to 10 m^3) Teflon (Atkinson et al. 1995), Pyrex (Niki et al. 1978; Behnke et al. 1988; Wallington and Japar 1989), quartz (Barnes et al. 1994), or stainless steel (Shetter et al. 1987) reaction chambers. Analysis of reactants and products typically is accomplished by in situ FTIR (Atkinson et al. 1995; Niki et al. 1978; Wallington and Japar 1989), differential optical absorption (Barnes et al. 1994) spectroscopic analysis systems, or conventional chemical-sampling methods that enable chemical analysis of gas mixtures using conventional means (gas chromatography, FTIR, mass spectrometry, etc.). The technique often used in such studies is to irradiate $CH_3ONO/NO/O_2/$ reactant/reference/air mixtures with the output from UV fluorescent "blacklamps." The reactions occurring in such systems are

$$CH_3ONO + hn \rightarrow CH_3O + NO \quad \text{(Reaction 2-14),}$$

$$CH_3O + O_2 \rightarrow HO_2 + HCHO \quad \text{(Reaction 2-15),}$$

$$HO_2 + NO \rightarrow OH + NO_2 \quad \text{(Reaction 2-5),}$$

$$OH + \text{reactant} \rightarrow \text{products} \quad \text{(Reaction 2-16),}$$

$$OH + \text{reference} \rightarrow \text{products} \quad \text{(Reaction 2-17).}$$

Experimental conditions are chosen such that the reactant and reference species are lost only via reaction with OH radicals and that neither reactant nor reference is reformed in any process. In these experiments the OH radicals are in steady state and the loss of reactant and reference are given by

$$\frac{-d[\text{reactant}]}{dt} = k_{16}[OH][\text{reactant}] \qquad \frac{-d[\text{reference}]}{dt} = k_{17}[OH][\text{reference}] \quad \text{(Equation 2-2).}$$

Integration gives

$$\ln\left(\frac{[\text{reactant}]_{to}}{[\text{reactant}]_{t}}\right) = k_{16}\int_{to}^{t}[OH]dt \qquad \ln\left(\frac{[\text{reference}]_{to}}{[\text{reference}]_{t}}\right) = k_{17}\int_{to}^{t}[OH]dt \quad \text{(Equation 2-3),}$$

where $[\text{reactant}]_{to}$, $[\text{reactant}]_t$, $[\text{reference}]_{to}$, $[\text{reference}]_t$, are the concentrations of reactant and reference at times "t_o" and "t," k_{16} and k_{17} are the rate constants for Reactions 2-16 and 2-17, and [OH] is the OH radical concentration. The reactant and reference have equal exposure to OH radicals, hence

$$\ln\left(\frac{[\text{reactant}]_{to}}{[\text{reactant}]_{t}}\right) = \frac{k_{16}}{k_{17}}\ln\left(\frac{[\text{reference}]_{to}}{[\text{reference}]_{t}}\right) \quad \text{(Equation 2-4).}$$

Plots of Equation 2-4 give the rate constant ratio k_{16}/k_{17}, which can be combined with the literature value for k_{17} to give k_{16}. Unlike absolute-rate techniques, there is no requirement in relative-rate studies for high reactant concentrations. Two research groups have gathered the majority of kinetic data for reactions of OH radicals with semi-volatile compounds using relative-rate methods. At the University of California, Riverside, experiments have been performed at room temperature in 740 Torr of air diluent using large (6400- to 8000-L) Teflon chambers with CH_3ONO photolysis as the OH source (Atkinson et al. 1984). At the University of Indiana, experiments have been performed at elevated temperatures (typically, 340 to 370 K) using a small (0.2-L) quartz reaction vessel with ozone photolysis used to produce $O(^1D)$ atoms, which then react with H_2O to give OH radicals (Anderson and Hites 1996a). A value of k_{OH} (298 K) is then calculated by extrapolation of the Arrhenius expression derived from the elevated temperature data. When using this approach, researchers should take care to verify that the mechanism of reaction does not change between room temperature and the temperature at which the measurements were made.

A new technique to study the reactions of OH radicals with low volatility organic compounds is under development (Palm et al. 1997). In this method, the rate of loss of organic compounds bound to aerosol particles (as a monolayer or less) is monitored in a large smog chamber in which OH radicals are produced via the photolysis of CH_3ONO (or another suitable precursor). The concentration of OH radicals in the system is calculated by monitoring the rate of loss of a gas-phase reference compound. The rate of reaction of the adsorbed organic is then derived from its observed loss rate and the calculated average OH radical concentration.

Very similar relative-rate techniques have proven useful in providing kinetic data for the reaction of NO_3 radicals with organic compounds (Atkinson 1991a). The most widely used source of NO_3 radicals in such experiments is N_2O_5, which exists in thermal equilibrium with NO_3 radicals and NO_2:

$$N_2O_5 + M \longleftrightarrow NO_3 + NO_2 + M \qquad \text{(Reaction 2-18).}$$

Unlike OH and NO_3 radicals, ozone is a closed-shell, stable molecule that can be routinely prepared and stored in large quantities. The kinetics of the reaction of ozone with POPs can be measured using 2 techniques. The simplest method is to prepare a mixture containing O_3, the POP of interest, and an OH radical scavenger, typically cyclohexane. Under conditions in which the concentration of ozone greatly exceeds that of the POP, the rate of POP loss follows pseudo–first-order kinetics. Pseudo–first-order rate constants for POP loss, k^{1st}, when plotted versus the ozone concentration yield linear plots whose slope is the bimolecular rate constant k_{19}

$$O_3 + POP \longrightarrow \text{products} \qquad \text{(Reaction 2-19).}$$

Relative-rate methods, such as those discussed above for the OH radical, can also be used to measure k_{19} (Greene and Atkinson 1992). In such experiments a reference compound is added and the relative rates of decay of the POP and the reference are used to derive k_{19}.

Kinetic data for persistent organic pollutants

Measured rate constants for the reactions of OH, NO_3, and ozone with compounds classified as POPs in the United Nations Economic Commission for Europe (UNECE) protocol are given in Table 2-1. The available kinetic database on which estimates of the atmospheric persistence of POPs are based is sparse. For example, kinetic data are available for only 6 of the 16 compounds, or classes of compounds, defined as POPs in the UNECE protocol. There are no data for aldrin, chlordane, chlordecone, dichlorodiphenyltrichloroethane (DDT), dieldrin, endrin, heptachlor, hexabromobiphenyl, mirex, or toxaphene. While the database concerning reactions of NO_3 radicals with candidate POPs is limited, these reactions are expected to be slow in most cases (aldrin is a possible exception). Reaction with NO_3 radicals and/ or O_3 is unlikely to be an important atmospheric removal mechanism for most candidate POPs.

Estimation of Rate Constants for Reactions of OH and NO_3 Radicals and O_3 with Organic Compounds

Estimation methods for OH radical reaction rate constants

A number of methods for the estimation of gas-phase OH radical reaction-rate constants for organic compounds have been proposed, ranging from estimation methods for single classes of organic compounds to generalized estimation methods (Gaffney and Levine 1979; Heicklen 1981; Zetzsch 1982; Güsten et al. 1984; Jolly et al. 1985; Atkinson 1986; Güsten and Klasinc 1986; Hodson 1988; Cooper et al. 1990; Klamt 1993, 1996; Medven et al. 1996; Sekušak et al. 1996; Chiorboli et al. 1996). Many of these estimation methods utilize molecular properties such as ionization energy (Gaffney and Levine 1979; Güsten et al. 1984; Güsten and Klasinc 1986), nuclear magnetic resonance (NMR) chemical shifts (Hodson 1988), molecular orbital energies (Cooper et al. 1990; Klamt 1993, 1996; Medven et al. 1996; Sekušak et al. 1996), bond-dissociation energies (Heicklen 1981) and infrared absorption frequencies (Jolly et al. 1985). Most of these methods are restricted in their use because of the limited database concerning the relevant molecular properties. An often-used estimation method is that developed and tested by Atkinson (Kwok and Atkinson 1995; Atkinson 2000), which uses structure-reactivity relationships and is the basis of the Syracuse Research Corporation's "Atmospheric Oxidation Program" (see Meylan and Howard [1993] for a discussion of an earlier version of the Atmospheric Oxidation Program).

The general approach of this estimation method has been described in detail previously (Kwok and Atkinson 1995; Atkinson 2000) and is based on the observations that gas-phase OH radical reactions with organic compounds proceed by four reaction pathways that are assumed to be additive: H-atom abstraction from C–H and O–H bonds; OH radical addition to >C=C< and -C≡C- bonds; OH radical

Table 2-1 Second-order room temperature rate constants (units of cm^3 $molecule^{-1}$ s^{-1}) for bimolecular reactions of OH and NO_3 radicals and ozone with compounds defined as POPs in the UNECE protocol

Organic reactant	k_{OH}	k_{NO3}	k_{O3}	Reference
2-chloro-biphenyl	2.9×10^{-12}			Atkinson and Aschmann 1985
	$2.7 \times 10^{-12*}$			Anderson and Hites 1996b
3-chloro-biphenyl	5.4×10^{-12}			Atkinson and Aschmann 1985
	$5.0 \times 10^{-12*}$			Anderson and Hites 1996b
4-chloro-biphenyl	3.9×10^{-12}			Atkinson and Aschmann 1985
	$3.4 \times 10^{-12*}$			Anderson and Hites 1996b
2,2'-dichloro-biphenyl	2.0×10^{-12}	$< 8 \times 10^{-16}$	$< 2 \times 10^{-20}$	Kwok et al. 1995
	$2.2 \times 10^{-12*}$			Anderson and Hites 1996b
3,3'-dichloro-biphenyl	4.1×10^{-12}			Kwok et al. 1995
3,5'-dichloro-biphenyl	4.2×10^{-12}			Kwok et al. 1995
4,4'-dichloro-biphenyl	$2.0 \times 10^{-12*}$			Anderson and Hites 1996b
2,4-dichloro-biphenyl	$2.6 \times 10^{-12*}$			Anderson and Hites 1996b
2,4,4'-trichloro-biphenyl	$1.1 \times 10^{-12*}$			Anderson and Hites 1996b
2,4,5-trichloro-biphenyl	$1.3 \times 10^{-12*}$			Anderson and Hites 1996b
2,4',5-trichloro-biphenyl	$1.2 \times 10^{-12*}$			Anderson and Hites 1996b
2',3,4-trichloro-biphenyl	$1.0 \times 10^{-12*}$			Anderson and Hites 1996b
2,2',3,5'-tetrachloro-biphenyl	$0.8 \times 10^{-12*}$			Anderson and Hites 1996b
2,2',4,4'-tetrachloro-biphenyl	$1.0 \times 10^{-12*}$			Anderson and Hites 1996b
2,2',3,5',6-pentachloro-biphenyl	$0.4 \times 10^{-12*}$			Anderson and Hites 1996b
2,3,3',4',6-pentachloro-biphenyl	$0.6 \times 10^{-12*}$			Anderson and Hites 1996b
2,3,4,5,6-pentachloro-biphenyl	$0.9 \times 10^{-12*}$			Anderson and Hites 1996b
Dibenzo-*p*-dioxin	1.5×10^{-11}	$< 8 \times 10^{-15}$	$<5 \times 10^{-20}$	Kwok et al. 1994
				Atkinson 1996
	$1.2 \times 10^{-11*}$			Brubaker and Hites 1998a
1-chlorodibenzo-*p*-dioxin	4.7×10^{-12}	$< 1.6 \times 10^{-14}$	$< 7 \times 10^{-20}$	Kwok et al. 1995
2,7-dichlorodibenzo-*p*-dioxin	$4.4 \times 10^{-12*}$			Brubaker and Hites 1998a
1,2,3,4-tetra-chlorodibenzo-*p*-dioxin	$8.5 \times 10^{-13*}$			Brubaker and Hites 1997
Dibenzofuran	3.9×10^{-12}	$< 1.6 \times 10^{-15}$	$< 8 \times 10^{-20}$	Kwok et al. 1994
				Atkinson 1996
	$3.5 \times 10^{-12*}$			Brubaker and Hites 1998a
2,8-dichlorodibenzofuran	$2.2 \times 10^{-12*}$			Brubaker and Hites 1998a
α-hexachlorocyclohexane	$1.4 \times 10^{-13*}$			Brubaker and Hites 1998b
γ-hexachlorocyclohexane (lindane)	$6.0 \times 10^{-13**}$			Zetzsch 1991
	$1.9 \times 10^{-13*}$			Brubaker and Hites 1998b
Hexachlorobenzene	$2.7 \times 10^{-14*}$			Brubaker and Hites 1998b

* extrapolated result based upon data measured at higher temperatures (typically 340–380 K)
** adsorbed on silica dust

addition to aromatic rings; and OH radical "interaction" with N-, S-, and P-atoms and with more complex structural units such as ->P=S, >NC(O)S- and >NC(O)O- groups. The total rate constant is assumed to be the sum of the rate constants for these four reaction pathways.

H-atom abstraction

Rate constants for H-atom abstraction from C–H and O–H bonds are based on the estimation of group rate constants for H-atom abstraction from $-CH_3$, $-CH_2$-, >CH– and –OH groups. The rate constants for H-atom abstraction from the C–H bonds of $-CH_3$, $-CH_2$- and >CH– groups depend on the identity of the substituents attached to these groups. k_{prim}, k_{sec} and k_{tert} are the group rate constants for H-atom abstraction from $-CH_3$, $-CH_2-$ and >CH– groups, respectively, for a standard substituent, and F(X), F(Y), and F(Z) are substituent factors for the substituent groups X, Y, and Z. The standard substituent group is chosen to be X (= Y = Z) = $-CH_3$, with $F(-CH_3)$ = 1.00. For example, the rate constant for H-atom abstraction from the $-CH_2-$ group in $X-CH_2-Y$ is

$$k(X-CH_2-Y) = k_{sec}\ F(X)\ F(Y) \quad \text{(Equation 2-5).}$$

It is assumed that the temperature-dependence of the substituent factors F(X) can be expressed in the form $F(X) = e^{Ex/T}$ and hence that the pre-exponential factors A (in the Arrhenius expression $k = A\ e^{-B/T}$ or in the alternative three-parameter expression $k = A'\ T^n\ e^{-D/T}$) for H-atom abstraction from C–H bonds are independent of the substituent groups. Atkinson (2000) gives the most recent tabulation of group rate constants, k_{prim}, k_{sec}, k_{tert}, and k_{OH}, substituent group factors, F(X), and ring-strain factors, F_{ring}.

Persistent organic pollutants for which the OH radical reactions are expected to proceed, at least in part, by H-atom abstraction from C–H bonds include aldrin, chlordane, DDT, dieldrin, endrin, heptachlor, toxaphene, hexachlorocyclohexanes, and the polycyclic aromatic hydrocarbons (PAHs) fluorene and acenaphthene.

OH radical addition to >C=C< and –C≡C– bonds

Rate constants for OH radical addition to carbon–carbon unsaturated bonds depends on the number, identity, and position of substituent groups around the >C=C< or –C≡C– bonds. Conjugated double-bond systems are dealt with by considering the entire conjugated >C=C-C=C< system as a single unit (Atkinson 1986) rather than as conjugated >C=C< subunits as done by Ohta (1983). The 298 K rate constants for addition to the CH_2=CH–, CH_2=C<, *cis*- and *trans*- –CH=CH–, –CH=C<, >C=C< non-conjugated carbon–carbon double-bond units, and the various –C≡C– and conjugated >C=C–C=C< units are given by Kwok and Atkinson (1995) and Atkinson (2000). There is no evidence for ring-strain effects on OH radical addition to cycloalkenes. For organic compounds containing >C=C< bonds with non-alkyl substituent groups, X, group substituent factors C(X) are used

[C(alkyl) = 1.00], and these group substituent factors are tabulated by Kwok and Atkinson (1995) and Atkinson (2000).

Persistent organic pollutants for which the OH radical reactions are expected to proceed, at least in part, by initial OH radical addition to >C=C< bonds include aldrin, chlordane, dieldrin, endrin, heptachlor, toxaphene, and certain PAHs such as acenaphthylene.

OH radical addition to aromatic rings

Rate constants for OH radical addition to aromatic rings are calculated by using the correlation between the OH radical addition rate constant and the sum of the electrophilic substituent constants $\Sigma\sigma^+$ (Zetzsch 1982). It is assumed that the OH radical adds to the position yielding the most negative value of $\Sigma\sigma^+$, that steric hindrance can be neglected, and that if all positions on the ring are occupied the *ipso* position is treated as a *meta* position (Zetzsch 1982). The correlation between the OH radical addition rate constants, k_{add}, and $\Sigma\sigma^+$ for monocyclic aromatic compounds and biphenyl is (Kwok and Atkinson 1995)

$$\log_{10} k_{add}\ (\text{cm}^3\ \text{molecule}^{-1}\ \text{s}^{-1}) = -11.71 - 1.34\ \Sigma\sigma^+ \qquad \text{(Equation 2-6).}$$

Electrophilic substituent factors are available in Brown and Okamoto (1958) and Atkinson (1996), the latter having estimated values of the $-C_6H_{5-x}Cl_x$ and $-OC_6H_{5-x}Cl_x$ groups needed to calculate the OH radical addition rate constants for PCBs, polychlorodibenzo-*p*-dioxins (PCDDs), and polychlorodibenzofurans (PCDFs). Rate constants for reactions of the OH radical with PAHs are quite uncertain for several PAHs (Brubaker and Hites 1998a), and at this time no reliable estimation of rate constants for PAHs can be carried out.

Persistent organic pollutants for which the OH radical reactions are known or expected to proceed, at least in part, by initial addition of the OH radical to aromatic rings include DDT, PCDDs, PCDFs, hexachlorobenzene, PAHs, PCBs, and hexabromobiphenyl.

OH radical interaction with N-, S-, and P-containing groups

The OH radical reactions with a number of N-, S-, and P-containing organic compounds appear to proceed, at least in part, by an initial addition reaction (Atkinson 1989), although the products observed may in some cases be those expected from H-atom abstraction. Rate constants are available for a number of N-, S-, and P-containing compounds (Atkinson 1989, 1994; Kwok et al. 1996; Koch et al. 1996, 1997). Group rate constants and substituent group factors are given by Kwok and Atkinson (1995) and Atkinson (2000). These group rate constants and substituent group factors should be used only for homologs of the compounds from which these factors were derived (Kwok and Atkinson 1995; Koch et al. 1996).

Based on the various parameters involved in the calculation of the OH radical reaction rate constants (Kwok and Atkinson 1995; Atkinson 2000), organic compounds that are predicted to be of low reactivity towards the OH radical are those containing F, Cl, or Br atoms and/or CF_3, CCl_3, CN, NO_2, and ONO_2 groups (i.e., haloalkanes, haloalkenes, halogen-substituted aromatic compounds, aliphatic nitrocompounds, organic nitrates, nitriles, and nitroaromatics). Many POPs contain chlorine, which is consistent with the expectation that the presence of halogens atoms or halogen-containing groups leads to low reactivity towards the OH radical.

Examples of calculations of OH radical reaction-rate constants

As examples of the calculation methods described briefly above, rate constants for the gas-phase reactions of the OH radical with aldrin and with 2',3,4-trichlorobiphenyl are given here.

Aldrin

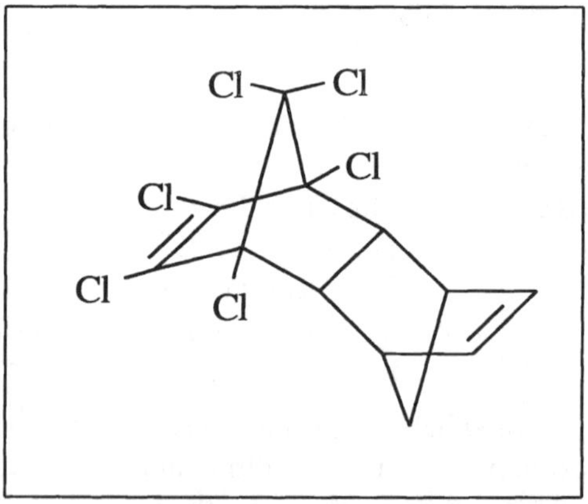

Aldrin contains a >C=C< bond with two Cl and two >CCl substituents; a *cis* -CH=CH– bond with two >CH substituents; two >CH– groups each with one >CCl and two >CH– substituent groups; two >CH– groups each with one >CH–, one >C=C<, and one $-CH_2-$ substituent groups; and a $-CH_2-$ group with two >CH– substituents. The calculated total rate constant for aldrin is given by

$$k_{total} = k(>C=C<)C(Cl)C(Cl)C(CCl)C(CCl) + k(cis\text{-}CH=CH\text{-})C(alkyl)C(alkyl) + 2k_{tert}F(>CCl)F(>CH\text{-})F(>CH\text{-})F_5F_5 + 2k_{tert}F(>CH\text{-})F(>C=C<)F(\text{-}CH_2\text{-})F_5F_5 + k_{sec}F(>CH\text{-})F(>CH\text{-})F_5F_5 \quad \text{(Equation 2-7)},$$

where F_5 is the ring-strain factor for a 5-membered ring. Using the group rate constants and substituent factors given by Kwok and Atkinson (1995) and Atkinson (2000), and assuming that C(CCl) = $C(CH_2Cl)$, results in a calculated rate constant for aldrin at room temperature of 6.3×10^{-11} cm^3 $molecule^{-1}$ s^{-1}, with the majority of the reaction predicted to occur via OH radical addition to the *cis* –CH=CH– bond.

No experimental data are available for aldrin or for a structurally similar Cl-containing compound.

2',3,4-Trichlorobiphenyl

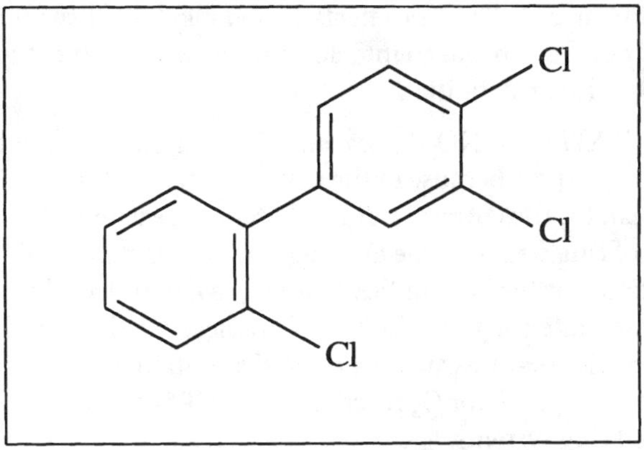

For the purpose of estimation of K_{OH}, 2',3,4-trichlorobiphenyl is viewed as consisting of 2 benzene rings, the first with Cl substituents at the 3 and 4 positions and a monochlorophenyl substituent at the 1 position, and the second with a dichlorophenyl substituent at the 1 position and a Cl substituent at the 2 position. Rate constants for OH radical addition to the two aromatic rings are calculated separately. For the ring with Cl substituents at the 3 and 4 positions, the most negative value of $\Sigma\sigma^+$ occurs with OH addition at the 2 position, and for the ring with 1 Cl substituent at the 2 position, the most negative value of $\Sigma\sigma^+$ occurs with OH radical addition at the 3 position. Using the electrophilic substituent constants listed by Brown and Okamoto (1958) and Atkinson (1996) leads to a calculated total rate constant of 9.9×10^{-13} cm^3 $molecule^{-1}$ s^{-1}, in agreement with the extrapolated experimental value of 1.0×10^{-12} cm^3 $molecule^{-1}$ s^{-1} (Anderson and Hites 1996b).

Estimation methods for NO_3 radical and O_3 reaction-rate constants

Analogous to the corresponding OH radical reactions, the room-temperature rate constants for the reactions of the NO_3 radical and O_3 with alkenes (acyclic alkenes in the case of the O_3 reactions, see below) depend on the number and position of the alkyl substituent groups around the >C=C< bond, the room-temperature rate constants generally increasing with the number of alkyl substituents. Group rate constants CH_2=CH-, CH_2=C<, *cis*- and *trans*- -CH=CH-, -CH=C<, and >C=C< can be derived from the rate constants for the methyl-substituted ethenes (Atkinson 2000). Analogous to the corresponding OH radical reactions, no effect of ring strain on the NO_3 radical reaction-rate constants for cycloalkenes is observed (Atkinson 1991a) and the group rate constants given by Atkinson (2000) can be used to estimate NO_3

radical reaction rate constants for non-6–membered ring cycloalkenes and non-conjugated cyclodienes. In contrast, there is a marked effect of ring strain on the O_3 reaction-rate constants for cycloalkenes (Atkinson and Carter 1984), and the group rate constants given by Atkinson (2000) for the O_3 reactions can be used only for acyclic alkenes and for cycloalkenes with 6-membered, essentially non-strained rings (Atkinson and Carter 1984). The effects of non-alkyl substituent groups around the >C=C< bond can be taken into account using substituent group factors C(X), with C(-alkyl) = 1.0 by definition.

However, values of C(X) for the NO_3 radical and O_3 reactions have not been tabulated in the literature, in part because of the smaller database for the NO_3 radical and O_3 reactions than for the corresponding OH radical reactions. Combined with the marked effects of ring strain on the O_3 reaction rate constants, calculated reaction-rate constants for the NO_3 radicals and O_3 reactions have higher uncertainties than do calculated rate constants for the OH radical reactions, and order-of-magnitude discrepancies may be expected for POPs containing non-alkyl substituted >C=C< bonds and, for O_3 reactions, for POPs containing >C=C< bonds in non-6–membered ring systems.

Persistent organic pollutants that are known, or expected, to react with NO_3 radicals and O_3 include aldrin, chlordane, dieldrin, endrin, heptachlor, toxaphene, and certain PAHs such as acenaphthylene. As for the OH radical reactions, the presence of halogen atoms or halogen-containing groups attached to the >C=C< double bond appears to decrease the reactivities of compounds containing >C=C< bonds towards NO_3 radicals and O_3 (Atkinson and Carter 1984; Atkinson 1991a).

Reliability of estimation methods for OH radical reaction-rate constants

As noted above, the majority of the estimation methods proposed to date have limitations because of the unavailability of specific properties of the molecule being considered. Furthermore, apart from certain of the molecular orbital calculations (Sekušak et al. 1996), the estimation methods proposed to date utilize the measured rate-constant database during their development. For example, the method proposed by Atkinson (1986) and described above requires a reliable rate-constant database for the various classes of organic compounds to derive the various group rate constants and substituent factors. As may be expected, this estimation technique is reasonably reliable when used within the database used in its derivation, and Kwok and Atkinson (1995) observed that the 298 K rate constants were predicted to within a factor of 2 of the experimental values for ~ 90% of the ~ 485 organic compounds that were considered and for which experimental data were available. Similarly, Version 1.8 of the Atmospheric Oxidation Program gives agreement between the calculated and experimental room-temperature rate constants to within a factor of 2 for 90% of 647 organic compounds, with a standard deviation of 0.242 log units (i.e., a factor of 1.75) (Meylan 1997).

Further and more relevant comparisons of the level of agreement between experimentally measured and estimated OH radical reaction-rate constants concern PCBs, PCDDs, PCDFs, hexachlorobenzene, and hexachlorocyclohexanes (Atkinson 1996; Anderson and Hites 1996a, 1996b; Brubaker and Hites 1997, 1998a, 1998b). Figure 2-3 compares the experimental versus calculated 298 K rate constants for the reactions of the OH radical with PCBs (Zetzsch 1982; Atkinson et al. 1984; Atkinson and Aschmann 1985; Klöpffer et al. 1986; Kwok et al. 1995; Anderson and Hites 1996a, 1996b), PCDDs (Kwok et al. 1994, 1995; Brubaker and Hites 1997, 1998a), PCDFs (Kwok et al. 1994; Brubaker and Hites 1998a) and hexachlorobenzene (Brubaker and Hites 1998b). The solid line denotes a 1:1 agreement, and it is evident that the agreement between experimental and calculated rate constants for PCBs, PCDDs, PCDFs and hexachlorobenzene is good, within a factor of 2 for all but one of these 25 POPs (i.e., 2,3,4,5,6-pentachlorobiphenyl, for which the experimental uncertainty is also a factor of 2).

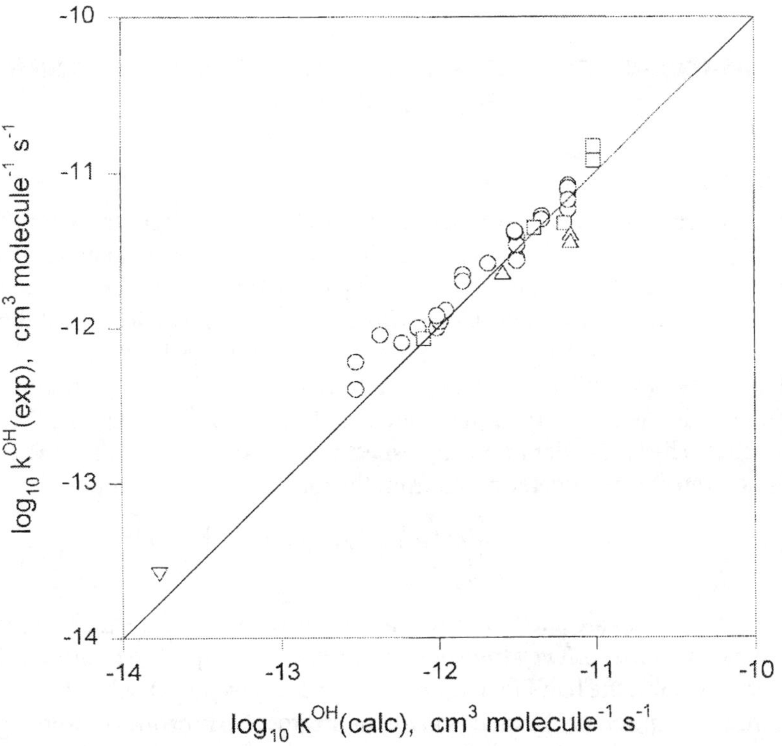

Figure 2-3 Plot of measured, k^{OH} (exp), and estimated, k^{OH}(calc), room-temperature rate constants for the gas-phase reactions of the OH radical with PCBs, including biphenyl (○), PCDDs, including dibenzo-*p*-dioxin (□), PCDFs, including dibenzofuran (Δ), and hexachlorobenzene (∇) The line shows a 1:1 agreement between measured and calculated rate constants.

However, this agreement is misleading because the experimental data of Atkinson and Aschmann (1985) and Kwok et al. (1994, 1995) for the monochlorobiphenyls dibenzo-*p*-dioxin, 1-chlorodibenzo-*p*-dioxin, and dibenzofuran were used to derive the electrophilic substituent constants necessary for the calculations of the rate constants for PCBs, PCDDs, and PCDFs (Atkinson 1996). In fact, an earlier version of the estimation method gave predicted rate constants for dibenzo-*p*-dioxin and dibenzofuran that were too high by factors of 2.5 and 8, respectively (Atkinson 1991b). For α- and γ-hexachlorocyclohexane, the calculated 298 K rate constants are identical for the two stereoisomers because the estimation method does not distinguish between stereoisomers, and the calculated values are higher than the extrapolated rate constants of Brubaker and Hites (1998b) by factors of 4 and 3, respectively.

As implied by the above discussion, extrapolation of this estimation method to compound classes not present in the database used for its development and testing results in a lack of assurance of its reliability and is not recommended.

Measurements of Ultraviolet Absorption Cross-sections and Photolysis Quantum Yields

Background

Direct photolysis is a mechanism that can contribute to the degradation of POPs in the atmosphere. The rate of atmospheric photolysis of a particular chemical species will be governed by 3 wavelength-dependent parameters: the absorption cross-section ($\sigma(\lambda)$, units of cm^2 $molecule^{-1}$), which quantifies the molecule's ability to absorb radiation at wavelength λ; the photolysis quantum yield ($\phi(\lambda)$, unitless), which gives the probability of chemical reaction occurring following photon absorption; and the actinic flux, ($I(\lambda)$, units of photons cm^{-2} s^{-1}) (ECETOC 1983; OECD 1992). The first-order rate with respect to photolysis (j, in units of s^{-1}) for the species of interest is then determined from the following relationship:

$$j = \int_{\lambda > 290nm} \sigma(\lambda)\, \phi(\lambda)\, I(\lambda)\, d\lambda \qquad \text{(Equation 2-8)},$$

where the integral is calculated over the range of wavelengths for which all three parameters are nonzero. The atmospheric lifetime with respect to photolysis is then the inverse of j, and the half-life in units of seconds is given by 0.693/j. Because of the intense absorption of UV radiation by the stratospheric ozone column, light reaching the troposphere is restricted to wavelengths greater than 290 nm, as shown in Figure 2-4. Solar-flux data as a function of wavelength for different latitudes and seasons are tabulated by OECD (1997).

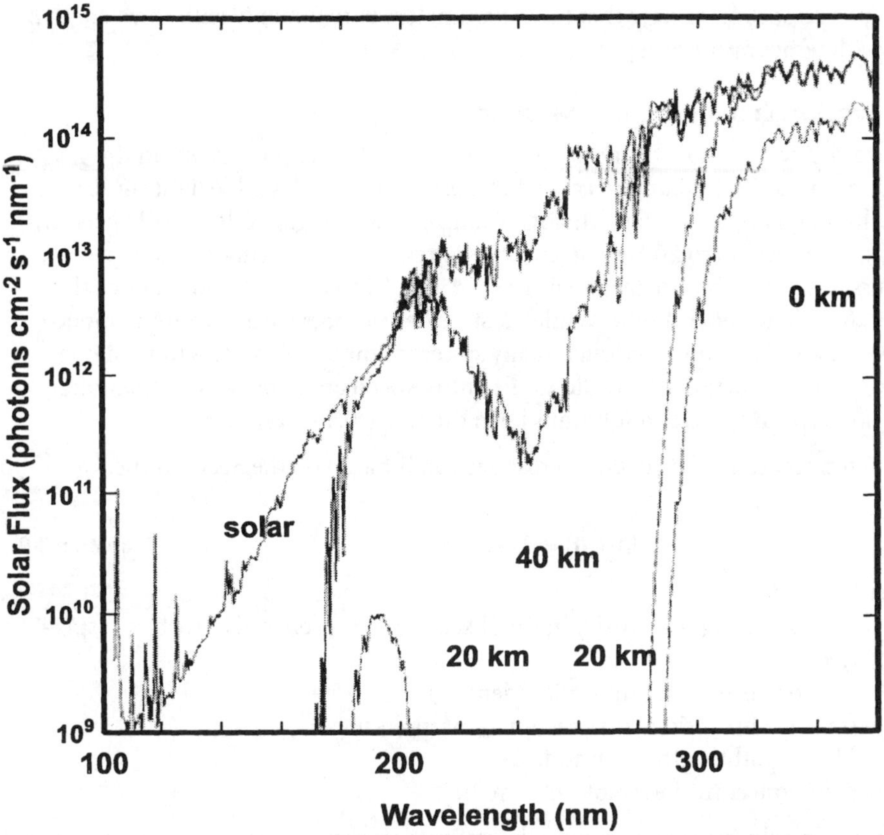

Figure 2-4 Solar flux (photons cm^{-2} s^{-1} nm^{-1}) as a function of wavelength at different altitudes in the atmosphere. Note the sharp onset of the surface flux near 290 nm. Figure provided by Dr. Ken Minschwaner (New Mexico Tech).

It is clear from the above discussion that in order for photolysis to be an important mechanism for the gas-phase destruction of a candidate POP, it must absorb to a significant degree beyond 290 nm. While absorption cross-sections and photolysis quantum yields cannot be reliably predicted at present, a first-order screening for the potential importance of photolysis can be done from a consideration of the fact that in order for the molecule to absorb to a significant degree beyond 290 nm, it must contain an appropriate chromophore, generally an unsaturated linkage. Thus, while such species as carbonyls and highly conjugated aromatic species, such as biphenyls, PAHs, and PCDD/Fs are all likely to absorb significantly in the region beyond 290 nm, photolysis can be ruled out as an important loss of saturated species such as the chloroalkanes. Once it is determined that a particular candidate POP may absorb beyond 290 nm (from a consideration of the molecular structure), absorption cross-sections and quantum yields need to be measured in the laboratory. It is important to note that these parameters vary not only with wavelength but also with temperature and pressure.

Absorption cross-sections: Methods

Laboratory systems for the measurement of absorption cross-sections in the gas phase are all quite similar and consist of a broad-band UV/visible light source (usually a D_2 lamp or a quartz/halogen lamp), an absorption cell (usually glass with quartz windows, 10 to 200 cm in length, and temperature-regulated), a light dispersal system, and a detector (ECETOC 1983; OECD 1992; Burkholder et al. 1993; Staffelbach et al. 1995). While most earlier systems used a scanning monochromator and a photomultiplier, many systems now use a spectrograph/diode array detector scheme, which allows the entire spectrum to be recorded simultaneously. Typical spectral resolutions lie in the range 0.1 to 1 nm.

The determination of absorption cross-section is based on the Beer-Lambert relation:

$$\ln(I_o/I) = A(\lambda) = \sigma \ell c \qquad \text{(Equation 2-9)},$$

where

I_o and I are the light intensity in the absence and presence of absorber, respectively;
A is the absorbance (or optical density);
σ is the absorption cross-section (cm^2 $molecule^{-1}$);
ℓ is the pathlength (cm); and
c the concentration (molecule cm^{-3}).

Note that in solution phase studies, the absorption cross-section, σ, is often replaced by the molar extinction coefficient (ϵ, units of M^{-1} cm^{-1}) defined by $(1/\ell c')\log_{10}(I_o/I)$, with c' in molar concentration units. The conversion factor from ϵ to σ is 3.84×10^{-21}. Typical procedures for absorption cross-section determinations involve the construction of plots of absorbance/pathlength versus concentration at each wavelength, the slope of these plots then yielding the absorption cross-section. Adherence to linearity in these plots allows one to conclude that such complications as stray light and dimerization are not occurring. For reasonable absorbance values (ranging from 0.01 to 2 for diode array systems, for example), accuracies of ± 5% are readily attainable.

While it would be best to make these measurements in the gas phase and at ambient temperatures, vapor-pressure limitations negate this possibility for most POPs, and data are available only at elevated temperatures in the gas phase or in solution (usually aqueous and/or organic solvent). It should be noted that solution-phase spectra can be shifted somewhat from the gas-phase spectra (see Finlayson-Pitts and Pitts 1986), as well as from solvent to solvent. While the solution-phase spectra provide a good first-order approximation in an assessment of the possibility that a candidate POP will photolyze in the gas phase, caution should be used in the adoption of solution-phase spectra for quantitative calculations, particularly when the absorption band of interest lies near the edge of the solar spectrum (i.e., near 290 nm, where the solar flux is a strong function of wavelength; see Figure 2-4).

Absorption spectra for POPs are also measurable in the adsorbed state by using photoacoustic spectroscopy (Alebić-Juretić et al. 1991). Examples of photoacoustic UV-absorption spectra of some POP species (Zetzsch 1991) are shown in Figure 2-5. Note that there is little difference between the spectra measured on various surfaces from those obtained in solution, though a red shift in the adsorbed state had previously been reported (Parlar 1990). Variations in the observed pentachlorophenol spectra can be explained in terms of the phenol/phenolate interconversion, which depends on the acid/base characteristics of the surface or solvent. Thus, in basic solution, the absorption is shifted to longer wavelengths.

Photolysis quantum yields: Methods

The determination of photolysis quantum yields for organic species in the gas phase is not an easy task. In fact, full wavelength-dependent quantum yields are known for only a very few species, and there is no single conventional experimental procedure in use for their study. Conceptually, the quantum yield for the photodecomposition of an organic species is determined from the measurement of the first-order rate of loss of that compound under its controlled exposure to UV radiation (ECETOC 1983; OECD 1992). To obtain an absolute quantum yield, the absolute intensity of the radiation must be determined, using a calibrated energy meter or through the use of a suitable chemical actinometer. While quantum yields are in general dependent on wavelength and pressure, seldom are such thorough investigations conducted. More commonly, measurements are made over a broad wavelength band representative of the solar flux to the troposphere or at single wavelengths within this region (e.g., excimer laser outputs at 308 and 351 nm, or the mercury emission line at 313 nm).

A good example of a quantum-yield determination for a gas-phase organic compound is provided by the work of Gierczak et al. (1997), who studied the photolysis of methylvinyl ketone and methacrolein. In this study, the compound of interest was photolyzed at 308 nm, 337 nm, or 351 nm with a pulsed laser, the energy of which was measured with a thermopile gauge. The first-order loss rate of the parent compound was monitored by gas chromatography and converted to a quantum yield using the known absorption cross-sections and the number and intensity of the laser pulses. Quantum yields (and thus atmospheric half-lives against photolysis) were determined with precisions of approximately ± 20%.

A much more convenient and inexpensive method for the determination of vapor-phase quantum-yield determinations of POPs considered in this report has been conducted by Bunce and co-workers (Bunce et al. 1988, 1989; Bunce and Nakai 1989). In these studies, a sample of the POP under study was vaporized in an

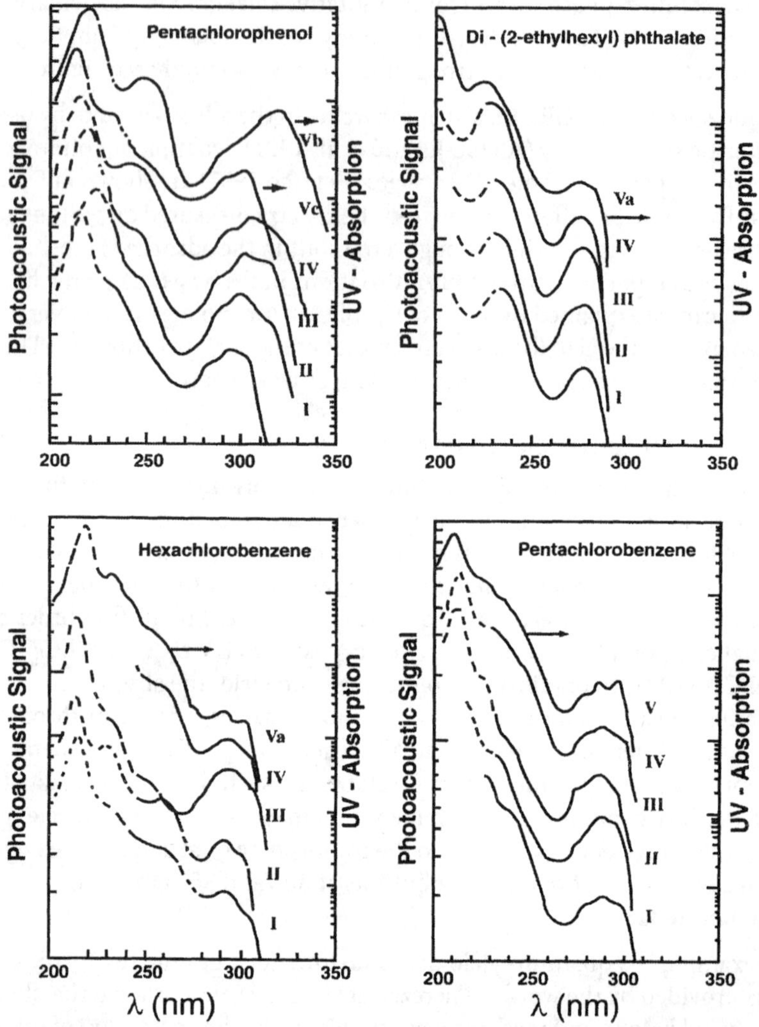

Figure 2-5 Photoacoustic UV spectra of aromatics adsorbed on various carriers (left-hand scale, I: Aerosil 200, Degussa, II: Silicagel 60r, Merck, III: Silicagel 60, Merck, IV: Alumina C, Degussa) and in solution (right-hand scale, Va: n-hexane, Vb: 1n NaOH, Vc: 1n HCl). Spectra II and III of pentachlorophenol are scaled up by a factor of 2 for clarity. Figure taken from Zetzsch 1991 and reprinted with permission from *Umweltwissenschaften und Schadstoff-Forschung*.

atmosphere of butane and irradiated with the 254-nm line of a low-pressure mercury lamp. The vessel containing the POP/butane mixture and the mercury lamp was surrounded with potassium ferrioxalate solution, used as the actinometer. Loss rates of the POP were measured by gas chromatography, and quantum yields determined using known absorption cross-sections and the light intensities determined by actinometry. Detailed tests have been conducted (Bunce et al. 1988) to ensure that surface effects did not influence the photolysis process. The major drawback is that 254-nm radiation was used in these studies, rather than a more tropospherically relevant wavelength, because the loss rates achieved with longer wavelength mercury emission lines were too slow to be measured accurately. Quantum yields at 254 nm are likely to exceed those at longer wavelengths (more energy is deposited by a 254 nm photon) and probably lead to overestimation of the tropospheric photolysis rates (see discussion of individual candidate POPs below).

Specific molecules

In general, little gas-phase information is available on the POPs under consideration here; thus, data from solution-phase studies also are presented below in some cases to provide an estimate of the potential importance of gas-phase photolysis.

Aldrin, chlordane, dieldrin, endrin, and heptachlor

The compounds aldrin, chlordane, dieldrin, and endrin, all of similar chemical structure, are unlikely to possess near-UV absorption features of sufficient magnitude to allow for significant photolysis to occur in the gas phase (Siek et al. 1976). Heptachlor, though also of similar chemical structure, does absorb beyond 290 nm, though quantum-yield data do not appear to be available to allow a quantitative assessment of its photolysis rate.

Dichlorodiphenyltrichlorethane

DDT does not appear to absorb to any significant degree in the wavelength region beyond 290 nm. Thus, it is unlikely that gas-phase photolysis will be an important loss process for this species (Gore et al. 1971).

Dioxins

The chlorinated dibenzo-*p*-dioxins all possess strong absorption features in the near UV, as can be seen from the gas-phase absorption spectra of 2,3,7,8-tetrachlorodibenzo-*p*-dioxin (2,3,7,8-TCDD), 1,2,3,7,8-pentachloro-*p*-dioxin and octachlorodibenzo-*p*-dioxin reported by Funk et al. (1995), and the solution-phase spectrum of dibenzo-*p*-dioxin (Kwok et al. 1994). Thus, photolysis is a potentially important process for their gas-phase removal. The only gas-phase quantum yield information available for PCDDs is from Orth et al. (1989), who determined a quantum yield of 0.033 ± 0.046 for 2,3,7,8-TCDD over the range 250 to 340 nm, sufficiently large to make the photolysis of this species rapid (half-life of about 1 h). Note, however, that this measurement is subject to a large uncertainty and, in addition, could have been influenced by photolysis occurring below 290 nm (i.e., at wavelengths shorter than those encountered in the troposphere), and thus may overestimate the rate of photolysis in the troposphere (Atkinson 1991b). Gas-phase photolysis of dibenzo-*p*-dioxin appears to be rather slow (Kwok et al. 1994). Aqueous-phase quantum yields for various tetra- through octachlorodibenzo-*p*-dioxins (Choudhry and Webster 1985, 1986; Dulin et al. 1986) range from about 2×10^{-3} to

1×10^{-5}, seeming to decrease with increasing chlorine content. This trend is also evident in gas-phase photodecomposition studies conducted by Sivils et al. (1994). These quantum yields, if valid in the gas phase, give photolysis half-lives of a few days for the TCDD species to a few years for the more highly chlorinated dioxins. Further studies of the gas-phase photolysis rates of these species are clearly warranted.

Furans

Like PCDDs, PCDFs all possess strong absorption features in the near UV, with absorption maxima between approximately 280 and 330 nm in solution and in the gas phase (Tysklind et al. 1993; Kwok et al. 1994; Funk et al. 1995). The intensities of these absorption features are such that quantum yields of even as low as 10^{-3} still would be sufficient to render photolysis an important process. To date, gas-phase quantum yields for the photolysis of the PCDF species have not been determined, though photolysis of dibenzodifuran itself appears to be slow. However, Dung and O'Keefe (1994) have shown that half-lives of 2,7-dichlorodibenzofuran, and of 2,3,7,8- and 1,2,7,8-TCDF in water exposed to natural sunlight were less than 1 d, shorter than the half-lives anticipated for their reaction with OH in the gas phase. Further studies of the gas-phase photolysis rates of these species are clearly warranted.

Hexachlorobenzene

While benzene itself does not possess absorption features beyond 290 nm, increasing chlorine content leads to red shifts of the absorption spectra and consequently greater overlap with the solar spectrum (Scharping et al. 1987). Hexachlorobenzene possesses an absorption band beyond 290 nm, with peak absorptions occurring near 291 and 301 nm (Weast 1974; Alebić-Juretić et al. 1991). While gas-phase quantum-yield data are not available, a quantum yield of 0.2 has been reported in solution by Bunce et al. (1983). Given the long half-life (about 2 y) (Brubaker and Hites 1998b) of this species with respect to reaction with the OH radical, photolysis may play a role in its gas-phase destruction. Further studies are warranted.

Chlorophenols

Much like chlorobenzenes, the absorption spectra for chlorophenols shift to longer wavelengths with increasing chlorine content, leading to greater overlap with the solar spectrum (Bunce and Nakai 1989). Thus, despite a decreasing photolysis quantum yield with increasing chlorine content (Bunce and Nakai 1989), photolysis rates for the more highly chlorinated phenols are larger than for the less substituted species. Coupled with a decreasing rate coefficient for reaction of OH with the more highly chlorinated species, the rate of photolysis has been estimated to exceed that of OH reaction for pentachlorophenol (Bunce and Nakai 1989). Note, however, that the vapor-phase quantum yield data reported by Bunce and Nakai (1989) correspond to a photolysis wavelength of 254 nm and may not be applicable to the troposphere ($\lambda > 290$ nm).

Polycyclic aromatic hydrocarbons

These species possess very intense near-UV absorption features and are known to photodecompose in the adsorbed state as a result of complex surface reactions possibly involving excited states of O_2 (Finlayson-Pitts and Pitts 1986). However, they probably do not photolyze to any appreciable extent in the gas phase because there is no simple bond-rupture process that is energetically accessible at the irradiation wavelengths applicable to the troposphere.

Polychlorinated biphenyls

These species may possess significant absorption features in the near UV, such that photolysis needs to be considered as a possible gas-phase degradation mechanism. Berman et al. (1992) have measured gas-phase absorption cross-sections for 3,3',4,4'-tetrachlorobiphenyl at elevated temperatures. Extrapolation of the measured spectra to 298 K results in a long-wavelength threshold near 293 nm and a marginal overlap of the absorption spectrum with the solar spectrum (Atkinson 1996). Bunce et al. (1989) have investigated the vapor-phase photolysis of 2-, 3-, and 4-chlorobiphenyl and concluded that the photolysis of 2- and 4-chlorobiphenyl would occur with half-lives of a few days, comparable to the time constants for OH reaction. Note again that these calculations were based on quantum yield measurements at 254 nm, and probably overestimate the quantum yields at atmospherically relevant wavelengths ($\lambda \geq 290$ nm).

Polybrominated biphenyls

Ruzo et al. (1976) have reported UV-absorption spectra in hexane solution for a number of polybrominated biphenyls. All species studied (ranging from dibromo- to octabromobiphenyls) showed significant absorption beyond 290 nm. Thus, photolysis must be considered as a loss process for these species. In general, photolysis of the brominated biphenyls will likely be more rapid than for their chlorinated analogs.

Atmospheric Persistence of Particle-borne Compounds and Impact of Adsorption/Uptake on Reactivity of Semi-volatiles

The atmospheric aerosol: Sources, characteristics, and behavior

The annual source strength of the atmospheric aerosol is estimated to range from 1 to 2 billion tons and exceeds the flux of the reactive organic trace gases (Warneck 1988). The total surface of the aerosol is several times larger than the surface of the globe (Güsten 1986). The atmospheric residence time of the aerosol is strongly dependent on the particle diameter, as shown in Figure 2-6 from a review by Jaenicke (1987). Particles with radii between 0.05 and 3 µm have tropospheric residence times longer than 1 week, corresponding to typical transport ranges of several thousand km (as demonstrated by Saharan dust and fallout from the Chernobyl accident). For particles between 0.05 and 1 µm, the residence time with respect to dry deposition reaches ≈ 85 d over grassland and oceans; wet deposition may shorten this to about 5 d (Warneck 1988). With increasing diameter, the particles settle faster, whereas smaller particles are also removed rapidly via Brownian diffusion and undergo coagulation by diffusion at high number densities, thus leading to particle growth.

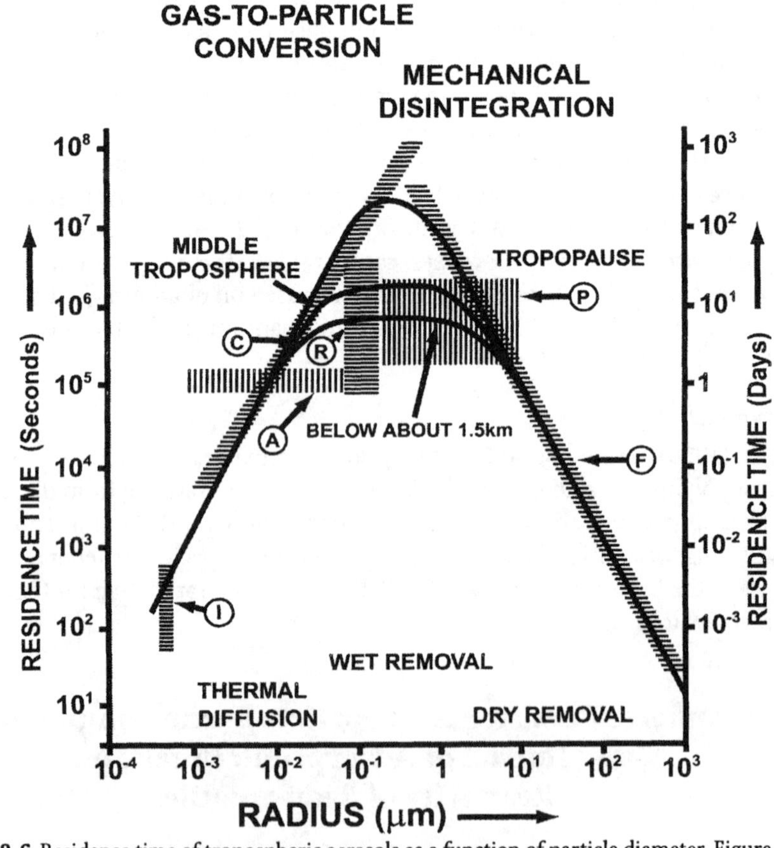

Figure 2-6 Residence time of tropospheric aerosols as a function of particle diameter. Figure taken from Jaenicke (1987) and reprinted with permission from Dechema e.V. The hatched areas represent published estimates for: small ions (I); Aitken particles (A); coagulation by thermal diffusion (C); radioactivity (R); precipitation (P); and sedimentation (F). The major production mechanisms and removal processes are indicated in the most important size ranges.

The largest portion of the atmospheric aerosol is NaCl from sea spray. Sulfuric and nitric acid are formed from gas-phase photooxidation of marine dimethyl sulfide and anthropogenic SO_2 and NO_x. Sulfuric acid has a low vapor pressure and condenses in the atmosphere; nitric acid may condense when neutralized by ammonia. Such processes form condensation nuclei and lead to the accumulation mode of the aerosol, forming ultrafine (< 0.1 μm diameter) and fine (0.1 to 2.5 μm diameter) particles. Figure 2-7a shows a size-resolved ionic composition of an aged marine aerosol (John et al. 1986). The sodium in the coarse mode above 1 μm is from sea-spray; the chloride has been substituted mostly by intrusion of HNO_3 and H_2SO_4. Growth curves of aerosol bulk samples for various sites as a function of humidity and other properties of the aerosol (significant growth occurs above 40% relative humidity) are reviewed by Warneck (1988).

A second major inorganic source of atmospheric aerosol originates from soil (about 200 Mt/y from the Sahara alone [D'Almeida 1986]), with potential photocatalytic properties leading to photochemical degradation of adsorbed organic volatiles and semi-volatiles (Güsten 1986; Parlar 1990). About 10% of the atmospheric mineral aerosol source strength is anthropogenic. Particles emitted from power plants, vehicle exhaust, and domestic heating are the predominant aerosol in the industrialized zones of the continents; photocatalysis and/or photooxidation in the aqueous phase of deliquescent particles may contribute to the oxidation of adsorbed semi-volatile organics, as found in experiments on silica gel, fly ashes, mica, synthetic clay, Fe_2O_3 and TiO_2 in fixed-bed and fluidized-bed photoreactors and in aerosol chamber experiments (Warneck et al. 1996). Although rapid photocatalytic degradation of several environmental chemicals adsorbed on silica gel—including extremely persistent compounds such as N_2O, CFCs, and CCl_4— has been reported (Parlar 1990), experiments on environmental aerosols, such as fly ash, show a large variability of degradation rates of PAHs, depending on unidentified properties of the carrier (Güsten 1986). Photocatalytic experiments on polychlorinated dibenzo-*p*-dioxins and dibenzofurans with 4 to 8 Cl substituents on a variety of fly ashes (from coal-fired power plants and waste incinerators) by Koester and Hites (1992) demonstrated unreactivity of the adsorbed compounds in the absence of any solvents, whereas dehalogenation was found to occur in organic solvents.

Soot from diesel engines and domestic heating is a source of elemental carbon in aerosols. Gas-phase oxidation processes convert reactive hydrocarbons (e.g., terpenes and aromatic compounds) into semi-volatile organic compounds that can then attach and contribute to aerosol particles, especially in polluted urban areas.

A schematic of an aged atmospheric aerosol particle is shown in Figure 2-7b (Weschler and Graedel 1982). An insoluble core, e.g., desert dust or fly ash, is embedded in an electrolyte solution that can be enclosed in an organic layer. The polar substituents of the molecules present in the organic layer are expected to be directed towards the electrolyte solution, thus masking a hydrophilic sea spray

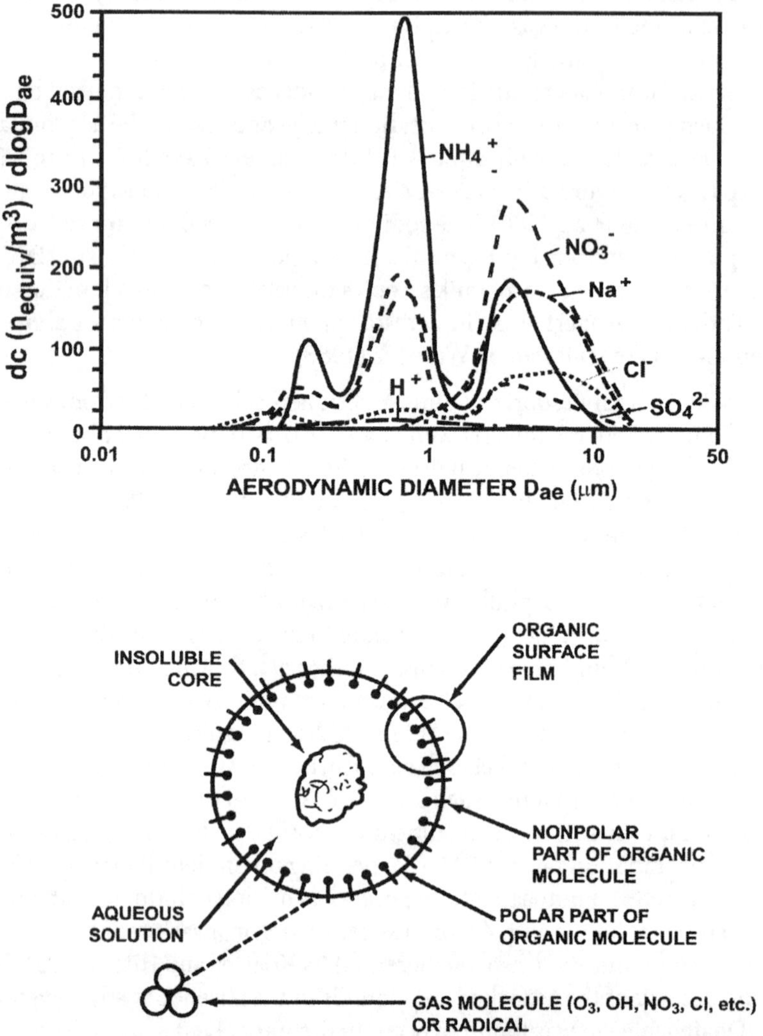

Figure 2-7 (a) Composition of inorganic ions in ambient atmospheric aerosol observed during an intercomparison field campaign at Claremont, California. Figure taken from John et al. (1986) and reprinted with permission (© 1986 Elsevier Science). (b) Schematic structure of an aged atmospheric aerosol particle and its interaction with the gas phase. Figure reprinted with permission from Weschler and Graedel (1982) ©1982 American Geophysical Union.

aerosol and converting its properties to a hydrophobic organic particle. Gas molecules will collide with the particles and can be taken up into the condensed phase, although the organic layer may decrease the sticking coefficient of hydrophilic and acidic gases. On the other hand, O_3, HO_2/RO_2, OH and NO_3 radicals, and other active species from the gas phase may react with the organic layer or penetrate into the droplet and lead to a degradation and partial oxidation of semi-volatile or dissolved particle-borne organics and POPs, such as PAHs, dioxins/furans, acidic/phenolic compounds, plasticizers, pesticides, and other industrial chemicals. Such oxidation processes may partly offset the hydrophobic properties of soot and other organic particles. Assuming a typical reaction probability of OH from the gas phase of about one-tenth of every collision with weak C–H bonds of macromolecules or C=C bonds of terpenoids, OH cannot penetrate deeper than a few molecular layers (Cooper and Abbatt 1996). The reactivity of POPs dissolved in the organic fraction of the aerosol will therefore be masked by more reactive constituents.

Uptake of H_2O_2 and HO_2 and heterogeneous reactions of O_3 and NO_X are expected to produce the radical precursors H_2O_2 (Heikes 1984) and HONO on the surface of aerosols. In cloud droplets, photolysis of H_2O_2 and photochemistry of Fe complexes and NO_3^- is expected to produce OH at levels around 10^7 to 10^8 molecule cm^{-3} (Zellner and Herrmann 1995), but levels in deliquescent aerosol at high ionic strength are unknown. Reaction-rate constants for OH with several POPs have been determined in the aqueous phase (see Sedlak and Andren 1991 for PCBs), and the contribution of deliquescent aerosols and clouds to atmospheric degradation needs to be assessed.

Experimental simulation of tropospheric chemistry in the presence of aerosols

Although the analytical chemistry of aerosols is now a well-developed area and a variety of compounds have been identified, it appears to be difficult if not impossible to draw conclusions regarding the reaction paths of individual semi-volatile compounds from the small portion of breakdown products detected so far.

Aerosol smog-chamber facilities employing natural sunlight and a large Teflon bag (Kamens et al. 1986, 1988, 1990) or artificial sunlight and a glass chamber (Behnke et al. 1988; Palm et al. 1997, 1999) have been described previously.

In the experimental facility of Behnke et al. (1988) and Palm et al. (1997, 1999), kinetic measurements are performed by coating highly dispersed inert powders, such as quartz, with a thin layer (less than a monolayer) of the test compound and by dispersing these powders as an aerosol into the smog chamber. The compounds are thereby exposed to well-defined concentrations of OH, which is produced by photolytic/photochemical methods, such as photolysis of HONO, H_2O_2, O_3 in the presence of H_2O, CH_3ONO, or NO_2, or in the dark by pernitric acid/NO, N_2H_4/O_3 or alkene/O_3. The OH concentration is calculated from the measured degradation rates of various added gaseous hydrocarbons and known rate constants. Under such experimental conditions, the mass fraction of the test compound in the aerosol is observed to disappear according to a pseudo–first-order rate law, the rate being proportional to the concentration of OH in the gas phase (Behnke et al. 1987a, 1987b). Figure 2-8 shows the decay rate of di(2-ethylhexyl)phthalate, a well-known plasticizer, adsorbed in monolayer thickness on SiO_2, Fe_2O_3, and Al_2O_3, as a function of the average concentration of OH in the smog chamber (Behnke et al. 1987c). The small decay rate in the absence of OH (intercept in Figure 2-8) is attributable to evaporation (Behnke et al. 1986) and the slope gives a rate constant of $k_{OH} = 1.5 \times 10^{-11}$ cm^3 molecule^{-1} s^{-1}, which is independent of carrier (i.e., SiO_2, Fe_2O_3, or Al_2O_3). The fact that the measured rate constant does not change with carrier suggests that the interaction between di(2-ethylhexyl)phthalate and the surface is weak. If this is indeed the case, the measured rate constant may provide a reasonable estimate for the rate constant of the gas-phase reaction. At this point it is worthwhile to note that, at ambient temperature, the vapor pressure of di(2-ethylhexyl)phthalate is about 1.8×10^{-7} Torr (2.3×10^{-10} atm) and the product (k_{OH} $\times$ vapor pressure) is 0.09 s^{-1}, which precludes any possibility of an absolute rate study of k_{OH} in the homogeneous gas phase at room temperature.

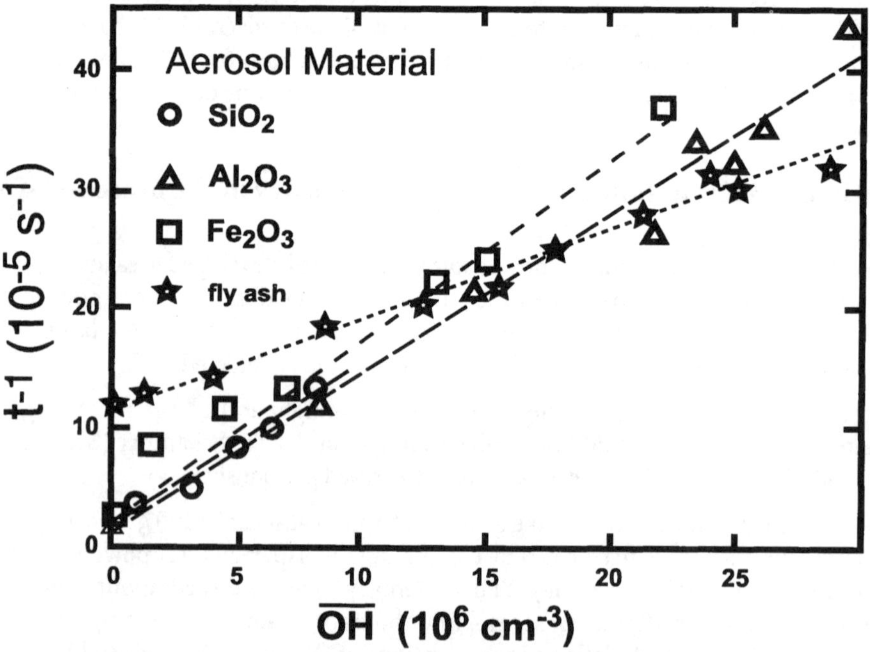

Figure 2-8 Decay rate of a monolayer of di(2-ethylhexyl)phthalate, adsorbed on various types of aerosols (SiO_2: Aerosil 200, DEGUSSA, Al_2O_3: type C, DEGUSSA, Fe_2O_3: Sicotram Orange, BASF) at various concentrations of OH in the smog chamber. Figure taken from Behnke et al. 1987c and reprinted with permission (©1987 Elsevier Science).

Several experimental problems with aerosol materials should be mentioned here: volatilization from particles, hydrolysis on the particles, and photocatalysis (Behnke et al. 1986, 1987a, 1987b, 1987c, 1987d, 1988; Zetzsch 1991; Palm et al. 1997, 1999).

Kamens and coworkers (Kamens et al. 1985, 1986, 1988, 1990, 1994, 1995; Fan et al. 1995; Fan, Kamens, Zhang et al. 1996; Fan, Kamens, Hu et al. 1996) have carried out series of experiments to investigate the decomposition of PAHs present in the particle or aerosol phase in the presence of sunlight and various reactive species and the formation and decomposition of nitro-PAHs and related compounds under similar conditions. Teflon chambers of various volumes ranging from 200-L stirred-flow reactors to 190-m^3 outdoor static chambers were used, and the results of these studies have been incorporated into computer models for the degradation of PAHs and the formation and decay of nitro-PAHs in the atmosphere (Kamens et al. 1994; Fan et al. 1995; Fan, Kamens, Zhang et al. 1996).

Atmospheric Deposition Processes

Dry deposition

Deposition of gases can be described as a sequence of 3 processes: 1) turbulent diffusion through the surface layer of the atmosphere, 2) diffusion through a laminar sublayer just above the surface, and 3) absorption (e.g., via dissolution), adsorption, or chemical reaction on the surface. The first two steps are a function of the atmosphere and terrain; they govern the transfer of the depositing chemical from the bulk atmosphere to the surface. The third step is a function of the surface (e.g., water, soil, vegetation) and chemical type (e.g., reactive, soluble).

Deposition of particles is also a function of the atmosphere and terrain, which affect the first two steps described above. In addition, it is a strong function of the size of the particle. Large particles are affected by gravitational settling and inertial impaction as the air flows around the surface. Very small particles are deposited via their Brownian motions near the surface. Particles in the size range of 0.1 to 1 μm deposit most slowly because they are not affected significantly by Brownian motion, inertial impaction, and gravitational forces.

Dry deposition is generally modeled using a dry deposition velocity, which is an effective parameter characterizing the chemical flux from the atmosphere to the surface. This deposition velocity can be specified by the modeler (e.g., using a default average value) or calculated by a deposition algorithm that takes into account the various steps described above and, therefore, accounts for the atmospheric characteristics, terrain features, surface type, and chemical or particle type.

Experimental data on the dry deposition of gases are limited. They are typically about 1 cm s^{-1} or more for soluble or reactive gases and less than 0.1 cm s^{-1} for gases that are not very soluble or reactive. If solubility (rather than chemical reaction) governs the removal of the chemical at the surface, the multimedia system can eventually reach equilibrium and dry deposition becomes zero. Assuming that dry deposition of gases is negligible (i.e., dry deposition velocity of zero) will provide an upper limit for the atmospheric residence time.

Experimental data for the dry deposition of particles are sparse. Measurements of dry deposition velocities of particles in the range of 0.1 to 1 μm in wind tunnels suggest values between 0.05 and 0.1 cm s^{-1} over rough surfaces and less than 0.01 cm s^{-1} over smooth surfaces (Seinfeld and Pandis 1997). Particles with diameters greater than 1 μm have deposition velocities greater than those of particles in the 0.1 to 1 μm range. Typically, deposition velocities of 0.1 to 0.2 cm s^{-1} have been used for POP-type chemicals.

The half-life ($t_{1/2}$) of a chemical for dry deposition can be estimated from the dry deposition velocity v_d by assuming some height, over which the chemical is well mixed:

$$t_{1/2} = 0.693 \times \text{height} / v_d \qquad \text{(Equation 2-10).}$$

If we assume that the bulk of the chemical mass is in the planetary boundary layer and that the average height of the planetary boundary layer is 1000 m, the half-life for a dry deposition velocity of 0.1 cm s^{-1} is 8 d.

Wet deposition

Wet deposition of gases occurs through rainout and washout. Rainout refers to the uptake of the gas molecule by a raindrop within the cloud, whereas washout refers to the uptake of the gas molecule by a raindrop below the cloud base. Gases also may be removed by precipitation of snow and ice, as well as settling of fog droplets, if they partition into the snowflake, ice particle, or fog droplet. Wet deposition of gases is, therefore, a function of their solubility in water. It can be characterized by their Henry's Law constant (i.e., ability to dissolve in dilute solutions); more soluble gases are removed more efficiently by wet deposition. If we assume equilibrium between the gas phase and the droplets, then the chemical concentration in the raindrops, c_W (in units of g per m^3 of water) can be calculated from the concentration in the air, c_A (in units of g per m^3 of air)

$$c_W = c_A / K_{AW} \qquad \text{(Equation 2-11),}$$

where K_{AW} is the temperature-dependent air–water partition coefficient. Such an approach has been shown to hold reasonably well for many organic compounds (Ligocki et al. 1985). K_{WA}, the reciprocal of K_{AW}, is the equivalent of a washout, or scavenging, ratio.

Wet deposition of particles occurs also through rainout (when particles are incorporated into cloud-water droplets) and washout (when particles collide with raindrops). In addition, some particles act as condensation nuclei for cloud droplet formation. The rate of removal by wet deposition is a strong function of the particle size. Particles in the size range of 0.1 to 1 μm are removed the slowest, for the same reasons as those mentioned above for removal by dry deposition. For particles in the range of 0.1 to 1 μm in diameter, the scavenging rate for a precipitation intensity of 1 mm h^{-1} is on the order of 2×10^{-6} s^{-1} for raindrops of 0.2 mm in diameter and 2 ×

10^{-8} s^{-1} for raindrops of 2 mm in diameter (Seinfeld and Pandis 1997). For larger particles, the scavenging rate increases with particle size; for example, a particle of 2 µm and raindrops of 0.2 mm in diameter has a scavenging rate of about 7×10^{-6} s^{-1}.

Wet deposition can be modeled using detailed models of cloud microphysics, but such approaches are computationally expensive and generally are not used in global models. Alternative approaches include the use of scavenging coefficients or washout ratios. The scavenging coefficient represents the rate of scavenging of the gas or particle by precipitation (expressed in $time^{-1}$) and is a function of the precipitation intensity. The concentration of the chemical in the air is then expressed as a function of time as follows:

$$c_A(t) = c_A(t = 0) \exp(-S\ t) \qquad \text{(Equation 2-12)},$$

where S is the scavenging coefficient and t is the time. The washout ratio represents the ratio of the chemical in precipitation and in the ambient air, and it is the reciprocal of the air–water partition coefficient. It implicitly assumes that there is equilibrium between the gas phase and the aqueous phase during the precipitation event. The washout ratio (W) is expressed as follows:

$$W = K_{WA} = c_W / c_A \qquad \text{(Equation 2-13)}.$$

For gases that are not very soluble, removal by wet deposition is slow.

The half-life of a chemical against wet deposition can be estimated from the washout ratio. If we use the washout ratio, W ($= K_{WA} = 1/K_{AW}$), then the half-life can be estimated by assuming some precipitation rate (P) and an average height (height) over which the chemical is assumed to be well mixed.

$$t_{1/2} = 0.693\ \text{height} / (W\ P) \qquad \text{(Equation 2-14)}.$$

For example, if we assume an annual precipitation amount of 1 m, a planetary boundary layer height of 1000 m, and a washout ratio of 10^3, the half-life for wet deposition is 250 d. For a washout ratio of 10^5, the half-life for wet deposition is 2.5 d.

If we use the scavenging rate, S, the half-life can be estimated by assuming the fraction of time with rain, f_r:

$$t_{1/2} = 0.693 / (S\ f_r) \qquad \text{(Equation 2-15)}.$$

For example, if we assume an annual precipitation amount of 1 m, the fraction of time with precipitation of 1 mm/h is 11%. For comparison, the fraction of time with precipitation is on the order of 3% in North America, suggesting precipitation intensities greater than 1 mm/h. Then, the half-life for a scavenging rate of 7×10^{-6} s^{-1} is 10 d.

Use of Kinetic/Photochemical Data and Deposition Data to Estimate Atmospheric Half-lives

Half-life and residence time

For a chemical that is input to the atmosphere by emission and/or chemical reactions and removed from the atmosphere via chemical reactions, dry deposition, and wet deposition, the rate of change of its atmospheric burden can be defined as follows:

$$\frac{dM}{dt} = P - R \qquad \text{(Equation 2-16)},$$

where

P is the atmospheric production rate (mass per time),
R is the atmospheric removal rate (mass per time),
M is the atmospheric mass of the chemical, and
t is time.

If steady state is achieved, then the production and removal rates are equal, and the chemical mass is constant:

$$P = R \quad \frac{dM}{dt} = 0 \quad \Rightarrow \quad M = constant \qquad \text{(Equation 2-17)}.$$

The residence time (t_r) of the chemical in the atmosphere is defined as follows (e.g., Aris 1969; Finlayson-Pitts and Pitts 1986; Seinfeld and Pandis 1997):

$$t_r = \frac{M}{P} = \frac{M}{R} \qquad \text{(Equation 2-18)}.$$

Other commonly used terms for t_r are lifetime and holding time. The residence time corresponds to the mean time spent by the chemical in the atmosphere (in reality, some chemical molecules will spend less time, and others will spend more time).

The removal rate can be expressed as the sum of first-order rates for chemical reaction, dry deposition, and wet deposition:

$$R = k_c M + k_d M + k_w M \qquad \text{(Equation 2-19)},$$

where k_c, k_d and k_w are the first-order rate parameters ($time^{-1}$) for chemical reaction, dry deposition, and wet deposition, respectively. Then, the residence time can be expressed as follows:

$$t_r = \frac{M}{k_c M + k_d M + k_w M} = \frac{1}{k_c + k_d + k_w} \qquad \text{(Equation 2-20)}.$$

We can define residence times specific to each removal process (i.e., the residence time corresponding to the scenario in which only that process removes the chemical from the atmosphere):

$$t_{rc} = \frac{1}{k_c} \qquad t_{rw} = \frac{1}{k_w} \qquad t_{rd} = \frac{1}{k_d} \qquad \text{(Equation 2-21)},$$

where t_{rc}, t_{rw}, and t_{rd} are the residence times of the chemical for chemical reaction, wet deposition, and dry deposition, respectively. The overall residence time can be expressed as a function of the process-specific residence times.

$$t_r = \frac{1}{\frac{1}{t_{rc}} + \frac{1}{t_{rd}} + \frac{1}{t_{rw}}} \qquad \text{(Equation 2-22)}.$$

One should note that the residence time differs from the half-life of the chemical. For example, the half-life of the chemical versus a first-order reaction is defined as follows (e.g., Aris 1969; Finlayson-Pitts and Pitts 1986; Seinfeld and Pandis 1997):

$$t_{1/2} = 0.693 / k_c \qquad \text{(Equation 2-23)}.$$

The half-life corresponds to the time required to deplete the initial chemical concentration to half of its value, in the absence of other production and removal processes. Half-lives can be defined similarly for dry deposition and wet deposition. We can also define an overall half-life for all atmospheric removal processes as follows:

$$t_{1/2} = 0.693\ t_r \qquad \text{(Equation 2-24)}.$$

If the chemical is partitioned between the gas and particulate phases, the overall half-life is then calculated as a weighted average of the half-lives of the gas-phase and particulate-phase fractions. Let ϕ be the fraction of the chemical mass that is in the particulate phase, $t_{1/2,g}$ the gas-phase half-life, and $t_{1/2,p}$ the particulate half-life, and the overall half-life is calculated as follows:

$$t_{1/2} = 1 / (((1 - \phi) / t_{1/2,g}) + (\phi / t_{1/2,p})) \qquad \text{(Equation 2-25)}.$$

Approaches to estimating half-lives

As described above, estimating the half-life of a chemical in the atmosphere requires estimating its removal rate via chemical reaction, dry deposition, and wet deposition. We discuss first the half-life corresponding to chemical reaction. The removal rate presented above is then expressed as follows:

$$R = k_c M \qquad \text{(Equation 2-26)},$$

where k_c is a pseudo–first-order rate parameter and M is the mass of chemical present in the atmosphere. In reality, the reaction of a chemical depends on concentrations of other reactants (i.e., oxidants such as OH radicals, O_3, and NO_3 radicals) and, for some chemicals, solar irradiation (i.e., actinic flux). Therefore, k_c is not constant but may be expressed as follows:

$$k_c = k_{c,OH}[OH] + k_{c,O3}[O_3] + k_{c,NO3}[NO_3] + k_{c,j} \quad \text{(Equation 2-27)},$$

where

$k_{c,OH}$, $k_{c,O3}$ and $k_{c,NO3}$ are the rate constants of the reaction of the chemical with OH, O_3, and NO_3, respectively (note that all these reactions may not be relevant for a given chemical);

[OH], $[O_3]$, and $[NO_3]$ are the atmospheric concentrations of these oxidants; and

$k_{c,j}$ is the photolytic rate parameter.

The mass of the organic chemical that is available for reaction with oxidants or photolysis may be only a fraction of the total atmospheric mass because adsorption to particulate matter may reduce chemical reactivity (Scheringer 1997). The rate parameters of the oxidation reactions are temperature-dependent; the oxidant concentrations are functions of the local atmospheric chemistry; and the photolytic rate parameter is a function of the solar zenith angle, ozone column, and cloud cover, since clouds/fog affect the solar flux. Therefore, all these parameters are functions of space and time.

Figure 2-9a depicts the spatial distribution of OH concentrations in the planetary boundary layer over the entire globe for the month of July as simulated with a three-dimensional chemical transport model (Spivakovsky et al. 1990). The model did not include reactions of non-methane hydrocarbons or the radiative reflection and scattering of aerosols, and the OH concentration results are likely to be uncertain by 10% to 30% on average. The values shown in Figure 2-9a are averaged over the month of July (i.e., day and night). We see that higher OH concentrations occur in the northern hemisphere than in the southern hemisphere because of higher solar radiation and more daylight hours in the northern hemisphere. Concentrations over land, where most anthropogenic pollution sources are located, are greater than concentrations over the oceans. For example, OH concentrations over North America and southern Europe are in the range of $(2–4) \times 10^6$ radicals cm^{-3}, whereas they are in the range of $(1–2) \times 10^6$ radicals cm^{-3} over the North Atlantic Ocean. The results are summarized in Table 2-2, where the OH concentration is averaged over all longitudes for selected latitudes.

Table 2-2 Longitudinally averaged OH radical concentrations (10^5 molecule cm^{-3}) calculated by using a 3-dimensional global model

Latitude	July	January
90 S (South Pole)	0	7.86
60 S	0.06	2.74
30 S	4.40	13.2
0 (Equator)	13.8	11.4
30 N	17.0	4.55
60 N	14.3	0.08
90 N (North Pole)	13.8	0
Global average	10.1	7.3

Figure 2-9b depicts the spatial distribution of OH concentrations in the plan-

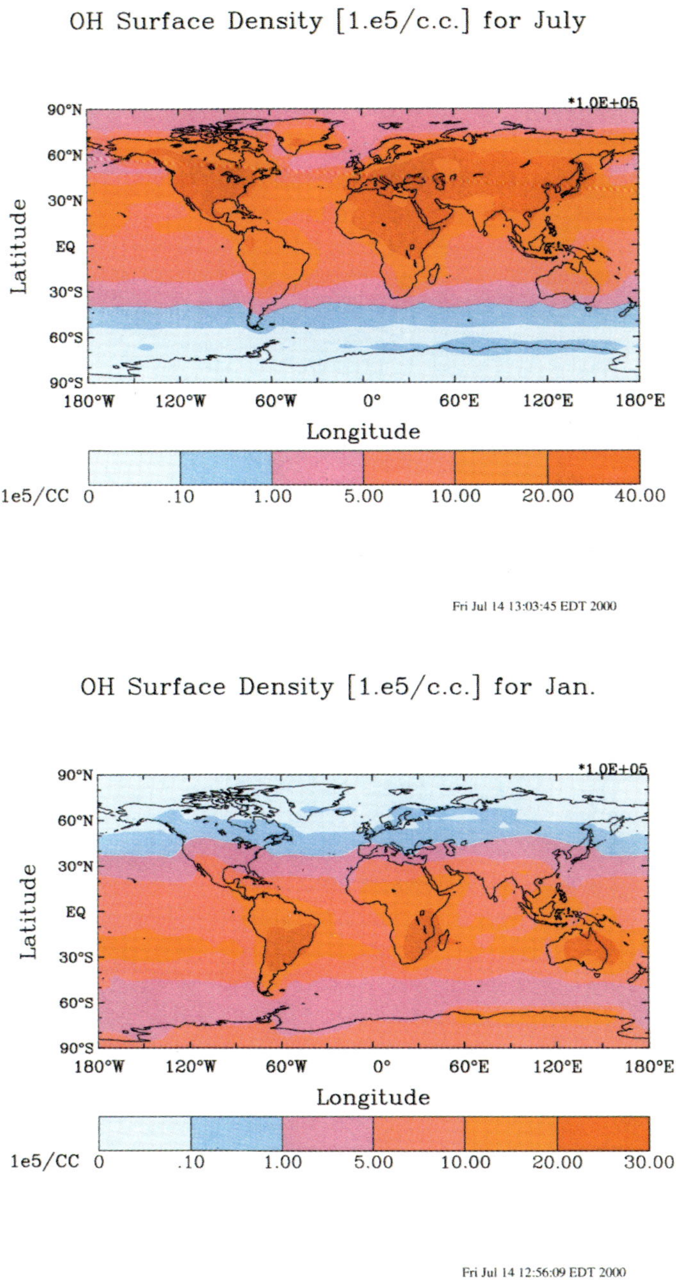

Figure 2-9 OH surface concentrations (10^5 molecule cm^{-3}) for (a) July and (b) January. Based on Spivakovsky et al. (1990).

etary boundary layer over the entire globe for the month of January as simulated by the same model. For comparison, Jefferson et al. (1998) measured OH concentrations in Antarctica in February 1994 in the range of (1–7) × 10^5 radicals cm^{-3}. The modeled values for January in the same region are in the range of (5–10) × 10^5 radicals cm^{-3}. If we compare Figure 2-9b with Figure 2-9a, we see that there is a strong seasonal variation in OH concentrations. In January, OH concentrations are higher in the southern hemisphere than in the northern hemisphere because of the higher solar radiation in the southern hemisphere. As for the July case, OH concentrations are greater over land than over the oceans. However, the overall OH concentrations are lower in January in the southern hemisphere than in July in the northern hemisphere because there are more anthropogenic sources of pollution in the latter.

Figure 2-9 clearly shows the strong spatial and temporal variations of OH concentrations in the atmosphere. Spatial and temporal variations also occur for the other oxidants, i.e., O_3 and NO_3.

Spatial and temporal variations also exist for dry and wet deposition processes: dry deposition depends on the atmospheric stability, terrain, and surface type; wet deposition depends on the type, intensity, and length of the precipitation event.

Ideally, the calculation of the half-life of a chemical in the atmosphere should take into account these variations of the chemical reaction rates and deposition processes with space and time. Such an approach, however, requires the spatially and temporally distributed concentrations of the oxidants and chemical studied and the spatial and temporal variations of the deposition rates. The calculation can be simplified if the rate parameters and oxidant concentrations are averaged in space, time, or both. We can group the various atmospheric modeling approaches into the following categories, in order of increasing complexity:

- Single-box model: This is the simplest approach. The chemical is assumed to be well mixed within the atmosphere (typically the troposphere, since there is little exchange between the troposphere and the stratosphere), and globally averaged concentrations and deposition rates are used for the atmospheric half-life calculations. Single-box models are suitable for chemicals that have an atmospheric half-life much greater than the inter-hemispheric exchange time of about 1 y.
- Compartmental model: The atmosphere is represented by several boxes (e.g., troposphere and stratosphere, northern and southern hemispheres) and the chemical is assumed to be well mixed within each of those boxes or compartments. Transport between the compartments must be specified to represent air flow or diffusion. Compartmental models are suitable for chemicals that have an atmospheric half-life greater than, but comparable to, the inter-hemispheric exchange time.

- One-dimensional model: The atmosphere is represented by a set of layers within which the chemical is well mixed. This type of model accounts for the fact that the chemical and the oxidant may not have the same vertical concentration profiles. Alternatively, a one-dimensional model can account for spatial variation of chemical concentrations in the longitudinal or latitudinal directions. One-dimensional models should be used for chemicals that have an atmospheric half-life much greater than the inter-hemispheric exchange time but have a chemical half-life that depends strongly on altitude (e.g., photolysis in the UV-B spectral region where the vertical distribution of tropospheric ozone may be important).
- Two-dimensional model: The atmosphere is represented by a two-dimensional grid system that provides vertical resolution as well as horizontal resolution along the north-south direction. Two-dimensional models use the north-south direction rather than the east-west direction to describe horizontal spatial gradients because wind flows mix chemicals faster in the east-west direction than in the north-south direction. Chemical concentrations are uniform within each grid cell, and transport in the vertical and horizontal directions is specified to represent atmospheric diffusion and wind flows. Two-dimensional models should be used for chemicals that have an atmospheric half-life between the time-scale for zonal mixing (i.e., in the longitudinal direction) and the inter-hemispheric exchange time. Time-scales for zonal mixing range from about 2 weeks in the vicinity of the jet stream to about 2 months near the surface.
- Three-dimensional model: A fully resolved three-dimensional representation of the atmosphere is used wherein the chemical concentrations are assumed to be uniform within each grid cell. The model spatial resolution is limited only by the size of the grid cells in the vertical and horizontal directions. The atmospheric motions (wind flow, turbulent diffusion) need to be specified to simulate the transport of the chemicals between the grid cells. Three-dimensional models should be used for chemicals that have a half-life less than the time-scale for zonal mixing.

For a screening study, a single-box model calculation will be conducted. Refined calculations can be carried out, if warranted, using more sophisticated approaches, ranging from compartmental models to three-dimensional models.

Summary of previous studies

Several modeling studies have been conducted over the past few years to estimate the residence times (or half-lives) of organic pollutants. Table 2-3 presents a summary of some of these studies, including the organic chemicals that were studied and the assumptions made in the calculations of the residence times or half-lives.

All studies used the OH reaction to calculate the residence time or half-life due to chemical reactions. In most cases, the kinetics of the OH reaction were estimated

Table 2-3 Summary of selected modeling studies of atmospheric residence times of organic pollutants

Reference	Reactions considered	Spatial/temporal characteristics	Organic compounds studied
Bunce et al. 1991a, 1991b	OH, photolysis	Space and time specific	Pentachlorophenol
Bunce et al. 1991	OH, photolysis	Space and time specific Global average[1]	8 chloroaromatics
Strand and Hov 1996	None	—	Hexachlorocyclohexane
Scheringer 1997	OH[2]	Global average	20 compounds
van Pul et al. 1998	OH	Global average[3]	30 compounds
Webster et al. 1998	OH	Space and time specific	Tetrachlorobiphenyl

[1] OH = 10^6 radicals cm^{-3}
[2] Fraction of organic compound adsorbed to particulate matter does not react
[3] OH = 6 x 10^5 radicals cm^{-3}

rather than obtained from experimental data. Only Bunce et al. (1991a, 1991b) considered solar photolysis in addition to the OH reaction. None of the studies considered reactions with O_3 or NO_3.

Bunce et al. (1991a, 1991b) and Webster et al. (1998) calculated atmospheric residence times (or half-lives) for specific times and locations, thereby explicitly acknowledging the spatial and temporal variation of chemical reaction rates. Bunce et al. (1991b) then calculated a global average concentration of OH (i.e., averaged spatially over the entire globe and temporally over an entire year) of 10^6 radicals cm^{-3} and used that value to calculate global residence times. It must be noted that this approach is rigorous only if the organic compound is well mixed in the global atmosphere, i.e., if its transport time scale is significantly shorter than its chemical reaction time scale. van Pul et al. (1998) also used a globally averaged OH concentration (6×10^5 radicals cm^{-3}) to calculate half-lives of 30 organic compounds.

Scheringer (1997) used globally averaged reaction rate parameters; he further assumed that the fraction of organic compound that is adsorbed on atmospheric particulate matter does not react. He used empirical values for the first-order rate parameters supposedly based on reaction rates with OH. Strand and Hov (1996) simulated hexachlorocyclohexane concentrations and assumed that chemical reaction was not the major process (compared to dry and wet deposition) governing the atmospheric residence time of that compound.

Case studies

α-hexachlorocyclohexane

In this case study, we present our screening calculation of the atmospheric half-life of α-hexachlorocyclohexane (α-HCH). The vapor pressure of this compound is sufficiently high that it is present in the atmosphere in the gas phase (Bidleman 1988). The measured kinetic rate parameter of the reaction with OH radicals is 1.4×10^{-13} cm^3 $molecule^{-1}$ s^{-1} (Brubaker and Hites 1998b). Therefore, assuming a globally averaged annual concentration of 10^6 radicals cm^{-3} for OH radicals, the half-life for chemical reaction is 57 d.

The Henry's Law constant is 3600 m^3 air/ m^3 water. If we assume that the great majority of the chemical mass is within a planetary boundary layer of 1000 m and that the annual precipitation amount is 1 m, then the half-life for wet deposition is 70 d. Dry deposition of a compound such as α-HCH can be assumed to occur because of its solubility in water and possible adsorption to soil. Therefore, it will depend on the concentration of the compound in the atmosphere as well as in the other medium (i.e., water or soil). If the chemical mass is totally in the atmosphere (i.e., concentrations in the other media are zero), then dry deposition will be an important removal pathway. If the chemical mass is distributed throughout all media and the multimedia system is near equilibrium, then dry deposition will be negligible. Strand and Hov (1996) estimated maximum velocities of 0.00035 and 0.68 cm s^{-1} for dry deposition over soil and sea water (i.e., for the case where all the chemical mass is in the atmosphere). Assuming that oceans cover two-thirds of the earth's surface area, the half-life for dry deposition is calculated to be 1.8 d when all the chemical mass is in the atmosphere. The half-life will increase as the multimedia system nears equilibrium and the net dry deposition rate approaches zero.

The overall atmospheric half-life can be calculated for the following 2 extreme scenarios: 1) all the chemical mass is in the atmosphere and 2) the multimedia system has reached equilibrium. For the first scenario, the atmospheric half-life is calculated to be 1.7 d and is governed by dry deposition; for the second scenario, the half-life is calculated to be 31 d and is governed by chemical reaction and wet deposition. For the case in which the calculated half-life is 1 month or more, the assumption that the chemical mass is primarily in the planetary boundary layer will not hold because vertical mixing will carry some of the chemical mass into the free troposphere and, at longer time-scales, into the stratosphere. If we assume that for the second scenario, α-HCH is uniformly mixed within the troposphere, then 15% of the mass is below 1000 m altitude. The half-lives for wet and dry deposition become 470 and 12 d, respectively. Then, the overall half-life is calculated to be 51 d and is governed by chemical reaction. For comparison, Strand and Hov (1996) used a comprehensive multimedia model to simulate the environmental behavior of this compound. They calculated an atmospheric residence time of 23 d, i.e., a half-life of

16 d. Their calculated half-life falls between the two extreme values obtained with our screening calculations (1.7 and 51 d).

2,3,7,8-tetrachlorodibenzodioxin

We present here an example of the screening calculation of the atmospheric half-life of 2,3,7,8-tetrachlorodibenzodioxin (TCDD), taking into account gas/particle partitioning, chemical reaction, dry deposition, and wet deposition.

The partitioning of TCDD between the gas and particulate phases can be estimated based on its particulate-vapor partitioning and particulate matter concentration or aerosol density (e.g., Bidleman 1988). Assuming an atmospheric particulate matter concentration of 65 $\mu g/m^3$, we estimated that 40% of TCDD was present in the particulate phase and 60% was present in the gas phase. The rate parameter for the reaction of TCDD with OH radicals was estimated to be 7.0×10^{-13} cm^3 $molecule^{-1}$ s^{-1} at 298 K (Atkinson 1996, 2000). We assumed that only TCDD in the gas phase reacts with OH. If we use a globally averaged annual concentration of 10^6 radicals cm^{-3} for OH radicals, this rate estimate leads to a half-life for chemical reaction of 11 d for gas-phase TCDD. Photolysis also may contribute to the degradation of TCDD in the atmosphere, in which case, the half-life of TCDD could be less than 11 d.

Dry deposition of gas-phase TCDD is assumed to be negligible. For particulate-bound TCDD, if we use a dry deposition velocity of 0.1 cm s^{-1} and a well mixed planetary boundary layer of 1000 m, the half-life for dry deposition is 8 d. The wet deposition of gas-phase TCDD can be assumed to be negligible because TCDD is not very soluble in water. If we use a scavenging rate of 7×10^{-6} s^{-1}, an average precipitation intensity of 1 mm/h, and an annual precipitation amount of 1 m, the half-life of particulate-bound TCDD for wet deposition is 10 d.

The overall half-life of the chemical in the atmosphere will then be 7 d. If we assume that particulate-bound TCDD reacts with OH radicals with the same kinetics as gas-phase TCDD, then the calculated atmospheric half-life is 5.5 d. van Pul et al. (1998) calculated an atmospheric half-life for TCDD of 1.5 d over land and 5.4 d over sea, consistent with our screening calculations.

Conclusions

Major atmospheric degradation and removal processes

- For most organic compounds, reaction with OH is the major chemical degradation process in the gas phase. The rate of reaction with O_3 and/or NO_3 should nevertheless be assessed, as should direct photolysis, because these processes may be significant or even predominant for some specific compounds.

- Semi-volatile organics, such as many of the listed or candidate POPs, may partition onto atmospheric particles and hence exist partly or even predominantly in the adsorbed state. This occurs in general for compounds with subcooled-liquid vapor pressures lower than about 10^{-2} Pa. The chemical degradation of species adsorbed on environmentally relevant aerosols therefore needs to be assessed, but there are severe limitations to our basic understanding of the impact of adsorption on degradation.
- Physical removal from the atmosphere, by wet and/or dry deposition, contributes to the overall disappearance rate of a given compound and hence to its atmospheric half-life. Wet and dry deposition must be assessed both for the gaseous compound and for the same species adsorbed on particles.

Experimental determination of rate constants in the gas phase

- A broad range of absolute- and relative-rate procedures can be used to determine rate data for homogeneous gas-phase reactions of the OH radical with relatively volatile compounds (i.e., for which $k_{OH} \times$ vapor pressure > $3\ s^{-1}$) to an accuracy of about ± 30%. Under ideal conditions, and after multiple independent studies, it is possible to establish the rate constants to an accuracy of ± 10%.
- For semivolatile compounds, rate constants can be determined at elevated temperatures and extrapolated back to ambient temperature, provided that the reaction mechanism does not change between the two temperature regimes.
- A possible alternative technique for semi-volatiles, still being developed and validated, involves determining the OH reactivity of the species adsorbed on inert solid aerosol particles. This procedure is believed to lead to rate constants comparable to those characteristic of the purely gas-phase reaction.
- Experimental procedures are available for determining rate constants for reaction of organics with O_3 and NO_3 in the gas phase, to an accuracy of about ± 30%. Techniques will have to be adapted or developed for compounds with very low volatility.

Estimation of rate constants in the gas phase

- To date, experimentally measured rate constants for the reactions of OH radicals, NO_3 radicals, and O_3 are available for only a few of the listed POPs. Reliable procedures for estimating these rate constants are therefore essential.
- The empirical estimation method of Atkinson (2000) allows the OH rate constants at 298 K of 90% of the approximately 500 to 600 organic compounds used in its development to be predicted to within a factor of 2 of the experimental values. Disagreements between calculated and measured rate constants most commonly occur for halogen-containing organic compounds

(i.e., the class of compounds that comprises many POP-type substances) and ethers.

- The limitations of present OH rate-constant estimation methods must be recognized. In particular, most estimation methods cannot be used with any degree of reliability for organic compounds outside of those compound classes actually used in the development of the estimation method. Compound classes for which experimental data are not available for at least one member of the homologous series could have uncertainties in the estimated rate constants for reaction with the OH radical of a factor of 5 or greater. In such cases, expert judgment should be sought. Atkinson's estimation technique has not been tested sufficiently to warrant its use for most N-, S-, and P-containing compounds.
- Further studies need to be carried out to extend Atkinson's estimation method to additional compound classes represented by POP-type compounds and to develop more direct methods for the calculation of OH radical reaction-rate constants.
- The accuracy of estimation methods for NO_3 and O_3 rate constants is generally worse than for OH rate constants, and order-of-magnitude uncertainties must be expected. However, degradation by reaction with O_3 or NO_3 is in most cases slower than in the case of OH, so the higher uncertainties are of minor importance.
- Rate constants for the atmospherically important reactions of POP-type compounds should be measured experimentally whenever possible, recognizing that experimental measurements are currently difficult for low-volatility chemicals.
- Finally, because only experimental determinations of k_{OH}, k_{O3}, and k_{NO3} can provide "certain" values, when the results of model calculations are published, it should be stated clearly whether estimated or experimental kinetic data were used.

UV/Visible absorption cross-sections and photolysis quantum yields

- The determination of the atmospheric photolysis rate of a particular compound requires that its absorption cross-sections and photolysis quantum yields be measured in the laboratory. These data are then coupled with solar flux to determine the half-life for photolysis.
- Satisfactory experimental techniques exist for determining UV/visible absorption cross-sections of relatively volatile compounds in the gas phase as a function of temperature and wavelength.
- For very low volatility species, which include many POP-type compounds, absorption cross-sections and quantum yields are difficult to measure and there are very few data currently available. For such compounds, it is often

necessary to resort to measuring cross-sections in an inert solvent. However, these spectra can be shifted somewhat relative to the gas-phase spectra, so extrapolation of the results to the gas phase requires expert judgment. This is particularly true when the absorption band under consideration is located at the short wavelength edge of the solar spectrum, i.e., near 290 nm.

- Determining quantum yields is in all cases a time-consuming process for which no standard methods exist. When results are available, they do not always refer to environmentally relevant wavelengths.
- No reliable methods exist for quantitative *prediction* of either absorption cross-sections or photolysis quantum yields. However, merely examining the molecular structure of a compound for the presence of known chromophores can indicate whether or not it is likely to absorb at environmentally relevant wavelengths and hence warrant experimental investigation.
- Atmospheric photolysis may be a significant loss process for certain classes of POPs (PCBs, PCDDs, PCDFs, chlorophenols, chlorobenzenes), but further studies are required to quantify the rates of photolysis.

Degradation on particles

- Despite the fact that many POP-type compounds are likely to be present in the atmosphere largely in the particle-borne state, our basic understanding of the impact of adsorption or absorption on the rate of degradation remains fragmentary.
- Reaction with OH may not cease when a substance is adsorbed on the outer surface of particles as a single molecular layer, as has sometimes been assumed in the literature. Indeed, the OH rate constant appears in this case to be comparable to that of the purely gas-phase reaction, but this conclusion has not yet been fully validated.
- Furthermore, some inorganic compounds present in aerosols (e.g., TiO_2 and Fe_2O_3) are known to photocatalyze degradation of adsorbed organics by production of OH and other reactive species. Such processes have nevertheless been neglected in this assessment because insufficient data are available to quantify their environmental relevance.
- Absorption of a gaseous species into an outer organic layer deposited on an inorganic particle may isolate the substance from gas-phase OH and hence impede its degradation. In a similar manner, adsorption onto the surface of pores whose particular shape hinders access by gaseous OH may lead to a reduced rate of degradation.
- It has been postulated in the literature that the UV/visible spectra of certain adsorbed species experience a red shift, enhancing their direct photolysis rates. However, recent studies on polychlorobenzenes, pentachlorophenol, etc., adsorbed on various powders, have failed to confirm this prediction.

Wet and dry deposition

- Chemicals are removed from the atmosphere via dry and wet deposition processes involving both gaseous and particle-borne species.
- Dry deposition rates depend on atmospheric stability, terrain, and surface type. For gases, their ability to absorb (e.g., by dissolution) into, adsorb on, or react with the surface strongly affects the deposition rate. For particles, the deposition rate is a strong function of particle size, with particles in the diameter range of 0.1 to 1.0 μm having the lowest deposition rate.
- Wet deposition rates depend on the deposition intensity, length of the precipitation event, precipitation type, and raindrop size distribution. Removal of gases increases with their solubility and chemical reactivity in the aqueous phase. Removal of particles is a strong function of particle size, with particles in the diameter range of 0.1 to 1.0 μm having the lowest precipitation scavenging rate.
- Deposition processes can be modeled with varying degrees of sophistication. Revolatilization must be considered.

Calculation of overall atmospheric half-life

- The calculated overall atmospheric half-life should account for 1) chemical destruction (gas-phase reaction with OH, O_3 and NO_3, gas-phase photolysis, and any heterogeneous chemical and photochemical reactions) and 2) physical removal by wet and dry deposition of species present both in the gaseous state and adsorbed on particles.
- At the screening level of the tiered approach, one may use a simple box model of the atmosphere and the earth's surface, assuming average values (global, regional, local) of certain input parameters (OH, O_3 and NO_3 concentrations; temperature; solar radiation intensity; aerosol densities; deposition velocities, etc.).
- At higher tiers of assessment, one should employ increasingly sophisticated models of the atmosphere, coupled to other media. These should take into account spatial and temporal variations in the parameters listed above and provide a more realistic description of atmospheric transport and inter-media partitioning.

Other conclusions

- It is important to note that the atmospheric concentration of the OH radical, on which the half-lives of most atmospheric pollutants are highly dependent, varies over orders of magnitude in time and space. Thus, in winter months at mid-latitudes, the OH concentration is so low (10^5 molecule cm^{-3}) that practically all organic compounds will have half-lives (for reaction with OH) greater than 2 d, a cutoff value that has been proposed by several regulatory bodies as a criterion for determining propensity for long-range transport.

- While persistence is probably much easier to quantify for the atmosphere than for other environmental compartments, there are still severe limitations to our predictive capability for purely gas-phase processes and to our basic understanding of chemical and physical processes involving substances adsorbed on, or absorbed by, particles.
- Despite the current environmental significance of POP-type compounds, the experimental database on their relevant properties is extremely limited.

References

Alebić-Juretić A, Güsten H, Zetzsch C. 1991. Absorption spectra of hexachlorobenzene adsorbed on SiO_2 powders. *Fresenius J Anal Chem* 340:380–383.

Anderson PN, Hites RA. 1996a. System to measure relative rate constants of semivolatile organic compounds with hydroxyl radicals. *Environ Sci Technol* 30:301–306.

Anderson PN, Hites RA. 1996b. OH radical reactions: the major removal pathway for polychlorinated biphenyls from the atmosphere. *Environ Sci Technol* 30:1756–1763.

Aris R. 1969. Elementary chemical reactor analysis. Englewood Cliffs NJ: Prentice Hall. p 158 and 215–217.

Atkinson R. 1986. Kinetics and mechanisms of the gas-phase reactions of the hydroxyl radical with organic compounds under atmospheric conditions. *Chem Rev* 86:69–201.

Atkinson R. 1989. Kinetics and mechanisms of the gas-phase reactions of the hydroxyl radical with organic compounds. *J Phys Chem Ref Data* Monograph 1:1–246.

Atkinson R. 1991a. Kinetics and mechanisms of the gas-phase reactions of the NO_3 radical with organic compounds. *J Phys Chem Ref Data* 20:459–507.

Atkinson R. 1991b. Atmospheric lifetimes of dibenzo-*p*-dioxins and dibenzofurans. *Sci Total Environ* 104:17–33.

Atkinson R. 1994. Gas-phase tropospheric chemistry of organic compounds. *J Phys Chem Ref Data*, Monograph 2:1–216.

Atkinson R. 1996. Atmospheric chemistry of PCBs, PCDDs and PCDFs. In: Hester RE, Harrison RM, editors. Issue 6, Issues in environmental science and technology. Cambridge, UK: The Royal Society of Chemistry. p 53–72.

Atkinson R. 2000. Atmospheric oxidation. In: Boethling RS Mackay D, editors. Handbook of property estimation methods for chemicals: environmental and health sciences. Boca Raton FL: Lewis. p 335–354.

Atkinson R, Carter WPL. 1984. Kinetics and mechanisms of the gas-phase reactions of ozone with organic compounds under atmospheric conditions. *Chem Rev* 84:437–470.

Atkinson R, Aschmann SM, Pitts Jr JN. 1984. Kinetics of the reactions of naphthalene and biphenyl with OH radicals and O_3 at 294 ± 1 K. *Environ Sci Technol* 18:110–113.

Atkinson R, Aschmann SM. 1985. Rate constants for the gas-phase reactions of hydroxyl radicals with biphenyl and the monochlorobiphenyls at 295 ± 1 K. *Environ Sci Technol* 19:462–464.

Atkinson R, Tuazon EC, Aschmann SM. 1995. Products of the gas-phase reactions of a series of 1-alkenes and 1-methylcyclohexene with the OH radical in the presence of NO. *Environ Sci Technol* 29:1674–1680.

Barnes I, Becker KH, Mihalopoulos N. 1994. An FTIR product study of the photooxidation of dimethyl sulfide. *J Atmos Chem* 18:267–289.

Behnke W, Nolting F, Zetzsch C. 1986. The heterogeneous reaction of di-(2-ethylhexyl-) phthalate, adsorbed on SiO_2 aerosol, with OH radicals originating from a photosmog. In: Israel G, editor. Aerosols: formation and reactivity. Oxford UK: Pergamon. p 879–881.

Behnke W, Nolting F, Zetzsch C. 1987a. An aerosol smog chamber for testing abiotic degradation of compounds with low volatility. In: Greenhalgh R, Roberts TR, editors. Pesticide science and biotechnology. Oxford UK: IUPAC/Blackwell. p. 401–404.

Behnke W, Nolting F, Zetzsch C. 1987b. A smog chamber study on the impact of aerosols on the photodegradation of chemicals in the troposphere. *J Aerosol Sci* 18:65–71.

Behnke W, Nolting F, Zetzsch C. 1987c. The atmospheric fate of di-(2-ethylhexyl)phthalate, adsorbed on various metal oxide model aerosols and on coal fly ash. *J Aerosol Sci* 18:849–852.

Behnke W, Nolting F, Zetzsch C. 1987d. The hydrolysis of a monolayer of di-(2-ethylhexyl)phthalate, adsorbed on various atmospheric model aerosol materials. *J Aerosol Sci* 18:853–856.

Behnke W, Holländer W, Koch W, Nolting F, Zetzsch C. 1988. A smog chamber for studies of the photochemical degradation of chemicals in the presence of aerosols. *Atmos Environ* 22:1113–1120.

Berman JM, Graham JL, Dellinger B. 1992. High temperature UV absorption characteristics of three environmentally sensitive compounds. *J Photochem Photobiol A: Chem* 68:353–362.

Bidleman TF. 1988. Atmospheric processes. Wet and dry deposition of organic compounds are controlled by their vapor-particle partitioning. *Environ Sci Technol* 22:361–367.

Brasseur GP, Orlando JJ, Tyndall GS. 1999. Atmospheric chemistry and global change. New York: Oxford University.

Brown HC, Okamoto Y. 1958. Electrophilic substituent constants. *J Am Chem Soc* 80:4979–4987.

Brubaker Jr WW, Hites RA. 1997. Polychlorinated dibenzo-*p*-dioxins and dibenzofurans: gas-phase hydroxyl radical reactions and related atmospheric removal. *Environ Sci Technol* 31:1805–1810.

Brubaker Jr WW, Hites RA. 1998a. OH reaction kinetics of polycyclic aromatic hydrocarbons and polychlorinated dibenzo-*p*-dioxins and dibenzofurans. *J Phys Chem A* 102:915–921.

Brubaker Jr WW, Hites RA. 1998b. OH reaction kinetics of gas-phase α- and δ-hexachlorocyclohexane and hexachlorobenzene. *Environ Sci Technol* 32:766–769.

Bunce NJ, Nakai JS. 1989. Atmospheric chemistry of chlorinated phenols. *J Air Pollut Control Assoc* 39:820–823.

Bunce NJ, Hayes PJ, Lemke ME. 1983. Photolysis of polychlorinated benzenes in cyclohexane solution. *Can J Chem* 61:1103–1104.

Bunce NJ, Langshaw J, Nakai JS. 1988. Gas phase photochemistry in round bottomed flasks. *J Photochem Photobiol A: Chem* 45:379–391.

Bunce NJ, Landers JP, Langshaw J, Nakai JS. 1989. An assessment of the importance of direct solar degradation of some simple chlorinated benzenes and biphenyl in the vapor phase. *Environ Sci Technol* 23:213–218.

Bunce NJ, Nakai JS, Yawching M. 1991a. A model for estimating the rate of chemical transformation of a VOC in the troposphere by two pathways: Photolysis by sunlight and hydroxyl radical attack. *Chemosphere* 22:305–315.

Bunce NJ, Nakai JS, Yawching M. 1991b. Estimates of the tropospheric lifetimes of short- and long-lived atmospheric pollutants. *J Photochem Photobiol A: Chem* 57:429–439.

Burkholder JB, Talukdar RK, Ravishankara AR, Solomon S. 1993. Temperature dependence of the HNO_3 UV absorption cross-sections. *J Geophys Res* D98:22937–22948.

Butler R, Snelson A. 1979. Kinetics of the homogeneous gas phase hydrolysis of CCl_3COCl, CCl_2HCOCl, $CH_2ClCOCl$ and $COCl_2$. *J Air Pollut Control Assoc* 29:833–837.

Chiorboli C, Piazza R, Carassiti V, Passerini L, Pino A, Tosato M-L, Riganelli D. 1996. A model for the tropospheric persistency of hydrohaloalkanes. *Gazz Chim Ital* 126:685–694.

Choudhry GG, Webster GRB. 1985. Quantum yields for the photodecomposition of polychlorinated dibenzo-p-dioxins (PCDDs) in water-acetonitrile solution. *Chemosphere* 14:893–896.

Choudhry GG, Webster GRB. 1986. Photochemical quantum yields and sunlight half-lives of polychlorodibenzo-p-dioxins in aquatic systems. *Chemosphere* 15:1935–1940.

Cooper DL, Allan NL, McCulloch A. 1990. Reactions of hydrofluorocarbons and hydrochlorofluorocarbons with the hydroxyl radical. *Atmos Environ* 24A:2417–2419.

Cooper PL, Abbatt JPD. 1996. Heterogeneous interactions of OH and HO_2 radicals with surfaces characteristic of atmospheric particulate matter. *J Phys Chem* 100:2249–2254.

Crutzen PJ. 1995. Ozone in the troposphere. In: Singh HB, editor. Composition, chemistry, and climate of the atmosphere. New York: Van Nostrand Reinhold.

D'Almeida GA. 1986. A model for saharan dust transport. *J Appl Meteor* 25:903–916.

Dulin D, Drossman H, Mill T. 1986. Products and quantum yields for the photolysis of chloroaromatics in water. *Environ Sci Technol* 20:72–77.

Dung MH, O'Keefe PW. 1994. Comparative rates of photolysis of polychlorinated dibenzofurans in organic solvents and in aqueous solutions. *Environ Sci Technol* 28:549–554.

[ECETOC] European Center for Ecotoxicology and Toxicology of Chemicals. 1983. Experimental assessment of the phototransformation of chemicals in the atmosphere. Brussels, Belgium: ECETOC. Technical report No. 7.

Fan Z, Chen D, Birla P, Kamens RM. 1995. Modeling of nitro-polycyclic aromatic hydrocarbon formation and decay in the atmosphere. *Atmos Environ* 29:1171–1181.

Fan Z, Kamens RM, Hu J, Zhang J, McDow S. 1996. Photostability of nitro-polycyclic aromatic hydrocarbons on combustion soot particles in sunlight. *Environ Sci Technol* 30:1358–1364.

Fan Z, Kamens RM, Zhang J, Hu J. 1996. Ozone-nitrogen dioxide-NPAH heterogeneous soot particle reactions and modeling NPAH in the atmosphere. *Environ Sci Technol* 30:2821–2827.

Finizio A, Mackay D, Bidleman T, Harner T. 1997. Octanol–air partition coefficient as a predictor of partitioning of semi-volatile organic chemicals to aerosols. *Atmos Environ* 31:2289–2296.

Finlayson-Pitts BJ, Pitts Jr JN. 1986. Atmospheric chemistry: fundamentals and experimental techniques. New York: Wiley.

Finlayson-Pitts BJ, Pitts Jr JN. 1999. Chemistry of the upper and lower atmosphere: Theory, experiments and applications. San Diego CA: Academic.

Funk DJ, Oldenborg RC, Dayton DP, Lacosse JP, Draves JA, Logan TJ. 1995. Gas-phase absorption and laser-induced fluorescence measurements of representative polychlorinated dibenzo-p-dioxins, polychlorinated dibenzofurans, and a polycyclic aromatic hydrocarbon. *Appl Spectrosc* 49:105–114.

Gaffney JS, Levine SZ. 1979. Predicting gas phase organic molecule reaction rates using linear free-energy correlations. I. $O(^3P)$ and OH addition and abstraction reactions. *Int J Chem Kinet* 11:1197–1209.

George LA, Hard TM, O'Brien RJ. 1999. Measurement of free radicals in OH and HO_2 in Los Angeles smog. *J Geophys Res* D104:11643–11655.

Gierczak T, Burkholder JB, Talukdar RK, Mellouki A, Barone SB, Ravishankara AR. 1997. Atmospheric fate of methyl vinyl ketone and methacrolein. *J Photochem Photobiol A: Chem* 110:1–10.

Gill PS, Graedel TE, Weschler CJ. 1983. Organic films on atmospheric particles, fog droplets, cloud droplets, raindrops and snowflakes. *Rev Geophys Space Phys* 21:903–920.

Gore RC, Hannah RW, Pattacini SC, Porro TJ. 1971. Infrared and ultraviolet spectra of seventy-six pesticides. *J Assoc of Anal Chem* 54:1040–1082.

Greene CR, Atkinson R. 1992. Rate constants for the gas-phase reactions of O_3 with a series of alkenes at 296 ± 2 K. *Int J Chem Kinet 24*:803–811

Güsten H. 1986. Photocatalytic degradation of atmospheric pollutants on the surface of metal oxides. In: Jaeschke W, editor. Vol. 6, Chemistry of multiphase atmospheric systems. NATO ASI Ser. G, Ecol. Sci. Heidelberg, Germany:Springer p. 567–594.

Güsten H, Klasinc L. 1986. Eine Voraussagemethode zum abiotischen Abbauverhalten von organischen Chemikalien in der Umwelt. *Naturwissen* 73:129–135.

Güsten H, Klasinc L, Marić D. 1984. Prediction of the abiotic degradability of organic compounds in the troposphere. *J Atmos Chem* 2:83–93.

Harner T, Bidleman TF. 1998. Octanol–air partition coefficient for describing particle/gas partitioning of aromatic compounds in urban air. *Environ Sci Technol* 32:1494–1502.

Heicklen J. 1981. The correlation of rate coefficients for H-atom abstraction by HO radicals with C–H bond dissociation enthalpies. *Int J Chem Kinet* 13:651–665.

Heikes BG. 1984. Aqueous H_2O_2 production from O_3 in impingers. *Atmos Environ* 18:1433–1445.

Hodson J. 1988. The estimation of the photodegradation of organic compounds by hydroxyl radical reaction rate constants obtained from nuclear magnetic resonance spectroscopy chemical shift data. *Chemosphere* 17:2339–2348.

Jaenicke R. 1987. About the atmospheric aerosol. In: Zellner R, editor. Formation, distribution and chemical transformation of air pollutants. Weinheim, Germany: VCH-Verlagsgesellschaft. p 127–146.

Jefferson A, Tanner DJ, Eisele FL, Davis DD, Chen G, Crawford J, Huey JW, Torres AL, Berresheim H. 1998. OH photochemistry and methane sulfonic acid formation in the coastal Antarctic boundary layer. *J Geophys Res* D103:1647–1656.

John W, Wall SM, Ondo JL, Wang HC. 1986. Acidic aerosol concentrations and size distributions in California. In: Israel GW, editor. Aerosols: formation and reactivity. Oxford, UK: Pergamon. p 25–28.

Jolly GS, Paraskevopoulos G, Singleton DL. 1985. Rates of OH radical reactions. XII. The reactions of OH with c-C_3H_6, c-C_5H_{10}, and c-C_7H_{14}. Correlation of hydroxyl rate constants with bond dissociation energies. *Int J Chem Kinet* 17:1–10.

Junge C. 1975. Basic considerations about trace constituents in the atmosphere as related to the fate of global pollutants. *Prepr Div Environ Chem Am Chem Soc* 15:4–5.

Kamens RM, Perry JM, Saucy DA, Bell DA, Newton DL, Brand B. 1985. Factors which influence polycyclic aromatic hydrocarbons decomposition on wood smoke particles. *Environ Int* 11:131–136.

Kamens RM, Fulcher JN, Zhishi G. 1986. Effects of temperature on wood soot PAH decay in atmospheres with sunlight and low NO_x. *Atmos Environ* 20:1579–1587.

Kamens RM, Gino Z, Falcher JN, Bell DA. 1988. Influence of humidity, sunlight, temperature on the daytime decay of polyaromatic hydrocarbons on atmospheric soot particles. *Environ SciTechnol* 22:103–108.

Kamens RM, Guo J, Guo Z, McDow SR. 1990. Polynuclear aromatic hydrocarbon degradation by heterogeneous reactions with N_2O_5 on atmospheric particles. *Atmos Environ* 24A:1161–1173.

Kamens RM, Fan Z-H, Yao Y, Chen D, Chen S, Vartianen M. 1994. A methodology for modeling the formation and decay of nitro-PAH in the atmosphere. *Chemosphere* 28:1623–1632.

Kamens RM, Odum J, Fan Z-H. 1995. Some observations on times to equilibrium for semivolatile polycyclic aromatic hydrocarbons. *Environ Sci Technol* 29:43–50.

Klamt A. 1993. Estimation of gas-phase hydroxyl radical rate constants of organic compounds from molecular orbital calculations. *Chemosphere* 26:1273–1289.

Klamt A. 1996. Estimation of gas-phase hydroxyl radical rate constants of oxygenated compounds based on molecular orbital calculations. *Chemosphere* 32:717–726.

Klöpffer W, Frank R, Kohl E-G, Haag F. 1986. Quantitative Erfassung der photochemischen Transformationsprozesse in der Troposphäre. *Chemiker-Zeitung* 110:57–61.

Koch R, Krüger H-U, Elend M, Palm W-U, Zetzsch C. 1996. Rate constants for the gas-phase reactions of OH with amines: *tert*-butyl amine, 2,2,2-trifluoroethyl amine, and 1,4-diazabicyclo[2.2.2]octane. *Int J Chem Kinet* 28:807–815.

Koch R, Palm W-U, Zetzsch C. 1997. First rate constants for reactions of OH radicals with amides. *Int J Chem Kinet* 29:81–87.

Koester CJ, Hites RA.1992. Photodegradation of polychlorinated dioxins and dibenzofurans adsorbed to fly ash. *Environ Sci Technol* 26:502–507.

Krol M, van Leeuwen PJ, Lelieveld J. 1998. Global OH trend inferred from methylchloroform measurements. *J Geophys Res* D103:10697–10711.

Kwok ESC, Arey J, Atkinson R. 1994. Gas-phase atmospheric chemistry of dibenzo-*p*-dioxin and dibenzofuran. *Environ Sci Technol* 28:528–533.

Kwok ESC, Atkinson R, Arey J. 1995. Rate constants for the gas-phase reactions of the OH radical with dichlorobiphenyls, 1-chlorodibenzo-*p*-dioxin, 1,2-dimethoxybenzene, and diphenyl ether: estimation of OH radical reaction rate constants for PCBs, PCDDs, and PCDFs. *Environ Sci Technol* 29:1591–1598.

Kwok ESC, Atkinson R. 1995. Estimation of hydroxyl radical reaction rate constants for gas-phase organic compounds using a structure-reactivity relationship: an update. *Atmos Environ* 29:1685–1695.

Kwok ESC, Aschmann SM, Atkinson R. 1996. Rate constants for the gas-phase reactions of the OH radical with selected carbamates and lactates. *Environ Sci Technol* 30:329–334.

Liang C, Pankow JF, Odum JR, Seinfeld JH. 1997. Gas/particle partitioning of semivolatile organic compounds to model inorganic, organic, and ambient smog aerosols. *Environ Sci Technol* 31:3086–3092.

Ligocki M, Leunenberger C, Pankow JF. 1985. Trace organic compounds in rain II: Gas scavenging of neutral organic compounds. *Atmos Environ* 19:1609–1617.

Lyman WJ, Reehl WF and Rosenblatt DH. 1990. Handbook of chemical property estimation methods: Environmental behavior of organic compounds. Washington DC: American Chemical Society.

Medven Z, Güsten H, Sabljić A. 1996. Comparative QSAR study on hydroxyl radical reactivity with unsaturated hydrocarbons: PLS versus MLR. *J Chemometrics* 10:135–147.

Meylan W. 1997. Estimation accuracy of the Atmospheric Oxidation Program Version 1.8. Syracuse NY: Syracuse Research Corporation.

Meylan WM, Howard PH. 1993. Computer estimation of the atmospheric gas-phase reaction rate of organic compounds with hydroxyl radicals and ozone. *Chemosphere* 26:2293–2299.

Miller BR, Huang J, Weiss RF, Prinn RG, Fraser PJ. 1998. Atmospheric trend of chlorodifluoromethane (HCFC-22) and the global tropospheric OH concentration. *J Geophys Res* D103:13237–13248.

[NASA] National Aeronautics and Space Administration. 1997. Chemical kinetics and photochemical data for use in stratospheric modeling, Evaluation Number 12. Pasadena CA: Jet Propulsion Laboratory. JPL Publication 97-4.

Niki H, Maker PD, Savage CM, Breitenbach LP. 1978. Relative rate constants for the reaction of hydroxyl radical with aldehydes. *J Phys Chem* 82:132–135.

[OECD] Organization for Economic Cooperation and Development. 1992. The rate of photochemical transformation of gaseous organic compounds in air under tropospheric conditions. Paris, France. Environment monograph No. 61. OCDE/GD(92)172.

[OECD] Organization for Economic Cooperation and Development. 1997. Guidance document on direct phototransformation of chemicals in water. Paris, France. Environmental Health and Safety Publications. Series on Testing and Assessment No. 7. OCDE/GD(97)21.

Ohta T. 1983. Rate constants for the reactions of diolefins with OH radicals in the gas phase. Estimate of the rate constants from those for monoolefins. *J Phys Chem* 87:1209–1213.

Orth RG, Ritchie C, Hileman F. 1989. Measurement of the photoinduced loss of vapor phase TCDD. *Chemosphere* 18:1275–1282.

Palm W-U, Elend M, Krüger H-U, Zetzsch C. 1997. OH reactivity of airborne terbuthylazine adsorbed on inert aerosol. *Environ Sci Technol* 31:3389–3396.

Palm W-U, Elend M, Krüger H-U, Zetzsch C. 1999. Atmospheric degradation of a semivolatile aerosol-borne pesticide: reaction of OH with Pyrifenox (an oxime-ether), adsorbed on SiO_2. *Chemosphere* 38:1241–1252.

Pankow JF. 1987. Review and comparative analysis of the theories on partitioning between the gas and aerosol particulate phases in the atmosphere. *Atmos Environ* 21:2275–2283.

Pankow JF. 1998. Further discussion of the octanol/air partitioning coefficient, K_{oa}, as a correlating parameter for gas/particle partition coefficients. *Atmos Environ* 32:1493–1497.

Parlar H. 1990. The role of photolysis in the fate of pesticides. In: Hutson DH, Roberts TR, editors. Volume 7, Progress in pesticide biochemistry and toxicology. Environmental fate of pesticides. Chichester, UK: Wiley. p. 245–276.

Pilling MJ, Seakins PW. 1995. Reaction kinetics. Oxford, UK: Oxford Science Publications.

Ruzo LO, Sundstrom G, Hutzinger O, Safe S. 1976. Photodegradation of polybromobiphenyls (PBB). *J Agric Food Chem* 24:1062–1065.

Scharping H, Zetzsch C, Dessouki HA. 1987. The uv absorption spectra of the trichlorobenzenes and the higher chlorinated benzenes in the gas phase and in n-hexane solution. *J Mol Spec* 123:382–391.

Scheringer M. 1997. Characterization of the environmental distribution behavior of organic chemicals by means of persistence and spatial range. *Environ Sci Technol* 31:2891–2897.

Sedlak DL, Andren AW. 1991. Aqueous-phase oxidation of polychlorinated biphenyls by hydroxyl radicals. *Environ Sci Technol* 25:1419–1426.

Seinfeld JH, Pandis SN. 1997. Atmospheric chemistry and physics: From air pollution to climate change. New York: Wiley-Interscience.

Sekušak S, Güsten H, Sabljić A. 1996. An *ab initio* study on reactivity of fluoroethane with hydroxyl radical: application of G2 theory. *J Phys Chem* 100:6212–6224.

Shetter RE, Davidson JA, Cantrell CA, Calvert JG. 1987. Temperature variable long path cell for absorption measurements. *Rev Sci Instrum* 58:1427–1428.

Siek TJ, Osiewicz RJ, Bath RJ. 1976. Identification of drugs and other toxic compounds from their ultraviolet spectra. Part III: Ultraviolet absorption properties of 22 structural groups. *J Forensic Sci* 21:525–551.

Sivils LD, Kapila S, Yan Q, Elseewi AA. 1994. Application of a two-dimensional chromatography system for gas-phase photodegradation studies of polychlorinated dibenzo-*p*-dioxins. *J Chromatog* A. 688:221–230.

Spivakovsky CM, Yevich R, Logan JA, Wofsy SC, McElroy MB. 1990. Tropospheric OH in a three-dimensional chemical tracer model: an assessment based on observations of CH_3CCl_3. *J Geophys Res* D95:18441–18471.

Staffelbach TA, Orlando JJ, Tyndall GS, Calvert JG. 1995. The absorption spectrum and photolysis quantum yields of methylglyoxal. *J Geophys Res* D100:14189–14198.

Strand A, Hovø. 1996. A model strategy for the simulation of chlorinated hydrocarbon distributions in the global environment. *Water Air Soil Pollut* 86:283–316.

Tomlin C, editor. 1995. The pesticide manual. Tenth edition. Surrey UK: The British Crop Protection Council.

Tully FP, Droege AT, Koszykowski ML, Melius CF. 1986. Hydrogen-abstraction from alkanes by OH. 2. Ethane. *J Phys Chem* 90:691–698.

Tysklind M, Lundgren K, Rappe C. 1993. Ultraviolet absorption characteristics of all tetra-to octachlorinated dibenzofurans. *Chemosphere* 27:535–546.

van Pul WAJ, de Leeuw FAAM, van Jaarsveld JA, van der Gaag MA, Sliggers CJ. 1998. The potential for long-range transboundary atmospheric transport. *Chemosphere* 37:113–141.

Wallington TJ, Japar SM. 1989. Fourier transform infrared kinetic studies of the reaction of HONO with HNO_3, NO_3 and N_2O_5 at 295 K . *J Atmos Chem* 9:399–409.

Wallington TJ, Schneider WF, Worsnop DR, Nielsen OJ, Sehested J, DeBruyn W, Shorter JA. 1994. Environmental impact of CFC replacements: HFCs and HCFCs. *Environ Sci Technol* 28:320A–326A.

Wallington TJ, Nielsen OJ. 1998. Measurement of rate constants for radical reactions in the gas phase. In: Alfassi Z, editor. General aspects of free radical chemistry. New York: Wiley.

Warneck P. 1988. Chemistry of the natural atmosphere. San Diego CA: Academic. International Geophysics Series Volume 41.

Warneck P. 1999. Chemistry of the natural atmosphere, 2nd edition. San Diego CA: Academic Press. International Geophysics Ser. Vol. 71.

Warneck P, Mirabel P, Salmon GA, Van Eldik R, Vinckier C, Wannowius KJ, Zetzsch C. 1996. Review of the activities and achievements of the EUROTRAC subproject HALIPP. In: Warneck P, editor. Transport and chemical transformation of pollutants in the troposphere, Volume 2: Heterogeneous and liquid phase processes. Heidelberg, Germany: Springer.

Weast RC, editor. 1974. Handbook of chemistry and physics, 55th edition. Cleveland OH: CRC.

Webster E, Mackay D, Wania F. 1998. Evaluating Environmental Persistence. *Environ Toxicol Chem* 17:2148–2158.

Weschler CJ, Graedel TE. 1982. Theoretical limitations on heterogeneous catalysis by transition metals in aqueous atmospheric aerosols. In: Schryer DR, editor. Heterogeneous atmospheric chemistry. Washington DC: American Geophysical Union. Geophysical Monograph 26. p 196–202.

Zellner R, Herrmann H. 1995. Free radical chemistry of the aqueous atmospheric phase. In: Clark RJH, Hester RE, editors. Advances in spectroscopy, Volume 24. Chichester, UK: Wiley. p 381–451.

Zetzsch C. 1982. Predicting the rate of OH-addition to aromatics using σ^+ -electrophilic substituent constants for mono- and polysubstituted benzene. Presented at the 15th Informal Conference on Photochemistry; June 27–July 1; Stanford CA.

Zetzsch C. 1983. Testing the atmospheric persistence of chemicals using pulsed vuv photolysis of H_2O and time resolved resonance fluorescence of OH. Proceedings of the 6th World Congress on Air Quality, Volume 1. Paris, France: SEPIC/IUAPPA. p 81–87.

Zetzsch C. 1991. Photochemischer Abbau in Aerosolphasen. *Umweltwissenschaften Schadstoff-Forschung-Z. Umweltchem Ökotox* 3:59–64.

CHAPTER 3

Quantification of Persistence in Soil, Water, and Sediments

Robert Larson, Larry Forney, Leslie Grady Jr., Gary M. Klečka, Shigeki Masunaga, Willie Peijnenburg, Lee Wolfe

Accurate predictions of environmental exposure, combined with an understanding of toxic effect levels, are important elements in evaluating the environmental safety associated with use of large-volume organic chemicals. Inherent in the prediction of exposure concentrations is the identification and characterization of sources and emissions, natural formation, and a variety of degradation processes that control steady-state levels in the environment. A number of biotic and abiotic reactions can be responsible for the degradation of organic chemicals in soil, water, and sediments. Biotic reactions, which include such processes as microbial transformation, biodegradation and phytodegradation, are catalyzed by living organisms. Abiotic reactions, such as hydrolysis, photolysis, ionization, oxidation-reduction, ligand exchange, and various kinds of partitioning reactions, also can be important. With respect to biotic reactions, microbial transformation typically results in modest changes in chemical structure, whereas biodegradation is often viewed as complete mineralization to CO_2 (Alexander 1994).

While a number of reactions can be involved in the degradation of a particular chemical, it is important to identify the major environmental compartments in which the chemical is likely to be found (from a mass-loading point of view) and the major processes (from a kinetic point of view) in order to maximize the usefulness of exposure information. This chapter focuses on developing generic approaches for quantifying persistence in the major environmental compartments (soils, surface waters, and sediments) in which chemicals are likely to reside, based on known or predicted transport and partitioning processes. Exposure concentrations in engineered systems (e.g., municipal wastewater treatment plants) are not explicitly

CHAPTER PREVIEW

Evaluation of Persistence and Long-Range Transport of Organic Chemicals in the Environment. Gary Klečka et al., editors.

addressed, although the principles and approaches described herein should be applicable to such systems (Cowan et al. 1995). To simplify the evaluation of persistence in soil, water, and sediments, the discussion of biotic transformation processes is limited to microbial biodegradation, with a brief consideration of phytodegradation. Likewise, the treatment of abiotic processes is limited to hydrolysis, direct and indirect photolysis, and oxidation/reduction reactions.

In developing a generic approach for quantifying persistence in soil, surface waters, and sediments, the dominant mechanism considered is microbial degradation. This approach is reasonable in view of the fact that microorganisms play a dominant role in recycling carbon and many other elements in the environment. Clearly, abiotic reactions (e.g., hydrolysis, photolysis) are important for certain chemicals in some environments (e.g., surface waters). Abiotic processes, however, generally result in only alterations of the chemical structure or changes in the distribution of a chemical between the different environmental compartments (e.g., sorption), not in mineralization. By contrast, biodegradation often leads to significant decreases in both the concentration and total mass of a chemical in the environment. Moreover, when complete biodegradation of an organic chemical occurs, the chemical looses its molecular identity and is converted into benign organic and inorganic end products.

The conceptual framework for quantifying persistence in soil, water, and sediments shown in Figure 3-1 is comprised of a tiered approach involving screening and detailed evaluation and resembles that for substance evaluation described in Chapter 8. This conceptual framework embraces common principles of sound science while incorporating the logic and spirit of the precautionary principle and allowing for the input of expert judgment and flexibility.

During the screening evaluation outlined in Figure 3-1, existing knowledge and partitioning information for a chemical are used to define which environmental compartments are relevant and whether the chemical is likely to be persistent in those compartments. When a chemical is deemed to be nonpersistent, it is set aside from further evaluation. As a simple example, chemicals that are expected to partition predominately to surface waters because they possess high water solubility and low vapor pressure would be considered nonpersistent and set aside, provided that they pass a ready biodegradation test. In contrast, chemicals that do not pass the screening level and could be persistent are carried through the process and evaluated further. At this level, additional data would be collected and a more detailed kinetic evaluation would be performed to determine whether the chemical could be set aside or warrants further study (risk assessment, etc.).

In the remainder of this chapter, a detailed discussion of the principles and methodology behind the conceptual framework for quantifying persistence in soil, water, and sediment environments will be presented. This discussion is designed to

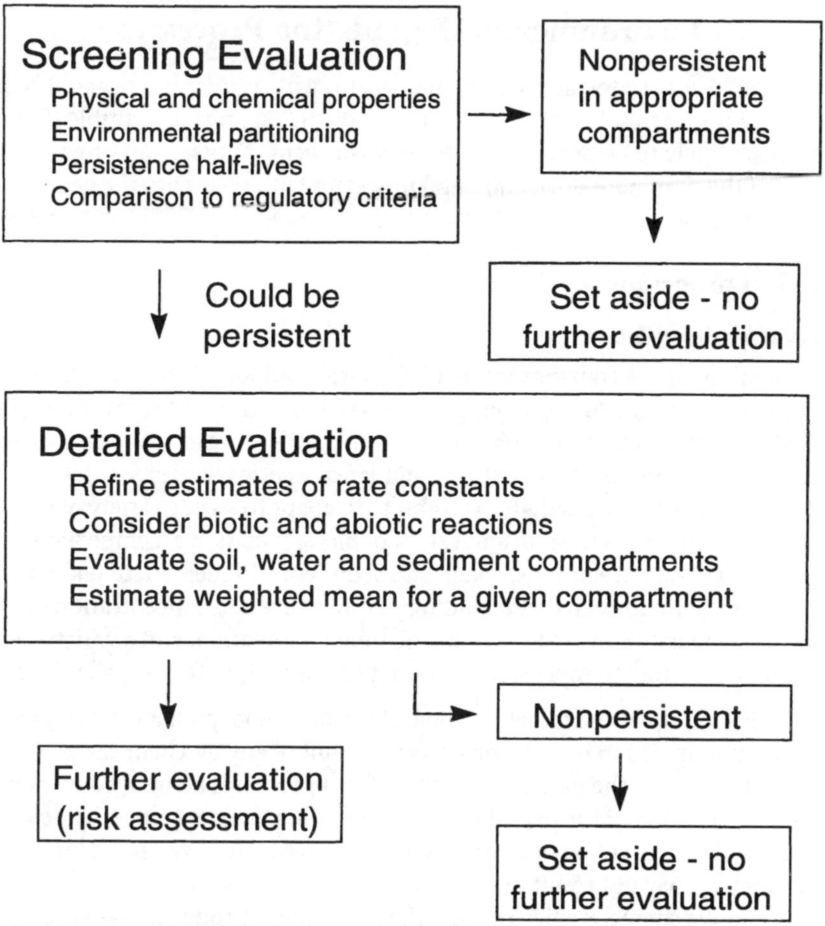

Figure 3-1 Conceptual framework for evaluating persistence in soil, water, and sediment

- provide a sound technical basis for regulatory decision-making that is both transparent and consistent and
- describe a simple generic framework for quantifying persistence that is broadly applicable to organic chemicals and provides a transparent conceptual basis for estimating environmental exposure concentrations in soil, water, and sediments.

The development of regulatory policies pertaining to the persistence and long-range transport of organic chemicals is rapidly evolving, and it is important that scientists contribute and participate in this process. To do so, we should make policymakers aware of what we know and believe to be true today and provide guidance as to how information can best be used, while recognizing that our scientific understanding and the quality of information will continue to advance in the future.

Environmental Degradation Processes

Degradation in soil, water, and sediment environments is mainly a result of biodegradation, hydrolysis, and photolysis. Compounds that are not susceptible to these processes are likely to be persistent in the environment. This section provides an overview of the dominant environmental processes with respect to the general types and kinetics of the reactions.

Biological processes

Microbial biodegradation

Biodegradation, or the transformation of chemical compounds by the action of living organisms, is one of the major processes that determines the fate of organic chemicals in aquatic and terrestrial environments. Microorganisms play a major role in biodegradation because of their abundance, species diversity, catabolic versatility, high metabolic activity, and ability to adapt to a wide variety of environmental conditions. Both the rate and extent of biodegradation are important in the evaluation of persistence in soils, water, and sediments. Biodegradation kinetics are dependent on numerous factors including the structure and concentration of the chemical, the density and activity of the microbial population, and a variety of environmental factors (temperature, pH, oxygen content, nutrient availability, etc.).

Enzymes that act upon the target compound mediate biodegradation. Primary biodegradation is said to have occurred as the result of any biochemical process that alters the structure of the parent compound. Such processes alone are sufficient to warrant classification of the parent compound as nonpersistent since, by definition, the identity of the parent compound has been changed. However, these biotransformations may or may not result in the loss of toxicity, persistence, or bioaccumulative characteristics because the metabolites produced (often referred to as daughter products) also could have persistent organic pollutant (POP) or persistent, bioaccumulative, and toxic (PBT) characteristics.

By contrast, it is also possible for a series of enzymes to act sequentially to mineralize an organic compound, forming carbon dioxide, methane, water, and other inorganic compounds. This process is sometimes referred to as ultimate biodegradation. Although ultimate biodegradation or mineralization are desirable, they are not necessary, provided that potential daughter products arising from the enzymatically catalyzed reactions are benign. If appropriate, metabolites may be evaluated in the same manner as the parent compound, or empirical evidence of mineralization can be collected directly through experimentation since the reaction rate for mineralization can never exceed the reaction rate for biotransformation of the parent compound. Consequently, experimental data from mineralization experiments provide a conservative estimate of the rate of biodegradation.

Kinetics of biodegradation

Several features commonly observed during the biodegradation of chemicals in the environment are illustrated by data on the biodegradation of a ^{14}C-labeled quaternary ammonium compound in activated sludge (Figure 3-2). Typically there is a lag phase before loss of the parent compound is observed. This lag phase (often referred to as an acclimation period) can arise because time is required for any or all of the following processes:

- induction or derepression of specific enzyme systems,
- adaptive evolution and natural selection of novel metabolic capabilities,
- increases in the numbers of microorganisms able to catalyze a particular transformation, and
- dilution of the compound to nontoxic levels.

The time required for acclimation to a particular compound in a specific environment is not predictable and varies with location, concentration of the compound, oxygen concentration, temperature, number and kinds of microbial populations, and so on (Alexander 1994). After the initial acclimation period, the lag phase generally is not seen again as long as subsequent additions of the compound occur within a reasonable time; this period may vary from weeks to more than a year (Alexander 1994). Following the onset of parent compound biotransformation, a transient accumulation of intermediates (hydrolysis products HP-1 and HP-2) may be observed, while substrate carbon is simultaneously converted to cell biomass and $^{14}CO_2$.

The rate of an enzyme-catalyzed reaction is proportional to the amount of enzyme present as well as the concentration of compound that serves as a substrate for the enzyme. Consequently, the kinetics of microbial degradation can be described by the generalized expression

$$(dS/dt)_b = f(E,S) \qquad \text{(Equation 3-1)},$$

wherein E is the concentration of enzyme and S is the concentration of the substrate being acted upon. If the substrate is metabolized intracellularly, the intracellular substrate concentration, not the extracellular substrate concentration, determines the reaction rate since the intracellular concentration of substrate is usually unknown. In these circumstances, it is assumed that substrate transport occurs via diffusion and, consequently, the concentration of substrate inside cells is in equilibrium with the soluble substrate outside the cells. It should further be noted that the concentration of soluble substrate outside the cells is not equivalent to the total substrate concentration in the compartment because some fraction of the substrate may be biologically unavailable for metabolism by virtue of sorption to organic matter, mineral surfaces, and other material in the compartment. Hence, it is important to estimate or measure the actual amount of substrate that is soluble and therefore biologically available to undergo biodegradation.

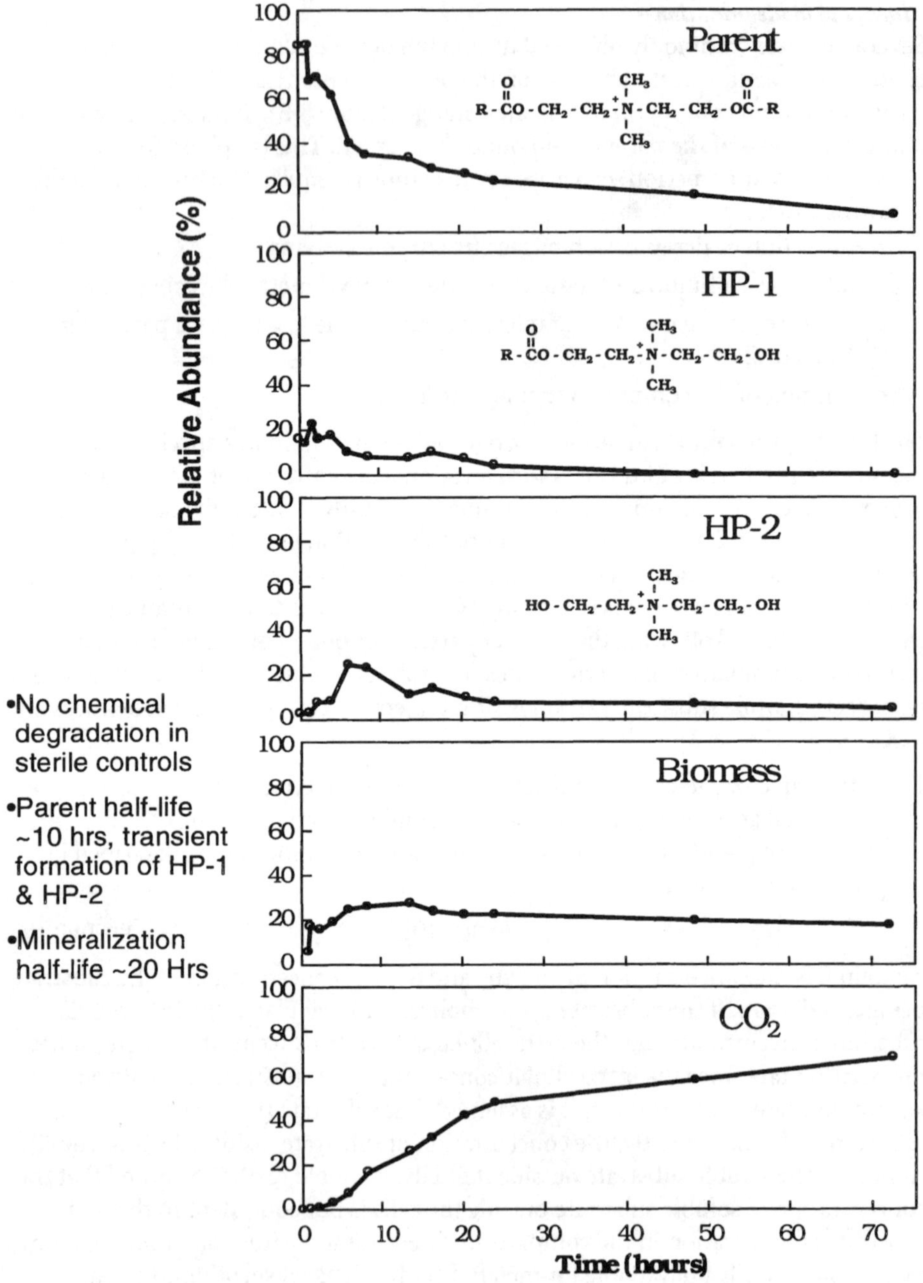

Figure 3-2 Kinetics of biotransformation and mineralization in activated sludge

Similarly, the concentration of enzyme involved in a biodegradation process cannot be readily determined, and a surrogate measure must be used. Because enzymes are produced by biomass, it can be assumed that the quantity of enzyme present is in some way proportional to the amount of biomass present. However, the exact nature of the proportionality depends on the circumstances under which biodegradation occurs. When a small but significant amount of a compound is discharged to a biological wastewater treatment system, certain microbial populations in the bioreactor may be able to use it as growth substrate because they have the genetic capability to synthesize the requisite enzymes. In other words, they are competent to perform the biotransformation. Typically, only a small fraction of the total biomass will be competent for the degradation of a particular substance because it constitutes a small fraction of the total organic matter entering the system (Magbanua et al. 1998). Consequently, the proportionality coefficient relating the quantity of enzyme to the quantity of biomass will be small.

In some cases, "cometabolism" of a compound may occur wherein the enzymatic transformation is simply fortuitous and provides no benefit to cells in terms of carbon or energy. In such cases, the fraction of biomass that is competent can be quite high or very low, causing the proportionality coefficient to vary proportionally. Obviously such differences can be important when attempting to extrapolate reaction-rate coefficients from one location to another.

On the basis of the previous discussion, the kinetics of biodegradation of an organic chemical in the environment can be expressed as the second-order relationship

$$(dS/dt)_b = -k_b[B][S] \qquad \text{(Equation 3-2)},$$

where k_b is the second-order biodegradation rate constant, B is the competent biomass concentration, and S is the concentration of the organic compound. If B and S are expressed as mass (M) per unit volume (L^3), then the units of k_b are $L^3M^{-1}t^{-1}$, giving the reaction rate units of $ML^{-3}t^{-1}$. The second-order rate constant (k_b) depends on a variety of environmental factors including pH and temperature, as well as properties of the microbial populations that carry out the biotransformation (Klečka 1985; Grady et al. 1996). Because the distribution and diversity of microbial populations are poorly understood, it is currently impossible to accurately predict how k_b might vary from location to location for a given set of environmental conditions (temperature, pH, and so on.). Moreover, the use of second-order kinetics to evaluate rates of biodegradation has been validated only for a limited range of organic compounds (Paris et al. 1981; Brown et al. 1990; Naziruddin et al. 1995).

Estimating rates of biodegradation by assuming pseudo–first-order kinetics

The effect of substrate concentration on enzyme-catalyzed reactions typically follows a saturation-type function, such as that described by the Michaelis-Menten equation (Alexander 1994). One characteristic of such a process is that it becomes first-order with respect to substrate at low concentrations and zero-order at higher substrate concentrations (Kovárová-Kovar and Egli 1998). Thus, when the target compound is present in the environment at low concentrations, it is reasonable to approximate the degradation rate by using kinetics that are first-order with respect to substrate concentration (Simkins and Alexander 1984). The concentration of substrate needed to allow approximation of the rate using first-order kinetics varies depending on characteristics of the compound being considered and on the microbial populations that are present, as well as their growth rate and exposure history. Generally it is assumed that first-order kinetics are applicable when the substrate concentration is in the low microgram-per-liter to milligram-per-liter range (Jannasch 1967; Simkins and Alexander 1984; Klečka and Maier 1985).

While a variety of kinetic expressions have been developed to describe the biodegradation of organic compounds in the environment (Paris et al. 1981; Simkins and Alexander 1984; Klečka 1985; Scow et al. 1986), the most commonly used approach is to incorporate the biomass concentration into the rate coefficient, making the rate of biodegradation pseudo–first-order overall

$$(dS/dt)_b = -k_{Bio}[S] \qquad \text{(Equation 3-3)},$$

where k_{Bio} is the pseudo–first-order biodegradation rate coefficient. For this expression to be valid, not only must the substrate concentration be low, making its effect first-order, but also the biomass concentration must not change appreciably as a result of growth so that it can be incorporated into the rate coefficient. This is a popular approach because

- pseudo–first-order rate coefficients are easy to calculate even when limited data are available,
- comparisons can be made with previously published data, and
- many programs used to model the environmental fate of chemicals employ first-order rate constants.

It is important to note that k_{Bio} is not an intrinsic property of a chemical. Rather, it is also a function of the number and activity of competent biomass present in an environment. Thus, when pseudo–first-order biodegradation coefficients are being extrapolated from one environment to another, one must account for potential differences in the amounts of competent biomass present in the two systems. For example, the competent biomass levels in a test that employs an activated sludge inoculum are typically much different from those in a river die-away test, and, accordingly, they yield very different pseudo–first-order rate coefficients. When pseudo–first-order biodegradation rate constants need to be standardized, division of the rate constant by biomass concentration may be done as a means to approximate a second-order rate constant, k_b (Paris et al. 1981). In principle, this approach should be used only when the level of competent biomass is known. However, in practice, biomass concentrations are commonly based on measurements of the total microbial biomass that are based on dry weight, viable cell plate counts, adenosine triphosphate concentrations, and so on. In addition, the levels of competent biomass can vary with location, nutrient status, and time. Thus, this approach can lead to overestimating the amount of competent biomass and introduce additional uncertainty into the kinetic analysis.

Phytodegradation

The transformation of organic compounds by green plants has received a considerable amount of recent attention (Schnoor et al. 1995). Phytodegradation is believed to be important in view of the fact that plants cover approximately 145,699,467 km^2 of the terrestrial surface (Loveland and Belward 1997). Although there are no large databases for these types of reactions, certain phytodegradation processes may represent important loss pathways for some classes of compounds. Halogenated hydrocarbons have been shown to be degraded in aqueous systems by plants (O'Neill et al. 1998). Jeffers and Wolfe (1997, 1998) have shown that volatile halogenated organic compounds can be degraded through uptake by foliage. Nitroaromatics have been shown to undergo reductive transformations with very short half-lives in terrestrial and aquatic plants (Hughes et al. 1997; Vanderford et al. 1997), and organophosphate pesticides can be rapidly oxidized and hydrolyzed by axenic plant tissue cultures (Gao et al. 1999). Similar studies have demonstrated that polycyclic aromatic hydrocarbon (PAHs) are susceptible to phytodegradation in plant/soil systems (Cunningham and Berti 1993; Banks et al. 1999). When phytodegradation is included in computer simulations of global persistence, the results suggest that plants can play a major role in the removal of certain organic compounds from the biosphere (Jeffers and Wolfe 1997, 1998).

Currently, phytotransformations are described by second-order kinetics, wherein

$$(dS/dt)_{phyto} = -k_{phyto}[P][S] \qquad \text{(Equation 3-4)},$$

where k_{phyto} is the second-order plant mediated disappearance rate coefficient with units of $L^3M^{-1}t^{-1}$, P is the plant biomass (M/L^3), and [S] is the substrate concentration (M/L^3).

In the future, phytotransformation should be included in assessing the persistence of chemicals in the environment as the database and understanding increases and models are expanded to include the process. As with microbial degradation, the second-order rate constant can be reduced to a pseudo–first-order rate coefficient by assuming a given amount of plant material and low substrate concentration (Jeffers and Wolfe 1998). This approach has been used in the global modeling of methyl bromide.

Abiotic processes

Hydrolysis

The reaction of organic compounds with water can be an important degradation mechanism for certain classes of chemicals in aquatic environments, moist soils, and ground water. Hydrolysis reactions can be divided into acid- and base-catalyzed reactions and neutral hydrolysis. Although "pure" hydrolysis can be considered to involve only the reactant species provided by water, i.e., H^+, OH^-, and H_2O, it would be incomplete to exclude from consideration the analogous reactions with other nucleophilic chemical species often present in the environment, such as sulfide ion in anaerobic bogs, chloride ion in sea water, and various ions that may be present in laboratory buffer solutions. The importance of the various reactants may complicate the assessment of hydrolysis because of the subtle variations in mechanisms reported for many compounds (Wolfe and Jeffers 2000). Since a given compound may be susceptible to one or more reaction mechanisms, a thorough evaluation often is warranted.

The various hydrolytic pathways can account for significant degradation of certain classes of organic chemicals, as well as selected members of other classes, while many organic chemicals are completely inert with respect to hydrolysis (Wolfe and Jeffers 2000). For example, esters can undergo neutral, base-catalyzed, and acid-catalyzed hydrolysis simultaneously. Compounds in the following classes, or that contain certain functional groups, may be susceptible to hydrolysis:

- alkyl halides,
- allyl and benzyl halides,
- polyhalogenated methanes,
- epoxides,
- aliphatic acid esters,
- aromatic acid esters,
- amides,
- carbamates,
- phosphoric acid esters,
- thiophosphoric acid esters,
- phosphoric acid halides,
- dialkylphosphonohalidates, and
- dialkylphosphorohalides (Lyman et al. 1990).

However, aromatic acid esters and amides only hydrolyze at a significant rate at very high pH values. Compilations of hydrolysis rate constants have been published in papers by Mabey and Mill (1978), Ellington et al. (1987), and Ellington (1989).

Neutral hydrolysis is the reaction of an organic compound with water. In this process, the parent carbon–X bond is broken and a carbon–oxygen bond is formed at the same time (Lyman et al. 1990). The rate of neutral hydrolysis is pseudo–first-order, as water is present in excess, and the rate changes only with the concentration of substrate present in the reaction mixture

$$(dS/dt)_{nh} = -\, k_{H_2O}\,[S] \qquad \text{(Equation 3-5)},$$

where k_{H_2O} is the rate constant for neutral hydrolysis and S is the concentration of the target compound.

In both acid- and base-catalyzed hydrolysis, hydrogen and hydroxide ion concentrations, respectively, as well as substrate concentration control the rate of reaction and thus the kinetics are second-order:

$$(dS/dt)_{ah} = -\, k_{H^+}\,[H^+][S] \qquad \text{(Equation 3-6)},$$

where k_{H+} is the rate constant for acid-catalyzed hydrolysis and H^+ is the hydrogen ion concentration, and

$$(dS/dt)_{bh} = -\, k_{OH^-}\,[OH^-][S] \qquad \text{(Equation 3-7)},$$

where k_{OH^-} is the rate constant for base-catalyzed hydrolysis and OH^- is the hydroxide ion concentration. A pseudo–first-order rate constant can be calculated from the acid- and base-catalyzed second-order rate equations (valid only at the selected pH) by multiplying the second-order rate constant with the hydrogen or hydroxide ion concentration at the measured pH value.

Taken together, hydrolysis can be described by the equation

$$-(dS/dt)_h = k_{H^+}\,[H^+][S] + k_{H2O}[S] + k_{OH^-}\,[OH^-][S] \qquad \text{(Equation 3-8)}.$$

It is important to note that all quantities in brackets represent concentrations; thus, H^+ or OH^- values deduced from pH measurement or buffer calculations must be converted from an activity to a true molar concentration. Furthermore, depending on the chemical under consideration, the susceptibility, and the presence of other reactants in the water, additional kinetic terms can be added to the generalized equation, including

$$k_{SH^-}[SH^-][S] \qquad k_{Cl^-}[Cl^-][S] \qquad k_{B^-}[B^-][S] \qquad \text{(Equation 3-9)},$$

where B^- represents other inorganic anions that function as nucleophiles and react with the substrate, and k_{SH^-}, k_{Cl^-}, and k_{B^-} represent second-order rate constants for reaction with sulfide, chloride, and other inorganic ions, respectively.

Hydrolysis in sediments is generally predicted on the basis of the aqueous-phase rate coefficients. In certain cases, alkaline hydrolysis in the solids phase can be attenuated relative to the aqueous phase, while acid hydrolysis may be accelerated. These effects, however, have not been extensively evaluated. Neutral hydrolysis rate coefficients are assumed to be the same in the sediment–solids sorbed phase as in the bulk aqueous phase. Because the global pH is assumed to be about 7.9 (Snoeyink and Jenkins 1980), neutral hydrolysis generally will dominate the reaction of organic compounds in sediment environments, and the contribution from acid-and base-mediated reactions can be omitted unless conditions dictate that they be included.

Hydrolysis in moist soils is not well-defined, and, unfortunately, many of the laboratory degradation studies that have been carried out in moist soils do not distinguish between abiotic and biotic degradation processes. Unless there are specific cases in which rate constants have been reported for a chemical, and the corresponding kinetic expressions have been derived, the hydrolysis of organic chemicals in moist soils will not be included in the persistence evaluations.

Photolysis

Photolytic transformations of organic compounds in the environment have been extensively studied. Photolysis in surface waters can contribute significantly to the attenuation of organic compounds under certain conditions. In contrast, the photolysis of organic compounds on soils and aquatic sediments generally is not considered to be of significance. The importance of photolysis in the degradation of organic compounds on plant surfaces is unclear, but, given the recent interest in the role of plants in attenuating the transport of organic compounds, further research may be warranted. As discussed in Chapter 2, photolytic degradation may be significant for certain classes of POPs (polychlorinated biphenyls [PCBs], polychlorinated dibenzo-*p*-dioxins [PCDDs], polychlorinated dibenzofurans [PCDFs], chlorophenols, chlorobenzenes, etc). Furthermore, Crosby and Wong (1977) reported that tetrachlorodibenzo-*p*-dioxin (TCDD) present in herbicide formulations on leaf surfaces was rapidly degraded within a few hours. Because TCDD losses were negligible in dark controls, the investigators attributed the disappearance to photochemical degradation and suggested that organic components on the leaf surfaces could potentially serve as hydrogen donors for the reaction.

In near-surface natural waters, photolysis can be addressed as 2 well-defined processes—direct and indirect (Zepp 1982). Direct photolysis is a first-order process, and the rate of the reaction for a given compound is governed by the quantum yield, the ultraviolet-visible absorption spectrum for the compound, and the sunlight intensity. Unfortunately, there are no reliable ways to predict the absorption spectra or quantum yields for most organic compounds. There are, however, standard experimental protocols for determining pseudo–first-order rate coefficients that are applicable to many types of organic compounds. Well-defined indirect processes include photosensitized reactions and secondary reactions of organic compounds with photo-derived reactive anionic and radical species.

The kinetics of photolysis of a compound in water is described by the equation

$$(dS/dt)_{photo} = -k_{photo}[S] \qquad \text{(Equation 3-10)},$$

where k_{photo} is the photolysis rate coefficient and S is the concentration of the target compound. In natural waters, k_{photo} is the sum of the direct and indirect photodegradation rate constants, respectively. As discussed below, the relative importance of these two processes is compound specific.

Direct photolysis

Certain compounds absorb energy and undergo direct photolysis when exposed to light, resulting in permanent alteration of the chemical structure. However, the ability to absorb light at environmental wavelengths (290 to 750 nm) is not sufficient evidence to indicate that a compound will undergo photodegradation. Many photoexcited compounds will release the absorbed energy without undergoing chemical transformation. Compounds containing unsaturated carbon–carbon bonds or aromatic rings—as well as nitrosamines, benzidines, chlorinated organics, and some metal complexes—are susceptible to direct photodegradation (Zepp 1982).

The rate of direct photolysis in water by sunlight is given by

$$(dS/dt)_{dir\ photo} = -k_{dir}\,[S] \qquad \text{(Equation 3-11)},$$

where k_{dir} is the direct photodegradation rate constant. The first-order rate expression is derived from the fact that the photolysis rates are generally proportional to the quantum efficiency and the rate of light absorption (Zepp 1982). This relationship can be written as

$$(dS/dt)_{dir\ photo} = -k_{abs}\,\phi\,[S] \qquad \text{(Equation 3-12)},$$

where k_{abs} is the specific rate of sunlight absorption and ϕ is the quantum yield.

Direct photolysis in distilled water is a first-order process at low concentrations of the compound; the first-order rate coefficient can be used after it is corrected for depth and sunlight intensity. The coefficient k_{abs} is calculated from the overlap between the absorption spectrum of the compound and the spectrum of the light source. The quantum yield, ϕ, is defined as the fraction of the species that is converted to an excited state that undergoes a particular fate. It is generally wavelength independent and is preferably determined in distilled water or in polar water/solvent mixtures for compounds of low water solubility. For many POPs, the quantum yield is low and thus even if the compound strongly absorbs sunlight its rate of transformation will be slow.

If photolysis studies are not available, spectral data can be used to indicate whether or not a compound may be susceptible to photodegradation. Because spectral data are usually generated in solvents other than water, the data should be corrected for solvent effects. A compound that adsorbs strongly in the range of 290 to 700 nm may potentially photolyze. Compounds that do not absorb light in this range are unlikely to photodegrade at environmentally relevant rates.

The computer program GCSOLAR is useful for calculating the rates of aqueous photolysis for organic compounds by sunlight as a function of latitude, time, and water-column depth (Zepp and Cline 1977). The program requires the quantum yield and the measured absorption spectra. The program has been shown to be useful for determining the contribution of photolysis in both marine and freshwater systems. Information about GCSOLAR is available on the Internet at the U.S. Environmental Protection Agency's Web site. It is available for electronic distribution with documentation at www.epa.gov/CEAM. The equations and scientific rationale for the model calculations have been described by Zepp and Cline (1977).

Indirect photolysis

Compounds that do not undergo direct photodegradation may be photolyzed indirectly via sensitized photolysis or photooxidation. Indirect sensitized photolysis can occur whenever a light-absorbing compound, referred to as the photosensitizer, absorbs light energy and then transfers the energy to the target compound. Photosensitizers in the environment include quinones, humic acid, flavins, trace metals, nitrate, nitrite, and hydrogen peroxide (Zepp 1991). Indirect photolysis is indicated when measured degradation rates are higher in natural waters then in distilled water or when there is a large variation in the rates measured in different natural waters.

Photooxidation also can play a role in degradation of compounds in aquatic environments. Light can be directly absorbed by dissolved organic matter naturally present in many freshwater systems and used to generate oxidants that then react with the organic compound. For example, singlet oxygen ($^1O_2$) is formed by the interaction of dissolved oxygen with an excited photosensitizer. Singlet oxygen then becomes the reactive species and can react with the target compound. Other oxidants commonly found in aquatic environments include the peroxy radical (RO·), hydroxy radical (HO·), $^1O_2$, ozone (O_3), and triplet diradicals (Mill 1989).

It is more complicated to calculate indirect photolysis rates than those of direct photolysis because the molar absorptivities and concentrations of the photosensitizers present in natural waters are usually unknown. Modeling the transformation of compounds due to sensitized reaction mechanisms also requires knowledge of the rate of light absorption by the sensitizer, the efficiency of the sensitizer triplet, and the efficiency of energy transfer from the sensitizer to the compound.

The reaction kinetics for indirect photolysis are second-order and are dependent on both the concentration of the sensitizer/oxidant and the target compound present in that environment (Mill and Mabey 1985; Mill 1999) as follows:

$$(dS/dt)_{sen} = -k_{sen}\,[Sen]\,[S] \qquad \text{(Equation 3-13),}$$

$$(dS/dt)_{photo\text{-}ox} = -k_{ox}\,[Ox]\,[S] \qquad \text{(Equation 3-14),}$$

where k_{sen} and k_{ox} are the second-order rate constants for photosensitization and photooxidation, respectively, and the concentrations of sensitizer and oxidant are given by [Sen] and [Ox], respectively. Average concentrations of the photosensitizers/oxidants present in natural waters have been published (Mill 1999); however, concentrations can vary considerably both temporally and spatially. In addition, there are other physical and chemical factors that affect sunlight intensity in the near surface-waters that are not well-defined. The kinetics can be reduced to first-order if it is assumed that the concentration of oxidant/sensitizer does not change during the study period. A single rate constant should be calculated for both sensitization and for photooxidation by summing the rate constants for each reacting species.

Oxidation/reduction reactions

Quantitative descriptions of oxidation/reduction (redox) reactions of organic compounds have been described for relatively few compounds in aquatic and sediment systems when compared to hydrolysis, but the reactions that have been studied can be very fast and occur across a diverse set of soil, water, sediment systems. Although there are limited kinetic data for abiotic reactions for organic compounds in soils, natural waters, and sediments, the database is expanding rapidly and abiotic oxidation/reduction is emerging as an important environmental degradation process. Oxidation/reduction transformations are divided into reductive and oxidative reactions. Reductive reactions are generally more important in sediments and oxidative reactions more important in surface waters.

Reductive transformations

Kinetic data are available for the abiotic reduction of selected functional groups on organic compounds in a variety of sediment systems. Rate coefficients are available for organic compounds containing nitro, cyano, halo, and azo functional groups and in several cases quantitative structure-reactivity relationships (QSRRs) can provide calculated rate coefficients.

In general, the rate of reductive transformation can be described by the second-order expression

$$(dS/dt)_{red} = -k_{red}[S][R] \qquad \text{(Equation 3-15)},$$

where [S] is the concentration of substrate and [R] is the concentration of reductant. Unfortunately, in many studies with environmental samples, the nature of the reductant is often not identified. Consequently, the kinetics of abiotic reductive transformations are generally reported using pseudo–first-order decay coefficients. However, it has been shown that for certain redox reactions, the reductant concentration/activity is proportional to the fraction of organic carbon (OC) in the sediments and the kinetic expression becomes

$$(dS/dt)_{redoc} = -k_{redoc}[S][OC] \qquad \text{(Equation 3-16)},$$

where k_{redoc} is the second-order reductive rate coefficient and [OC] is the organic carbon content.

Peijnenburg, ‘t Hart et al. (1992) have shown that to correctly describe reductive dehalogenation transformations, the algorithm must take into account sorption, because only the fraction of the compound present in the aqueous phase of the sediment–water system reacts. Their study suggested that OC not only provides the source of the reactant, but also decreases the overall reaction rate by contributing to sorption of the target compound. The corrected rate expression is given by

$$(dS/dt)_{red(corr)} = -k_{redoc}[S](1/(1 + \rho K_d)) \quad \text{(Equation 3-17)},$$

where [S] denotes the substrate concentration, ρ is the sediment–water mass ratio (M/M), and K_d is the sediment–water distribution coefficient.

Oxidative transformations

The abiotic oxidation of certain organic compounds has been shown to occur in surface waters. One of the most extensively studied set of reactions involves the absorption of solar ultraviolet radiation by nitrate and nitrite in water with subsequent homolysis to produce free radicals (Larson and Weber 1994):

$$NO_3^- + H_2O + h\nu \rightarrow \cdot NO_2 + HO\cdot + OH^- \quad \text{(Equation 3-18)},$$

$$NO_3^- + h\nu \rightarrow NO_2^- + \cdot O\cdot \quad \text{(Equation 3-19)},$$

$$NO_2^- + H_2O + h\nu \rightarrow \cdot NO + HO\cdot + OH^- \quad \text{(Equation 3-20)}.$$

The rate of reaction of organic compounds with hydroxy radicals in near surface waters is given by the equation

$$(dS/dt)_{hydrox} = -k_{OH\cdot}\,[S]\,[\cdot OH\cdot] \quad \text{(Equation 3-21)}.$$

Zepp et al. (1987) and Mopper and Zhou (1990) have suggested that nitrate photolysis is the dominant process responsible for generation of hydroxy radicals in natural waters. While these free-radical reactions have been well described, the importance of oxidative transformations in the removal of organic compounds is unclear, because natural organic matter and sediment particles, which are present in much higher concentrations, will compete for the reactive species. Using known measurements of hydroxy radicals and measured rate coefficients, it is possible to model this oxidative reaction pathway (Mill 1999).

Factors Influencing Transformation Reactions

As discussed in the previous section, biotic and abiotic reactions for a variety of transformation processes can be described by second-order relationships. The generic rate equations for these second-order reactions are

$$(dS/dt)_b = -k_b[B]\,[S] \quad \text{(Equation 3-22)}$$

$$(dS/dt)_a = -k_a[A]\,[S] \quad \text{(Equation 3-23)},$$

where

k_b and k_a are the second-order biotic and abiotic transformation coefficients, respectively,

[B] is the competent biomass concentration for biotic reactions,

[A] is the reactant concentration for abiotic reactions, and

[S] is the aqueous phase concentration of the target chemical in the environmental compartment under consideration.

Numerous physical, chemical, and environmental factors are known to effect k_b, k_a, [B], [A], and [S] (Klečka 1985; Grady et al. 1996). The most important factors and the parameters they affect are summarized in Table 3-1. The following section discusses these factors and their qualitative and quantitative impacts on transformation rates in soil, water, and sediment compartments in more detail.

Table 3-1 Factors influencing transformation reactions

Factor	Units	Compartment and parameter		
		Water	Soil	Sediment
Temperature	Degrees C	k_a, k_b	k_a, k_b	k_a, k_b
pH	Units	k_a, k_b, A	k_a, k_b	k_a, k_b
Redox potential	mV	k_a, k_b	k_a, k_b	k_a, k_b
Light wavelength	nm	k_a		
Water activity			B	
# active cells	cells/L	B	B	B
Water solubility	mg/L	S	S	S
K_d/K_{oc}			S	S

Factors affecting k_a and k_b

Temperature

All biotic and abiotic reactions are affected by temperature. The effect of temperature on biotic and abiotic rate coefficients can be described thermodynamically by the Arrhenius equation

$$k = Ae^{-Ea/RT} \quad \text{(Equation 3-24)},$$

where

A is the frequency factor (time^{-1}),

Ea is the activation energy (cal mol^{-1}),

R is the gas constant (1.986 cal deg^{-1} mol^{-1}), and

T is the absolute temperature (°K).

For most reactions in which Ea is in the range of 8000 to 40,000 cal/mol, the Arrhenius equation can be simplified to a Q_{10} relationship, i.e., the change in rate coefficient for every 10 °C temperature change. For many biotic and abiotic reactions, the Q_{10} is in the range of 2 to 3, i.e., an approximate doubling of the rate coefficient for every 10 °C temperature rise. Biotic reactions are also susceptible to temperature effects that are not strictly thermodynamic, i.e., effects that are dictated by the temperature optima for different microorganisms carrying out the biodegradation reaction. Psychrophilic, mesophilic, and thermophilic microbial populations tolerate fairly narrow temperature changes (~ 10 to 15 °C), and biotic reactions requiring a specific microorganism or group of microorganisms will be disproportionately affected by temperature if it is outside of the tolerance zone for a particular group. Generally speaking, however, temperature effects on biotic reactions in natural systems are relatively small and predictable, particularly if temperature changes are gradual and sufficient time is available for the selection of populations with the appropriate temperature optima.

Hydrogen ion concentration (pH)

Abiotic reactions can occur over a wide range of pH values (2 to 12), and some abiotic reactions (e.g., alkaline hydrolysis) involve hydroxyl ions as one of the reactants. While the effect of pH on abiotic reactions can be substantial, it is usually predictable and quantifiable for specific reactions. By contrast, biotic reactions are typically restricted to a much narrower range of pH values (5 to 9), except in the case of extremely acidic or alkaline environments (e.g., thermal vents, hot springs). The effect of pH on biotic rate coefficients tends to be small and is comparable to values observed for temperature effects, i.e., a factor of 2 to 3 over a 5 to 9 pH range. As with temperature, however, biotic reactions requiring specific microbial populations will be disproportionately affected by pH if it is outside of the tolerance zone for a particular group.

Oxidation/reduction potential

As a general rule, aerobic microorganisms tend to degrade organic pollutants more efficiently from a thermodynamic standpoint than do fermentative or obligately anaerobic microorganisms. Furthermore, the concentration of electron acceptors (nitrate, sulfate, carbonate, iron) in such environments can be rate-limiting. As a result, biotic rate coefficients in anaerobic or reducing environments (–400 mV) are typically on the order of 10-fold lower than in aerobic environments. However, it is important to note that certain biotransformation processes, such as biological reductive dehalogenation of highly halogen-substituted compounds, occur almost exclusively in anaerobic environments. Examples of obligately anaerobic or reducing environments include methanogenic and sulfide-reducing environments like anaerobic digesters, organic rich sediments or highly eutrophic surface waters.

Radiant energy (light)

Light intensity plays a major role in abiotic reactions like photodegradation. Light generally is not considered to be an important factor in biotic reactions unless photosynthetic microorganisms (such as algae or photosynthetic bacteria) are involved.

Factors affecting competent biomass concentration, B

Microbial adaptation

Previous exposure of a natural microbial population to a particular compound has been shown to influence rates of degradation upon subsequent exposures (Spain et al. 1980; Spain and Van Veld 1983). The ways in which a microbial population can adapt to transform compounds more rapidly include

- induction or derepression of specific enzyme systems for metabolism or transport of a compound,
- the selection of new metabolic capabilities as a result of genetic changes,
- acquisition of catabolic genes through horizontal gene transfer, and
- an increase in the numbers of organisms able to catalyze a particular transformation through growth of a population.

Thus, the continual introduction of an organic compound into the environment may result in an increased number and kinds of competent bacteria in the microbial community.

Competent biomass

By far, the single most important variable controlling biotic reactions is the number of active, competent cells. The number, and therefore activity of competent cells is the key factor controlling biodegradation rates and, more than any other factor, dictates the accuracy of extrapolation of biotic rate coefficients from one compartment to another. There are numerous techniques for characterizing the number, diversity, and activity of microorganisms in laboratory and natural systems, and any discussion of this topic is clearly outside the scope of this chapter. The key point is that the amount of competent biomass must be known or estimated in a given compartment before it is possible to extrapolate pseudo–first-order rate coefficients among and between compartments with confidence. This extrapolation assumes that other factors like nutrient concentration and electron acceptors are not rate-limiting for cell growth or metabolism and that the activity per cell does not vary substantially between various microbial populations.

Water activity and ionic strength

Moisture content or water activity is a key factor affecting biological activity in soil. Soil moisture content must generally be greater than 10% for biotic reactions to occur. Soils with low water activity also may not support certain abiotic reactions (e.g., hydrolysis).

Ionic strength is also known to affect microbial activity. Microorganisms have adapted to grow in aquatic environments with high ionic strength, and it now appears that marine organisms have a specific requirement for certain ions. Alternatively, the activity of freshwater organisms may be affected in marine environments. The influence of ionic strength on microbial degradation of organic compounds has been studied, and there is some evidence that biodegradation rates in marine systems differ from those in freshwater environments (Bourquin and Przybyszewski 1977; Ward and Brock 1978). Moreover, ionic strength may affect the pathway of biodegradation, as recently described by Fu et al. (2000) for the reductive dehalogenation of chlorinated dibenzo-*p*-dioxins in freshwater and estuarine sediments.

Factors affecting substrate concentration, S

Water solubility and partitioning

Because abiotic and biotic reactions tend to act on soluble species, water solubility can be a key factor controlling reaction rates for many organic compounds. All other factors being equal, biotic and abiotic reaction rates for poorly soluble materials will be directly proportional to substrate concentration (i.e., first order) at concentrations below the solubility limit, provided that the rate of dissolution is not rate-limiting. In contrast, the degradation of slightly soluble materials will be zero-order, or independent of substrate concentration, at concentrations above their solubility limit. Therefore, the observed first-order rate coefficient reaches a maximum value, and may actually decrease, as the substrate concentration increases beyond the solubility limit.

Partitioning behavior, i.e., sorption of compounds to surfaces, directly affects the rates of biotic and abiotic reactions by reducing the concentration of soluble substrate. The extent of sorption that occurs is a function of the concentration of substrate and sorbent and the OC content of the sorbent. Sorption or partitioning coefficients are best determined experimentally and reported as a partition coefficient, K_d, defined as the concentration on solids (M/M) divided by the concentration in water (M/L^3). The K_d value, which typically is used to quantify sorption in soil systems, is a very useful parameter that allows the fraction of substrate sorbed to particulate matter to be estimated over a range of solids and substrate concentrations. If experimental data are not available, sorption can be estimated from octanol–water partitioning coefficients (K_{OW}).

In general, K_{OW} extrapolations are well-established for neutral hydrophobic materials and can be used with a high degree of certainty (Chiou et al. 1983; Baker et al. 1997). However, little research has been done on the extrapolation of K_{OW} values for charged hydrophilic materials, and experimental measurements are preferable. Both K_{OW} and K_d can be converted to K_{OC} values by normalizing for the OC content of the sorbent. Either K_d or K_{OC} can be used to calculate the fraction of substrate available for specific reactions under a given set of conditions.

As mentioned earlier, substrate concentrations used to estimate rate coefficients for biotic and abiotic reactions are for the soluble species only.

Sources of Information on the Extent and Rate of Degradation in Soil, Water, and Sediment

As discussed in previous sections of this chapter, the assessment of degradation typically should rely on empirical data with known reliability and validity. Kinetic information on the biotic and abiotic degradation of organic chemicals in soils, water, and sediments may be obtained from either field studies, laboratory simulation tests performed with grab-samples collected in the field, or standardized screening tests routinely conducted in the laboratory. When available, results of field studies are preferred for the purpose of deriving environmentally realistic transformation rates. For most chemicals, however, this type of data will be scarce, which will usually necessitate the use of data obtained from laboratory testing of grab-samples or screening tests. When quantitative structure-activity relationships (QSARs) have been properly validated, they can be used to predict the environmental fate of specific chemicals as well.

Field studies

Field studies generally involve monitoring the disappearance of a chemical in a soil plot, pond, stream, or groundwater aquifer. The experimental design typically specifies the location and number of monitoring sites, and computer models are often employed to analyze the data to obtain kinetic information. Because the extent and rate of degradation are derived indirectly from the monitoring data, care should be taken to insure that non-degradative losses (e.g., dilution, volatilization, leaching, etc.) do not result in the overestimation of degradation rates. This is often addressed in the experimental design by including the use of stable compounds with physical properties similar to those of the test chemical as tracers to correct for non-degradative processes.

Because the test conditions in field studies cannot be rigorously controlled, it is often difficult to reproduce the same conditions at other times or at other sites. Hence, with field studies, it is more difficult to anticipate and account for abiotic degradation or transport processes that may reduce the chemical substrate concentration. However, when properly controlled, field studies can provide the most realistic measurements of degradation rates. When considering the use of field studies for deriving rates of compound degradation, it is important to validate the measurement techniques used and the detection limits of the analytical methods, as well as to ensure that the data are representative of the environmental compartment and that relevant environmental conditions are well documented. Because of the cost and logistical complications of such studies, field data on biotic and abiotic degradation will be scarce for most chemicals.

Laboratory studies

The majority of information on the persistence of organic chemicals in soils, water, and sediments has been gathered through laboratory experimentation. Internationally accepted guidelines for evaluating both the abiotic and biotic degradation of organic compounds have been developed and updated under the umbrellas of the Organization for Economic Cooperation and Development (OECD 1996) and the International Organization for Standardization (ISO 1994). Additional testing protocols have been published by the USEPA (USEPA 1998) and the Society of Environmental Toxicology and Chemistry (Lynch 1995). The use of standardized test methods will facilitate interpretation and extrapolation of the resulting test data and the assessment of their reliability. Test conditions should be well-documented in order to assist in the assessment of the impact of environmental factors on rates of abiotic transformation.

Biodegradation

Biodegradation studies may be conducted using either pure or mixed cultures of microorganisms. Pure-culture studies are useful for characterizing potential degradation pathways, but they should not be used when evaluating environmental persistence because the organisms employed may not be active or important in the environment. Because biodegradation in nature generally involves the action of several different microbial populations within the community, it is preferable to use naturally occurring mixtures of organisms (e.g., grab-samples) as the inoculum for such tests.

Mixed-culture laboratory biodegradation studies can be categorized into the following types:

- screening tests,
- biological treatment simulations, and
- grab-sample tests.

OECD (1996) and USEPA (1998) guidelines are described for a variety of both "ready" and "inherent" biodegradability screening tests. In these tests, the rate of degradation usually is determined indirectly by measuring oxygen consumption, CO_2 evolution, consumption of chemical oxygen demand (COD), or disappearance of dissolved organic carbon (DOC). While the use of indirect analytical methods allows application of screening tests to a wide variety of chemicals, indirect methods provide less information about the loss of the chemical substrate. Often, variable results are obtained with screening tests; such variability often can be attributed to the source of the inoculum, the opportunity for acclimation during the test period (Cowan et al. 1996), the amount of biomass used in the test, and the physiological state of the microorganisms (Grady et al. 1996).

Biological treatment simulations consist of either semi-continuous (SCAS) (USEPA 1998) or continuous (CAS) (OECD 1996) activated sludge tests that use either synthetic or natural sewage. The sewage and the inoculum for such tests are usually obtained from a domestic wastewater treatment plant in order to reduce variability and to simulate treatment of consumer products. For compounds that exhibit greater than 90% degradation in semi-continuous tests, the results have been reported to be similar to those obtained with CAS tests. The OECD Coupled Units Test (OECD 1996) has been used more frequently, especially in Europe, and provides more reliable prediction of treatment efficiency in wastewater treatment systems than do results of SCAS tests if degradation of the compound in the SCAS test is < 90% (Boethling et al. 1997).

Grab-sample tests consist of monitoring the disappearance of the test chemical in a natural sample of soil, aquifer material, ground water, aquifer material/groundwater slurries, surface water, or surface water/sediment slurries. These studies often use low concentrations of the test chemical to more accurately simulate the environmental fate and degradation kinetics. Because natural samples often have background levels of organic materials and the test-chemical concentrations are low, indirect measurement techniques cannot be used; generally, disappearance of the test chemical is monitored using sensitive and specific analytical methods or by using radioactively (^{14}C or other) labeled test chemicals and radioanalytical techniques.

Guidelines for determining aerobic biodegradation in freshwater (USEPA 1998) and seawater (OECD 1996) environments have been described. Standard protocols are also available for testing the aerobic biodegradation of organic chemicals in soils (ISO 1994; Lynch 1995; OECD 1996). In addition, guidelines for determining the anaerobic biodegradation of organic chemicals in soil are available (ISO 1998). These guidelines will be generally applicable for the development of laboratory data on the degradation of POPs/PBTs. However, the duration of the incubation period may need to be adjusted for chemicals that degrade slowly, in order for half-lives in the range of 2 to 6 months to be determined.

The results of laboratory biodegradation tests are highly dependent upon the test protocol. This is especially true for the screening tests. For example, most screening tests do not include an acclimation period prior to initiating the test, and the incubation periods are of short duration which may not be long enough for acclimation to occur. In addition, the reproducibility of individual tests can be poor, especially between laboratories and, in some cases, even within the same laboratory.

Hydrolysis

Standard methods for measuring the aqueous hydrolysis of organic compounds as a function of pH have been described by the USEPA (1998), OECD (1996), and Lynch (1995). With these methods, the changes in the aqueous concentration of the test substance are monitored as a function of time in reactions conducted at a specific temperature and a specific pH. The experiments generally are performed for at least three pH values, enabling the calculation of the contributions of acid- and base-catalyzed, as well as neutral hydrolysis, processes. Essentially the same basic procedures can be used to measure the reaction of organic compounds with other reactants (SH^-, Cl^-, etc.) in aqueous solution.

Photolysis

Standard methods for measuring the photolysis of organic compounds in water and on soil have been described by the USEPA (1998), OECD (1997) and Lynch (1995). With photochemical transformation processes, the amount of light absorbed by the test chemical and the intensity of the light-source as a function of wavelength are the main factors responsible for determining reaction rates. In all the methods for measuring rates of photolysis described in the literature, these factors are taken into careful consideration. Use is often made of a reference compound with well-known absorption characteristics for which the quantum yield has been assessed as a function of wavelength (Zepp 1982). Methods to derive rate coefficients for indirect photolysis and oxidation in the aquatic environment were reported by Mill (1980) and Haag and Yao (1992).

Published sources of information (databases/handbooks)

Information on the rate and extent of degradation of specific organic chemicals may be obtained from various sources, including published results of independent studies, compilations of information reported in handbooks, and on-line databases that are accessible via the Internet.

A number of handbooks are available that provide information on the relationship between chemical structure and degradability, as well as tabulations of specific degradation data (e.g., Verschueren 1983; EurEco 1990; Pitter and Chudoba 1990; Alexander 1994; Mackay et al. 1992a, 1992b, 1992c, 1995). Biotic and abiotic transformation data for many pesticides are given in the Pesticide Manual (Tomlin 1997). Information on abiotic and biotic half-lives for many organic compounds in air, soil, water, and sediment can be found in the environmental handbook by Howard et al. (1991). The Japanese Chemicals Inspection and Testing Institute (CITI) has also published a large compilation of biodegradation and bioaccumulation data obtained under their Chemical Substances Control law (CSCL-Japan). The biodegradation data consist of 2- and 4-week results from the MITI-I ready biodegradability test (CITI 1992).

Sources of environmental fate information can be accessed via electronic databases as well as by on-line searches as summarized in Table 3-2. Many of the sources contain redundant information. The Hazardous Substances Databank (HSDB), Environmental Fate Data Base (EFDB), and Chemical Abstracts, in combination, are suggested as the most generally productive and convenient sources of information. If environmental fate information is not available for a particular chemical, substructure searches may be necessary to identify structurally similar chemicals. Among the most comprehensive sources of biodegradation information are the

Table 3-2 Databases available for environmental fate information (from Aronson and Howard 1999)

Database name	Description
AGRICOLA	Over 3 million bibliographic citations for U.S. agricultural and life sciences information
AOP	Database of 766 experimental hydroxyl radical, nitrate radical, and ozone rate constants within the AOP (Atmospheric Oxidation Program) estimation program.
BIODEG file (EFDB)	BIODEG contains the experimental results from biodegradation studies for approximately 800 chemicals. Experimental details, such as chemical concentration and rate of degradation, are included.
BIOLOG file (EFDB)	BIOLOG is a bibliographic database with publications indexed by oxygen conditions, inoculum source, mixed or pure culture used, etc. (currently 63,000 records for 7,800 compounds).
BIOSIS Previews	Electronic version of Biological Abstracts, largest printed reference publication for life sciences research information.
Chemical Abstracts	Contains over 14 million records. Covers all references which contain information on chemicals.
CHEMFATE file (EFDB)	CHEMFATE contains numerical values for 23 categories of environmental fate and physical/chemical property information on commercially-important chemical compounds. Actual experimental values (rate constants, experimental conditions, physical properties, etc.) are abstracted and retained in the file for 1700 chemicals.
CITI	Chemicals Inspection and Testing Institute (1992) reports biodegradation screening test data for approximately 1000 compounds in either the MITI (I) (ready biodegradability) or the MITI (II) (inherent biodegradability) test.
DATALOG file (EFDB)	DATALOG contains bibliographic citations to physical/chemical properties, biodegradation data, atmospheric data, field studies, monitoring information (currently 324,000 catalogued records for 16,000 compounds).
ECDIN	Environmental Chemicals Data and Information Network, compiled by the EC Joint Research Center. Contains environmental monitoring and fate data.
EFDB	The Environmental Fate Data Base contains four separate databases: DATALOG, BIOLOG, CHEMFATE, and BIODEG. Accessible online at esc.syrres.com
HSDB	Hazardous Substances DataBank provides information on over 4400 hazardous compounds including physical/chemical properties and environmental fate data. Accessible online at http://toxnet.nlm.nih.gov/servlets/simple-search.
IUCLID	International Uniform Chemical Information Database provides data on 1408 HPVC (High Production Volume Chemicals) from the European Union. Includes environmental fate information.
NTIS	National Technical Information Service provides bibliographic and abstract information on U.S. government and world-wide government-sponsored research. Over 700,000 citations.
PhysProp	An ISISBASE database which contains evaluated physical/chemical properties, including gas-phase hydroxyl radical reaction rates, for 13,000 compounds (also available as a book; Howard and Meylan, 1997).
TSCATS	Toxic Substances Control Act Test Submissions database contains health effects, environmental effects, and environmental fate information for over 7,500 chemicals. Accessible online at esc.syrres.com

BIOLOG (~ 6000 chemicals) and BIODEG (~ 800 chemicals) files of the Environmental Fate Database (Howard et al. 1986). Other databases of interest include ASTER (1994), ARS (1995), CS (1996), Enviro-Net (1996), JRC (1996a, 1996b), MedChem (1996), University of Akron (1996), and USEPA (1996).

Overview of estimation methods for predicting abiotic and biotic degradation

Although the assessment of degradation typically should rely on empirical data, the ability to accurately predict degradability based on chemical structure has long been regarded as a desirable alternative to costly and time-consuming laboratory experiments. The use of QSARs may save time and financial resources when appropriate methods are used. Furthermore, QSARs may provide additional information to complement laboratory results on the degradability of a specific chemical.

A variety of methods have been developed to estimate specific degradation processes for specific chemicals or specific classes of chemicals. Some are designed for the non-expert user and provide reasonable documentation. Overviews of models are presented by Mulder et al. (1993), Peijnenburg (1994), and Howard (2000).

Biodegradation

While experimental data are considered to be the best indication of degradation in the environment, there are no data available for many compounds that will be screened for POP/PBT persistence characteristics, and it may not be economically feasible or practical to complete such studies. In these cases, it may be reasonable to employ estimation programs such as QSARs in the initial stages of POP/PBT identification to predict the potential for degradation, and thereby categorize compounds as "little potential concern" (model predicts rapid degradation), "possible concern" (intermediate probability of degradation) and "persistent" (model predicts resistance to degradation). Laboratory or field testing would then be needed for compounds placed in the last two categories to assess whether they in fact exceed POP/PBT criteria for persistence.

Steady progress has been made toward the development of means to accurately predict the environmental fate of organic compounds. Characteristics intrinsic to the compound itself, such as high molecular weight, low water solubility, presence of three or more aromatic rings, branching in the molecule, and presence of halogen atoms (aerobic conditions only), have all been implicated as properties indicating that a compound will have increased resistance to biodegradation (Alexander 1994; Howard 2000). However, the biodegradation kinetics of a particular compound is dependent on environmental conditions, the presence of other compounds, the availability and concentration of the studied compound, and the level of competent biomass present.

The prediction of biodegradability has been hampered because of the lack of reproducibility of biodegradation data and because of the numerous protocols that have been used for biodegradation tests. For these reasons it is difficult to predict the rate of biodegradation of a compound based solely on its structure; consequently, nearly all of the available modeling programs provide only qualitative or semi-quantitative data. While numerous algorithms for predicting biodegradation have been published, most are suitable only for specific classes of compounds and are not appropriate for screening the biodegradation potential of large numbers of compounds that may have quite different structures and structural fragments. As a result, quantitative prediction of biodegradation rates has been attempted on only a very limited number of structurally related chemicals.

Despite these limitations on predicting biodegradation rates, a number of comprehensive approaches using fragment constants have been attempted to qualitatively predict the extent of biodegradability. Many of the models are based upon structural fragments whose occurrences are correlated with high or low rates of degradation. Taking a somewhat different approach, Punch et al. (1997) have developed methodologies that take the metabolic pathways into account by simulating the biodegradation of synthetic chemicals through the sequential application of plausible biochemical reactions, thus allowing for the prediction of the formation of possible toxic metabolites. Klopman et al. (1993, 1995) have utilized metabolic information and a fragment approach based on biophores to generate biodegradation predictions. Although both methodologies are very promising, they are still under development and are not yet applicable for generating predictions for large numbers of diverse chemicals. The system developed by Punch et al. (1997), however, does contain over 3300 rules for predicting biodegradation pathways and has recently been rewritten as an Internet application to facilitate use and expansion of the rulebase by multiple users. A similar project is the PREDICT-BT program, under development at the University of Minnesota. When complete, the PREDICT-BT program will predict biodegradation pathways and products based on rules compiled by experts and will be linked with the University of Minnesota Biocatalysis/Biodegradation Database (UM-BBD), which has detailed information on ~ 100 pathways. A workshop aimed at capturing expert knowledge as a system of programmable rules was conducted in 1998 (www.labmed.umn.edu/umbbd/predictbt/).

At present, models based on structural fragments appear to be best suited for assessing the biodegradability potential of specific chemicals. These include the multiple linear and nonlinear regression models incorporated in the Biodegradation Probability Program (BIODEG or BPP) (Howard et al. 1992), 2 of the models reported by Degner et al. (1993), and the Partial Least Squares (PLS) model developed by Loonen et al. (1999). BPP1 and BPP2 of the BIODEG program give qualitative estimates of the probability of slow or fast aerobic biodegradation of an organic chemical in the presence of mixed populations of microorganisms. The recommended models of Degner et al. (1993) include one for predicting biodegradation of acyclic aliphatic compounds and one for monocyclic aromatics. The multivariate model developed by Loonen et al. (1999) uses a large set of test data obtained by means of the MITI-I test (OECD 1996) and a large set of predefined structural fragments to predict whether or not a compound is to be classified as being degradable. The model is applicable to a wide range of compounds and has been thoroughly validated, both internally and externally. An external validation carried out by Rorije et al. (1999) showed that the PLS model performed best of all of the broadly applicable biodegradation models included in the study; on average about 83 % of the predictions made correctly predicted biodegradability. Since that study, another modeling exercise based on the MITI-I data, in this case using multiple linear and nonlinear regression against a modified set of BIODEG fragments, has been completed (Tunkel et al. 2000). This most recent study achieved comparable accuracy and was independently validated.

The characteristics of comprehensive biodegradability models in current use are summarized in Table 3-3 (MITI-based data) and Table 3-4 (models not based on the MITI data). Of the existing comprehensive models for predicting biodegradability, only those based on biodegradation expert survey data (Boethling et al. 1994) are capable of providing any sort of rate prediction. The BPP3 (ultimate degradation) and BPP4 (primary degradation) models in the BIODEG program estimate the approximate total length of time that might be needed to degrade a substance, expressed semi-quantitatively in general terms like days, weeks, or months. The models were developed from data collected in a survey of experts, in much the same fashion as in the development of PREDICT-BT. The experts were asked to estimate biodegradation rates for a set of 200 chemicals and to express their answers using the same general terms. These models (for primary and ultimate biodegradation) may be used to estimate half-lives for input to multimedia models after assigning specific numbers to the general terms, assuming pseudo–first-order kinetics, and specifying how many half-lives may reasonably constitute complete degradation, since that was the specified survey endpoint.

Table 3-3 Comprehensive biodegradability estimation methods based on the MITI data

		Accuracy (% correct)	
Reference	Methodology	Training	Validation
Degner et al. 1993	Model 75: acyclics + tert-amines + phosphoric acids; 7 descriptors	92 (n = 65)	91 (n = 222)
	Model 78: monocyclic aromatics; 9 descriptors	85 (n = 60)	84 (n = 57)
Loonen et al. 1999	127 Structural fragments	85 (n = 894)	83*
Tunkel et al. 2000	42 Structural fragments and MW; linear model	83 (n = 589)	83 (n = 295)
	Same variables; nonlinear	84 (n = 589)	84 (n = 295)

*Average of 4; leave 25% out

Table 3-4 Comprehensive estimation methods for biodegradability (excluding MITI methods)

Reference	Methodology	Endpoint/Training set
Niemi et al. 1987	28 structural fragments or classes	Biodegradable yes or no/ BOD data
Gombar and Enslein 1991	Aromatics model: 22 fragments + 5 MCI or charge calculations Aliphatics model: 21 fragments + 1 MCI	Biodegradable yes or no/ BOD data
Howard et al. 1992 (BPP1 and 2)	35 structural fragments	Probability of rapid biodeg/ BIODEG SUM summary evaluations
Klopman et al. 1993	18 biophores and 6 biophobes	Probability of rapid biodeg/ BIODEG SUM eval
Boethling et al. 1994 (BPP1 and 2, revised)	36 Structural fragments and MW	Same as above
Boethling et al. 1994 (BPP3 and 4)	Same as above	Estimated time for complete biodeg/ estimates from expert survey, using terms days, weeks, months, etc.
Devillers et al. 1996	Atomic and group contributions	Same as above

Abiotic transformations

Most of the models for predicting specific abiotic transformation pathways can be used to generate quantitative predictions of rate coefficients. The majority of the models are based upon multiple linear regression techniques. At present, procedures are available for estimating the rates of hydrolysis, photolysis, and redox transformations for specific classes of organic chemicals.

Models for predicting hydrolysis reactions are generally based upon Hammett and Taft constants. As summarized by Mulder et al. (1993), 5 models can be recommended for estimating hydrolysis rates for the following specific classes of chemicals:

- brominated alkanes (Vogel and Reinhard 1986);
- esters (Drossman et al. 1988);
- carbamates, including hydrogen substitution (Drossman et al. 1988);
- carbamates, excluding hydrogen substitution (Drossman et al. 1988); and
- benzonitriles (Masunaga et al. 1995).

A structure-based aqueous hydrolysis estimation program (HYDROWIN) (Meylan and Howard 1995) can be used to estimate rate constants for acid- and base-catalyzed hydrolysis (Mill et al. 1987). Neutral hydrolysis is not predicted by the program. Chemical classes for which hydrolysis can be predicted by HYDROWIN include esters, carbamates, epoxides, halomethanes, and selected alkyl halides. Other structures such as isocyanates, acyl halides, cyclic esters, ureas, amides, and carbonate esters are highlighted by the program (either as reacting extremely rapidly, e.g., isocyanates and acyl halides, or as reacting very slowly, e.g., cyclic esters or amides) although a rate constant is not determined. Not all structures that can hydrolyze are included in this estimation program (e.g., benzyl and allyl halides). The program Abiotic Transformations of Halogenated hydrocarbons In Aqueous Solution (ATHIAS) can be used to obtain base-catalyzed and neutral hydrolysis rates at various pH values and temperature conditions for 58 different halogenated aliphatic compounds (alkyl halides) (Ellenrieder and Reinhard 1988).

Models for photolysis of substituted phenols by reaction with singlet oxygen were presented by Tratnyek and Hoigné (1991), whereas quantum yields for the photohydrolysis of substituted aromatic halides may be calculated on the basis of the equations given by Peijnenburg, de Beer et al. (1992).

Models for predicting abiotic reduction reactions are currently available for

- reductive dehalogenation of aromatic hydrocarbons (Peijnenburg, 't Hart et al. 1992),
- reductive dehalogenation of aliphatic hydrocarbons (Rorije et al. 1997),
- hydrolytic transformation of substituted benzonitriles (Peijnenburg et al. 1993),
- para-substituted benzonitriles (Masunaga et al. 1993), and
- nitroaromatics (Schwarzenbach et al. 1990).

Assessing the reliability and variability of transformation data

In order to evaluate the reliability and variability of environmental persistence data derived from laboratory or field studies, it is important to have sufficient information to allow for interpretation of the data reported. For example, information on the environmental factors known to influence reactivity (like pH, temperature, etc.) should be reported to enable proper assessment of their impact and to facilitate extrapolation to the other environmental conditions. In addition, sufficient information should be available to permit assessment of the quality of the data. This would include information on the following:

- analytical methods used, in terms of sensitivity, detection limits, overall recovery, product interference, etc.;
- methodologies employed, in terms of mass balances, temperature control, pH control, impurities, etc.; and
- underlying statistics of the reported data, in terms of number of replicates, standard errors, correlation coefficients, etc.

It is not possible to assign quantitative criteria for deciding which data are valid. However, when data are collected according to standardized test procedures conducted under Good Laboratory Practice (GLP) guidelines, they will generally be accompanied by sufficient information to assess their quality. Consequently, data collected according to a standardized test method are generally preferred over data gathered by means of a non-standardized test method.

Quantitative structure-activity models should be used only after a thorough evaluation. It is possible to define to only a limited extent a strict set of criteria for determining whether a QSAR model will be acceptable for use, because such criteria will depend on factors such as the endpoint under consideration, the method used to generate the QSAR, and the domain of application of the QSAR. However, when generating predictions using a QSAR model, the following factors should be considered (Hermens et al. 1995):

- the exact degradation process modeled;
- specification of the experimental conditions (test methods);
- definition of the domain of the model (i.e., specification of the model descriptors and the groups of substances for which the model is valid, including the corresponding structural rules);
- the statistics related to the model, in terms of the number of chemicals used for deriving the model, the value of the correlation coefficient, and the estimated standard deviation of predictions;
- identification of outliers, preferably with a mechanistic explanation; and
- external validation of the model.

In addition, for pragmatic reasons, the models should be given explicitly and the descriptors should be easily obtainable.

Uncertainty in data estimation using QSAR models often results from the appropriateness of the model and the reliability of the model (e.g., standard error of estimate, which is usually available). In particular, the former is of primary importance and illustrates the need for defining the domain of a model. Expert judgment may be needed to address this matter.

In the case of biodegradation, assessing the reliability of QSAR models based upon structural fragments is more complicated. In principle, it can be done by both internal and external validation. To this point, only Loonen et al. (1996) have presented the results of an internal validation of their model. External validation of several structural fragment-based models was carried out by Langenberg et al. (1996) and Rorije et al. (1999) using a set of MITI-I test data. The percentages of correct predictions generated with each model according to the study of Rorije et al. (1999) are presented below. In all cases the first number relates to the percentage of correct predictions for compounds that are classified as being readily biodegradable on the basis of the MITI-I test results; the second number relates to the percentage of correct predictions for compounds that are not readily biodegradable in the MITI-I test.

- Biodegradation Probability Program (BIODEG) (Howard et al. 1992): 68.1% and 76.0%
- MultiCASE (Klopman et al. 1993, 1995): 72.0% and 80.8%
- Models of Degner et al. (1993): 81.9% and 88.2%
- Multivariate model of Loonen et al. (1996): 80.0% and 84.8%

From these results it is apparent that current QSAR models are better at predicting chemicals that are not readily biodegradable. This is because our knowledge of chemical functional groups that consistently inhibit biodegradation is much better than our knowledge of functional groups that enhance the process. In addition it should be noted that although the model developed by Degner et al. (1993) appeared to give the best performance, it is not broadly applicable.

Evaluation of Persistence at the Screening Level

As articulated previously in this chapter, a full, detailed assessment of all industrially significant chemicals for persistence in soil, water, and sediments would be prohibitively time-consuming and expensive. It would also be a waste of resources because a large database already exists concerning the fate of many chemicals. Logic requires that maximum use be made of the available data and current scientific understanding to eliminate from further consideration those chemicals that are clearly nonpersistent. Obviously, only those data that have been deemed to be valid should be used for these evaluations.

The purpose of this section is to describe in more detail the steps involved in "screening level evaluations" as outlined in Figure 3-1. The process involves an initial examination of physicochemical properties, which often can provide an indication of whether the compound is susceptible to degradation in the environment. Subsequently, the environmental partitioning of the compound is evaluated to identify whether the compound is likely to partition predominately to one, or more than one, environmental compartment. Knowledge of the potential environmental distribution of the compound allows the researcher to focus efforts to determine environmental half-lives in the relevant compartments. Guidance is provided in this section on methodology for determining half-lives for the dominant biotic and abiotic degradation processes from the results of field and laboratory studies. When measured values are unavailable, additional guidance is provided on how predictive models (QSARs) can be used in the evaluation.

Consideration of physicochemical properties

Although environmental transformation data are clearly desirable for screening-level evaluations, certain physicochemical properties can provide insight to the possible persistence of chemicals. In the absence of fate data, the properties described in Table 3-5 can be used to provide a preliminary indication of the potential for a chemical to persist in the environment.

Henry's Law constant (H) can be used as a qualitative indicator of the importance of volatilization; chemicals with H values less than 1×10^{-7} atm-m^3/mol would have little tendency to volatilize from water to the atmosphere (Howard 1989). To determine whether a compound will be available for degradation, the propensity of

Table 3-5 Preliminary indicators of the potential for persistence in soil, water, and sediments

Property	Criterion	Relevance
Henry's law constant	$< 1 \times 10^{-7}$ atm-m^3/mol	Partitioning to air
Log K_{ow}	< 2	Availability as solute
Molecular weight	< 600	Transport across biological membranes
Presence of functional groups or chemical structure	Functional groups, e.g., -OH, -COOH, etc. Chemical structure, e.g., aliphatic, aromatic, etc	Susceptibility to enzymatic attack or abiotic hydrolysis

a compound to adsorb to soil, sediments, or particulate matter can also be considered. Because it is widely recognized that for non-ionized organic chemicals, sorption to particulate matter is governed by sorption to the OC present in the system, the OC–water partition coefficient, K_{OC}, provides a good indication of the propensity for adsorption. Further, Sabljic et al. (1995) have shown that K_{OC} is directly related to K_{OW}, and hence an index based on K_{OW} may be used to evaluate whether a chemical will be sufficiently available for degradation. Similarly, the molecular weight of a chemical provides an indication of the ability for the compound to cross a biological membrane and hence to be susceptible for microbial degradation reactions to occur. Consideration of the chemical structure of a compound also can reveal information as to the degradation potential. For example, the presence of certain functional groups on the molecule often provides an indication of the potential for biodegradation reactions. Alternatively, certain classes of compounds are known to be susceptible to hydrolysis.

Evaluation of environmental partitioning

While it is possible to evaluate environmental persistence on the basis of half-lives determined for all environmental media (air, soil, water, sediment, etc.), the more logical approach for screening-level evaluations recognizes that a compound released to the environment often has a tendency to accumulate in one medium more than in others and that transformation rates for the compound will differ in each medium. Environmental partitioning can be evaluated using a variety of multimedia models as described in Chapter 5. Degradation processes in the dominant medium (i.e., where the compound preferentially partitions) can be expected to have more effect on the overall persistence of the compound than do degradation processes in the other media (Webster et al. 1998). Thus, when the majority of a chemical (e.g., 90%) is predicted to partition into a single compartment, it is sufficient to judge the overall persistence of the chemical by consideration of the degradation rates in that compartment. Alternatively, when a chemical is predicted to partition into more than one compartment, the overall persistence can be calculated on the basis of the fraction of the material present and the degradation rates in the appropriate environmental media.

Determination of persistence half-lives

Once the relevant environmental media have been identified, the next phase of the evaluation involves collecting the available information on the rates of the dominant degradation reactions. For soil, water, and sediments, this would include kinetic information for both biotic (biodegradation) and abiotic (hydrolysis, photolysis) reactions. Because multiple processes may be involved in the degradation of a particular chemical, the overall persistence in a given environmental medium can be described by the following equation:

$$k_{Per} = k_{Bio} + k_{Hyd} + k_{Photo} \qquad \text{(Equation 3-25)},$$

where

k_{Per} is the overall persistence first-order rate constant,
k_{Bio} is the first-order biodegradation rate constant,
k_{Hyd} is the first-order hydrolysis rate constant, and
k_{Photo} is the first-order photolysis rate constant.

Environmental half-lives for a given compartment can then be calculated using the equation:

$$t_{1/2} = \frac{0.6931}{k_{Per}} \qquad \text{(Equation 3-26)},$$

where $t_{1/2}$ is the half-life and k_{Per} is the overall first-order rate constant. Once half-lives are obtained, they can be compared with the published POP/PBT criteria in order to determine whether or not a compound should be classified as persistent.

The following sections present methodologies for determination of first-order rate constants for the predominant environmental degradation processes. Preference is given to the use of published experimental data in such evaluations. When experimental data are not available, one can consider estimating first-order rate constants using predictive models (i.e., QSARs) or experimental data for structurally related compounds.

Determination of first-order biodegradation rate constants

This section provides an overview of the different types of biodegradation studies reported in the literature, examines their strengths and weaknesses, and provides methods for determining first-order rate constants or half-lives. Biodegradation tests include pure-culture studies, enrichment studies using a pre-acclimated inoculum, screening studies, grab-sample studies, and field studies (Howard and Banerjee 1984; Howard et al. 1987).

Examination of the available literature for a particular chemical often reveals considerable variability in biodegradation rates. Such variation is often the result of non-standardized experimental methods as well as heterogeneity and the natural inherent variability in the environment. Consequently, the choice of a single "best" value is often difficult for compounds with large quantities of biodegradation data.

If possible, studies measuring mineralization versus primary biodegradation should be separated and evaluated independently for each study type and environmental medium. Depending on the amount of data available, rate constants may be presented as a frequency distribution, a range, or as a median or mean value.

When calculating environmental half-lives, results of environmentally realistic tests should be given higher priority during the screening evaluation. Both field and grab-sample studies provide data that are environmentally relevant and give the best indication of the potential for biodegradation in the natural environment. Results of field studies should be examined first, as they are likely to provide the most environmentally relevant data available. Grab-sample studies are given second priority, followed by ready biodegradability screening tests and inherent biodegradability screening tests.

While it is difficult to state a priori what would constitute a sufficient dataset, it is possible to list those factors that are desirable. First, experimental data should be available for the relevant environmental compartments. Second, several datasets should be available from a variety of locales within any given compartment. Ideally, the locales should be widely separated geographically, although this may be difficult given the lack of data from developing countries. Third, data should be available from different types of environments within a compartment. For example, data should be available from both eutrophic and oligotrophic water bodies, or from several soil types. Fourth, when possible, the data should be reduced to a common base for comparison. And finally, when available, information should be obtained that will enable assessment of the factors that modulate the availability of the chemical for degradation, the impact of the environmental factors modulating reactivity, and the potential for uptake of the chemical by microorganisms.

Field studies

As field studies directly measure the loss of a compound in the actual environment (usually as a dissipation half-life [DT_{50}]), they are the most realistic source of information for the persistence of a particular chemical. However, careful analysis of the data are warranted, because loss mechanisms in the field may include both degradation and transport processes and distinguishing between these sometimes can be difficult. Transport mechanisms include leaching and volatilization from soil, adsorption to suspended solids/sediment, and dilution and volatilization from water.

Techniques that are used in field studies to account for non-degradative losses include mass-balance modeling and the use of tracers to account for dispersion, sorption, and volatilization processes. Mass balance modeling requires that the system be at equilibrium, which may not be the case for certain chemicals, particularly for persistent compounds that are often water-insoluble and adsorptive. Considerations for selecting an appropriate tracer are based on the biodegradability (should be non-degradable under the study conditions) and sorption/volatility of the tracer compound (should be similar to that of the studied compound) in that system. A recent study by Kim et al. (1995) measured the loss of toluene in a stream in situ using tracers to measure transport processes. In this study, the importance of transport versus degradation as a loss mechanism changed with the season. Biodegradation was the dominant process during the warmer months; however, when temperatures dropped in winter, volatilization became a significant loss mechanism (Kim et al. 1995). Field studies in soil often measure compound residues at different depths or use soil cores to measure losses that are due to leaching (Lichtenstein and Schulz 1959; Stewart and Chisholm 1971). Physical properties such as the Henry's Law constant, log K_{OW}, and water solubility can be used to estimate K_{OC} and volatilization effects in order to predict whether loss due to transport could be significant when this information is not published.

Differences in environmental parameters such as soil type, pH, OC content, temperature, nutrient status, time of year, and moisture between different locations can result in considerable variation among degradation rates derived from field studies. However, when field data are available with adequate controls for transport effects (leaching, volatilization, dilution), such information should be given the highest priority. First-order rate constants for loss in a field study can be directly converted to overall persistence half-lives, after correcting for transport loss, as degradation is due to both biotic and abiotic processes.

Grab-sample studies

If field data are not available, or are questionable due to the lack of adequate controls, then grab-sample data should be used. Grab-sample tests (also called "batch tests") are frequently performed to measure the loss of a compound in an environmental sample (soil, water, or sediment) in the laboratory. The most commonly encountered tests are river die-away and soil metabolism tests. In these tests, an environmental sample is collected, treated with the test compound, and the loss of the added chemical (and often formation of metabolites) is monitored with time. Greater control of environmental parameters (e.g., substrate concentration, temperature, moisture content, presence of other compounds) is possible in these tests. Because biodegradation can be considered separately from other degradation and transport processes in grab-sample tests, a more accurate measurement of the biodegradability of the compound within that system is possible in comparison to that measured in a field test. In addition, metabolites can be identified by common laboratory procedures, definitively showing the transformation of the test compound by the microbial community.

Grab-sample tests in which surfactants are added to increase bioavailability or a specialized microbial population or enrichment culture is added to increase rates of degradation should be evaluated carefully. In some cases, the surfactant can be used as a carbon source that inhibits biodegradation rates of the insoluble compound, or it may be toxic to the microbial population. Acclimated cultures can give data that might be relevant to biodegradation under conditions in which a compound is continuously released, but they cannot be applied to situations in which the compound is intermittently released. Many of these types of studies are more appropriately used to evaluate the bioremediation potential of a chemical rather than to determine its intrinsic biodegradability.

When grab-sample studies are conducted with appropriate controls (to account for other loss mechanisms), results of such studies can be used to calculate first-order biodegradation rate constants for use in the determination of overall persistence. When non-biological losses are indicated, control data should be analyzed to separate out the other processes, and a first-order biodegradation rate constant should be calculated. For example, in laboratory tests with aquatic samples, dark controls can be used to separate photolysis reactions, while sterile controls can be used to differentiate between hydrolysis and biodegradation. If separation of the different degradation mechanisms is not possible based on the controls but non-degradation losses (sorption, volatilization, etc.) can be accounted for, an overall persistence first-order rate constant can be calculated after correcting for any non-degradation loss.

Rates of transformation obtained in the laboratory are generally believed to be more rapid than those reported in the field (Aronson and Howard 1997; Aronson et al. 1999), at least for compounds that are not readily degraded (Barker et al. 1987). Thus, rates reported from grab-sample studies may be higher than the actual transformation rate that would be seen in the environment. However, for many compounds, laboratory experiments represent the most reasonable approach for quantifying potential biodegradation rates in the environment.

Conversion factors are available to extrapolate biodegradation data between different environmental media. Federle et al. (1997) measured the aerobic mineralization rates in sewage-amended river water and sludge-amended soil for 9 diverse compounds and obtained a mean scaling factor of 1.1; scaling factors ranged from 0.14 to 3.24. In addition to the variation in the scaling factors, the correlation between the different rates in soil and water was not strong ($r = 0.53$). The scaling factor reported by Federle et al. (1997) is very similar to that derived by Boethling et al. (1995). These authors measured the mean ratio of aerobic soil half-lives to aerobic water half-lives (looking at primary biodegradation and mineralization separately) that had been published for 20 chemicals and obtained a scaling factor of roughly 1, again indicating that half-lives reported in soil are reasonably similar to those in water. Scaling factors ranged from 0.13 to 5.22 for the primary biodegradation data and from 0.049 to 5.66 for the mineralization data.

The published POP/PBT criteria do not distinguish between aerobic and anaerobic conditions, although it is assumed that aerobic conditions are implied. Aerobic conditions are expected to predominate in drained soil, surficial sediments, and fresh and marine waters, while anaerobic conditions may be found in sediment, the hypolimnia of stratified lakes, waterlogged soils and wetlands, and in some groundwater aquifers. Studies reporting the degradation of a compound under anaerobic conditions may be used to estimate degradation rates under aerobic conditions (Boethling et al. 1995). Published half-lives in surface soil were compared to half-lives in flooded soils for 25 different compounds. Rates of degradation in aerobic soil were determined to be approximately 3 to 4 times faster than those found in flooded soil (presumed anaerobic), whether primary degradation or mineralization was the measured endpoint. However, expert judgment should be consulted when using this approach, since compounds containing either multiple halogen atoms or nitroaromatics can degrade much more rapidly under anaerobic conditions than under aerobic conditions.

Boethling et al. (1995) compared biodegradation half-lives in marine water versus fresh water for 11 compounds. While it is generally accepted that rates of degradation are slower in salt water than in fresh water (Bartholomew and Pfaender 1983), these authors reported that a scaling factor cannot be predicted based on the available data.

Screening tests

When data from field or grab-sample studies are not available, screening studies can be conditionally used to estimate the potential for biodegradation in the environment. Screening studies measure the biodegradation of a compound using a small quantity of inoculum (sewage, activated sludge, surface water, soil, or sediment) added to a defined medium. These studies were developed to identify compounds that would rapidly undergo ultimate biodegradation in the environment and are the most commonly reported biodegradation tests found in the literature. However, these tests do not mimic conditions in the environment for the following reasons:

- they use a mixed-culture inoculum and a mineral salts medium and are run for relatively short periods of time (< 1 month),
- the inoculum is often from sources such as sewage that are quite different from surface waters and soil and may also be pre-acclimated to the studied compound,
- the microbial biomass used in the test can be quite high compared to that found in some environments,
- these tests are often run at high concentrations of the test chemical (often necessitated by analytical methods), which may affect the degradation kinetics as well as result in toxicity to the microorganisms, and
- other carbon sources are limited to facilitate analysis of biodegradation of the test chemical.

Thus, the calculation of half-lives from screening tests in order to predict biodegradation in the environment must recognize and consider these caveats.

The Organization of Economic Cooperation and Development divides screening test methods into ready and inherent biodegradability tests (OECD 1996). Ready biodegradability tests (e.g., modified OECD screening test, CO_2 evolution test, manometric respirometry test, DOC die-away test, closed bottle test, and the MITI[I] test) are stringent tests that provide limited opportunity for biodegradation and acclimatization to occur. Such tests measure mineralization over a 28-d period using a low concentration of a sewage or activated sludge inoculum, and a high concentration of the test compound. A compound is considered to be "readily biodegradable" if 60% to 70% biodegradation has occurred in 28 d (including the lag phase) and the time between the start of mineralization (defined as 10% of the theoretical) and the plateau does not exceed 10-d (the "10-d window") (OECD 1996). It may be assumed that a chemical giving a positive result in a ready test will rapidly biodegrade in the environment (Hales et al. 1997).

Inherent biodegradability tests (e.g., Zahn-Wellens test, SCAS test, MITI[II] test) allow prolonged exposure of the test compound to acclimated microorganisms, and thus the likelihood of biodegradation is increased in these tests (Struijs and van den Berg 1995). Inherent tests are run with high microbial population densities, again using a sewage or activated sludge inoculum. As with the ready biodegradability test, mineralization is the measured endpoint and the same criteria for degradability are used. A compound giving a positive result in a test of this type may be classified as "inherently biodegradable," but its rapid and reliable biodegradation may not be assumed in the environment (Hales et al. 1997). In contrast, a compound that "fails" the inherent biodegradability test generally is considered to be non-biodegradable; however, further studies would be warranted to confirm that the compound is indeed persistent.

The stringent conditions applied in these screening tests (i.e., mineralization as the measured endpoint, incorporation of the lag phase into the biodegradation time frame) indicate that the results can be useful in predicting whether or not a chemical is likely to biodegrade in the environment. The high concentration of compound added to many of these tests, however, means that results from the test can be affected by the solubility or toxicity of the test compound and often yield false negative results (Federle et al. 1997). This is less of a concern in grab-sample and field studies, which use lower substrate concentrations.

Kinetic data from screening tests have been used to predict biodegradation rates in the environment (Federle et al. 1997; Struijs and van den Berg 1995). Federle et al. (1997) measured mineralization rates in the modified Sturm test (ready CO_2 test) and in die-away tests using river water and sludge-amended soil for nine diverse compounds. Mineralization rates in the environmental media were not correlated to rates reported in the screening test ($r = 0.04$ and 0.12 in river water and soil, respectively), despite the addition of a microbial inoculum from the screening test to "standardize" the microbial population in the other environmental media. The rate of biodegradation in the screening test was, however, strongly correlated to the solubility and the log K_{OW} of the added compound. The authors concluded that while prediction of biodegradation kinetics in environmental media was not possible from screening-test results, the ready tests are a very useful indicator of the ability of a compound to undergo complete biodegradation.

Struijs and van den Berg (1995) attempted to use results reported in standardized OECD biodegradation tests to assign first-order biodegradation rate constants for degradation in water and soil. This was done by referencing the first-order rate constant to the expected population density (biomass) in the studied medium. A compound that passes the ready-test criteria (including the 10-d window) has a lower limit of 0.14 d^{-1}. First-order rate constants for water and soil are assigned as follows:

- Surface waters: A rate constant of 0.14 d^{-1} is given to compounds passing the ready-test criteria. This is based on the assumption that the microbial density in surface water is similar to that found in the ready biodegradability tests.
- Soils: Rate constant of 14 d^{-1} is used to describe biodegradation in the soil pore water for compounds that pass the ready-test criteria. This is based on the assumption that the microbial density in soil is 100 times greater than that found in the ready biodegradability test and that the compound is found completely in the soil pore water. If a compound is not expected to reside completely in the pore water, i.e., adsorption to the soil occurs, a correction can be made to the rate constant as described by the authors, using the soil partition coefficient (K_d); in this method, no biodegradation is assumed when a compound is adsorbed to soil/sediment.

These rate constants were validated using only a small sample of compounds (14 for surface water, 4 for soil) and the authors report that the extrapolation of these rate constants to more persistent compounds appears to result in an overestimation of biodegradation in the environment. In contrast, the work of Federle et al. (1997) showed that biodegradation rates were not strongly correlated with the microbial population density, thereby questioning the assumptions used by Struijs and van den Berg (1995).

The EU Technical Guidance Document (EC 1994, 1996) assigns rate constants based on ready- and inherent-test data; these are not measured rate constants but are "guesstimates" based on expert judgment. Default values for extrapolation of screening test results to surface water are summarized in Table 3-6.

Table 3-6 Default half-lives for biodegradation in surface water based on screening tests (EC 1994, 1996)

Test result	Half-life (d)
Readily biodegradable	15
Readily biodegradable, but failing 10-d window	50
Inherently biodegradable	150
Not biodegradable	∞

Chemicals that pass the ready biodegradability test are assigned a half-life of 15 or 50 d, depending on whether the 10-d window criteria are met. The default half-life of 150 d is assigned to inherently degradable compounds. Any other result (e.g., if the compound fails the ready test and no inherent-test data are available, or if it fails the inherent test and no ready-test data is available) is assigned a half-life of infinity.

Because other factors not related to the intrinsic biodegradability of a chemical, e.g., low water solubility, toxicity to microorganisms, etc., may cause a chemical to fail a screening test, the default half-lives assigned by the EU Technical Guidance Document have been criticized as excessively stringent. A less-conservative approach that is currently under consideration by the USEPA for use in several regulatory initiatives, including the High Production Volume Chemical (HPVC) project, is summarized in Table 3-7.

Table 3-7 Half-lives for biodegradation in surface water based on results of standardized biodegradation tests

Ready-test result	Inherent-test result	Rate constant (d^{-1})	Half-life (d)
Passes test	—	0.14	5
Does not pass, but ≥ 40%	—	0.069	10
Does not pass, ≥ 20 < 40%	≥ 70%	0.023	30
—	≥ 20 < 70%	0.0069	100
Does not pass, < 20%	< 20%	0	10,000 or other default for no biodegradation

In situations in which there are conflicting data between ready- and inherent-test results, if a compound exceeds 40% degradation in a ready test, the results preempt the inherent data. Conversely, the inherent results preempt the results of a ready test if the latter are < 40%.

Because sorption to soil may decrease the bioavailability of some compounds (mainly those with high soil–water partition coefficients) soil half-lives were assigned by the Technical Guidance Document based on both screening-test results and sorption characteristics (EC 1994, 1996). Default values for extrapolation of screening test results to soils are summarized in Table 3-8.

Table 3-8 Half-lives for bulk soil based on results from standardized biodegradation tests (EC 1994, 1996)

	Half-life in soil (d)		
K_d soil (L/kg)*	< 100	100–1000	1000–10,000
Test result:			
Readily biodegradable	30	300	3000
Readily biodegradable, but failing 10-d window	90	900	9000
Inherently biodegradable	300	3000	30,000

*K_d: soil–water partition coefficient ≅ organic carbon content × octanol–water partition coefficient

The scheme outlined in Table 3-8 is based on generic microbial population densities, which are assumed to be 100 times higher for soil pore water than for surface water. Equilibrium partitioning, which requires additional information on solids/water distribution, is then used to adjust the predicted biodegradation rate for distribution of the chemical between sediment and pore water (i.e., for sorption), and it is assumed that no degradation occurs in the bound phase. While it has been suggested that the approach yields reasonable estimates of soil biodegradation rates (Struijs and van den Berg 1995), the number of chemicals included in the analysis was small. It is important to note that on the basis of these conservative default assumptions, chemicals with K_d values greater than 100 L/kg will exceed many of the proposed POP/PBT criteria for soil even if the compound is shown to be biodegradable in screening-level tests.

Another approach is to use the recommended default values for water and then extrapolate the results to the other media using scaling factors. A "scaling factor" is simply a number that, when multiplied by a degradation rate constant or half-life for one set of environmental or test conditions, yields a rate for a second, different set of conditions. Scaling factors of approximately 1 for the water/soil extrapolation were suggested in two independent studies using quite different methodologies (Boethling et al. 1995; Federle et al. 1997). Federle et al. (1997) also found that, contrary to earlier results (Struijs and van den Berg 1995), biodegradation rates in river water verses soil did not scale on the basis of microbial density, even if the different biodegradative availabilities of sorbed versus free chemical were factored into the comparison.

Although results of biodegradation screening tests have been used to estimate biodegradation rate constants when field or grab-sample studies were unavailable, this approach does have scientific limitations. For the reasons previously outlined, biodegradation screening tests do not simulate the conditions in soils and surface waters. A more logical approach would be to use the screening-test data as a qualitative indication of whether or not the chemical is persistent. Further, because screening tests are well known for having considerable variability, a weight-of-evidence approach could be adopted as was done in the development of the BIODEG file for the Environmental Fate Data Base (Howard et al. 1987). A suggested qualitative approach is summarized in Table 3-9.

As summarized in Table 3-9, if a chemical consistently reaches a pass level in one or several ready biodegradation or inherent biodegradation tests, the chemical should be considered nonpersistent. When there is more than one test result, the potential for variable or conflicting data increases, especially for chemicals that are less biodegradable. If there are conflicting data, but the chemical reaches a pass level in one ready biodegradation test, the chemical should still be considered nonpersistent. If the chemical passes an inherent biodegradation test and 1) there is no other data or 2) if it does not pass the ready test, the chemical should still be considered

Table 3-9 Qualitative interpretation of environmental persistence based on evaluation of biodegradation screening tests

Type of test	Criteria	Conclusion	Comment
Ready biodegradation test	Passes at least one	Nonpersistent	Rate in environment should be rapid
Inherent biodegradation test	Passes at least one	Nonpersistent	Rate in environment may be slow
Ready biodegradation test	Does not pass, but degradation is > 20%	Nonpersistent	Most likely nonpersistent; would be desirable to conduct additional testing
Inherent biodegradation test	Does not pass, but degradation is > 20%	Persistent	May be nonpersistent, but requires further testing
Ready biodegradation test	Does not pass, and degradation is < 20%	Persistent	May be nonpersistent, but requires further testing
Inherent biodegradation test	Does not pass, and degradation is < 20%	Persistent	Likely to be persistent, but requires further testing

nonpersistent, especially if the criteria are a half-life of 6 months, although the rate in the environment may be somewhat slow (half-life of months).

If the chemical does not reach a pass level in any of the screening tests, the actual results should be examined. If no biodegradation is observed in any of the tests, the chemical may be persistent, although further research is warranted (e.g., grab-sample tests). If some biodegradation (> 20% degradation) occurs in any of the ready-test methods, the chemical is unlikely to be highly persistent; it is likely that such chemicals will pass the 6- to 12-month soil/sediment half-life criteria and may pass the 2-month water criteria if tested in a grab-sample test. Therefore, chemicals in this class should also be considered nonpersistent. Similarly, if a chemical exhibits some biodegradation (>20% degradation) in an inherent biodegradation test, the chemical may not be persistent, but further studies are warranted to confirm this.

Note that when complex mixtures (e.g., creosote, PCBs, toxaphene) are being tested, screening-test results will need to be carefully assessed. There may be portions of the mixture that are readily biodegraded (e.g., mono- and dichlorobiphenyls within the PCB mixture), which may allow small percentages of degradation to occur while the majority of the material may be persistent.

Pure-culture studies

Pure-culture studies, in which a compound is added to a mineral salts medium and inoculated with a pure culture of bacteria or fungi, cannot be used to determine the biodegradation kinetics of a compound in the environment. Their main utility is in identifying metabolites and thus possible pathways of biodegradation (Howard et al. 1987). These studies generally are characterized by high substrate and nutrient concentrations. The extent of breakdown of a chemical shown in a pure culture study may not be representative of that found in the environment, since biodegradation in the environment often proceeds through the action of several different microbial populations. Negative results in a pure culture study may result from toxicity to the microorganisms (due to the high substrate concentrations used) or may simply reflect the inability of that particular bacterial strain to biodegrade the test compound.

Estimation using quantitative structure-activity relationship

Currently, extrapolation of biodegradation estimations to actual default rate-constant values is not possible. Some authors have used qualitative estimation data, however, to classify compounds as either persistent or readily degraded. (USEPA 1994; Tyle and Niemela 1999). Until a more detailed comparison of known persistent compounds with the currently available biodegradation estimation programs is completed, it is not possible to scientifically select probability values to be used as kinetic predictors. Ideally, with further experience, values could be selected and used to classify compounds as readily biodegraded and persistent and as having intermediate biodegradability. Because qualitative estimates cannot be used to determine an overall persistence half-life—unless experimental data from hydrolysis or photolysis studies meet the proposed criteria—estimations of biodegradation should be used only to rank compounds with a similar lack of data.

It should be reiterated that biodegradation-estimation programs are not a substitute for experimental data and should be used with caution. They can be used to readily screen a large number of compounds and may be useful in the initial classification of compounds without experimental data. However, a compound that is shown to have intermediate persistence or to be persistent based on current estimation methods needs further laboratory or field testing before it is definitively identified as a persistent compound according to POP/PBT criteria. If current modeling programs are not appropriate for the chemical structure under consideration (e.g., if it contains unique structural fragments that are not included in current biodegradation models), then experimental degradation data for structurally analogous compounds should be located and expert judgment employed to obtain a scientifically valid estimation of degradation for the unknown compound.

Determination of first-order abiotic degradation rate constants

Hydrolysis

As previously discussed, certain chemical structures are known to be susceptible to abiotic degradation. Of these processes, hydrolysis and photolysis should be considered at the screening level when evaluating persistence in the environment. In comparison to microbial degradation, abiotic degradation generally results in smaller alterations of the parent compound and is often slow or specific to a particular pH range or light condition (which may not occur where the compound is released to the environment). However, it is important to note that for some chemicals, abiotic processes can convert a biologically recalcitrant molecule into an intermediate that is more easily biodegraded.

Rate constants for hydrolysis of chemicals in aqueous solution can be determined in laboratory studies as previously described. Such information often can be directly incorporated into calculations of overall environmental persistence, provided that the rate constants are corrected for environmental factors such as pH and temperature.

The abiotic neutral hydrolysis of compounds in sediment–water systems has been shown to be comparable to that reported in water alone (Wolfe et al. 1989). Hydrophobic esters, however, exhibited a decreased rate of base-catalyzed hydrolysis in the presence of sediment (Macalady and Wolfe 1985). As the effect of ionic strength on the hydrolysis reaction rate of any given compound cannot be predicted (Mabey and Mill 1978), hydrolysis rates calculated in salt or brackish water cannot be used without qualification to estimate freshwater hydrolysis and vice versa. Perdue (1983) reports that, based on limited data, the rate of base-catalyzed hydrolysis is inhibited while the rate of acid-catalyzed hydrolysis is increased when susceptible compounds are incubated in the presence of humic acids. Humic acids are commonly found in natural waters.

Hydrolysis data obtained in aqueous systems must be used carefully to describe hydrolysis in moist soil; organic carbon and clay content can affect the sorption of a compound to the soil matrix, soil temperatures can reach higher values than those in an aquatic environment, and varying soil moisture can act to either increase or decrease the rate of hydrolysis in comparison to that measured in an aqueous system. The effect from these differing conditions is not currently predictable (Wolfe et al. 1989). Soil hydrolysis data can sometimes be inferred from soil grab-sample studies where biodegradation loss is corrected for losses due to hydrolysis.

Photolysis in aquatic environments

In aquatic photolysis studies, photolysis rates in natural versus distilled water have been shown to vary because of the presence of particulates in natural waters. Levels of particulate matter in surface waters may increase in summer months as a result of algal and bacterial growth. Furthermore, the presence of dissolved substances (e.g., humic acids) in natural waters may 1) attenuate the available sunlight, 2) promote indirect photoreactions, or 3) interact either chemically or physically to make the compound less available for photolysis (Hwang et al. 1987; Zepp and Wolfe 1987). Consequently, experiments that are conducted under realistic environmental conditions are most likely to yield the most relevant rate constant for a particular body of water. Laboratory experiments conducted in water or water–solvent mixtures can suggest that a compound may photolyze in the environment; however, data from these experiments should be used carefully when calculating environmental rate constants. For example, data on 13 different PAHs (calculated at latitude 40

°N at midday in midsummer) suggest that these compounds photolyze rapidly with half-lives, as a result of direct photolysis alone, ranging from 0.13 h for 9-methylanthracene to 4.4 h for chrysene to 21 h for fluoranthene (Zepp and Schlotzhauer 1979). However, several of these compounds can be quite persistent in the environment, because they partition to particulate material in the water column and end up in the sediment layer, in which photolysis is unlikely.

Photolysis on soils

Photolysis may occur on soil surfaces, although caution is warranted when estimating rate constants from available literature. For example, laboratory studies of the photolytic degradation of DDT on soil reported a half-life of 125 d (Vollner and Klotz 1994); however, field studies in soil showed that this compound had substantially longer degradation half-life values, ranging from 2 to more than 15 y (Lichtenstein and Schultz 1959; Stewart and Chisholm 1971). Some compounds may leach below the soil surface and become unavailable or, over time, compounds may become strongly adsorbed to the soil matrix, resulting in stabilization to photolytic degradation. Photooxidation also is assumed to occur on soil surfaces; however, very little is published on this subject because of the complexity of the system (Mill 1989).

Rate constants derived from aquatic systems should not be used to estimate the kinetics of photolysis on soil surfaces; therefore, a first-order photolysis rate constant should be calculated separately for both soil and water, if possible. Photolysis in sediment is not expected to be a major degradation process (Zepp and Schlotzhauer 1979). Estimation programs that can estimate either direct or indirect photolysis rates in water or soil are not currently available for screening purposes (Mill 1989, 1999).

Screening-level evaluation of multimedia chemicals

While the analysis of the overall persistence half-life in a single compartment is appropriate at the screening level for chemicals that are shown to partition predominately into a single environmental medium, a multimedia approach is necessary for chemicals that are shown to partition into more than one compartment (Mackay 1991).

The overall persistence half-life for a multimedia chemical can be calculated on the basis of the fraction of the material present and the degradation kinetics in the appropriate environmental media, by using the following equation:

$$t_{1/2,\text{ overall}} = t_{1/2,\text{ soil}} \times f_{\text{soil}} + t_{1/2,\text{ water}} \times f_{\text{water}} + t_{1/2,\text{ sediment}} \times f_{\text{sediment}} \quad \text{(Equation 3-27)},$$

where, $t_{1/2,\text{ c}}$ is the degradation half-life and f_c is the fraction of chemical present in the environmental compartment, c, respectively. When calculating the overall half-life for soil, water, and sediment, the same approaches for determining overall half-lives from first-order biotic and abiotic rate constants apply in the evaluation of multimedia chemicals. The fraction of the chemical present in the compartment can be obtained by using multimedia models as described elsewhere in these proceedings.

Environmental Heterogeneity and Its Impact on Quantifying Persistence

One of the most significant problems confronted in efforts to understand the fate of organic compounds in the environment is the heterogeneity of the biosphere. While this complicates the efforts to predict the fates of those compounds, it does not negate the need to make such predictions. Rather, environmental diversity must be recognized and dealt with in a rational and logical way. The purpose of this section is to explicitly consider the impacts of environmental diversity, thereby setting the stage for the prediction of persistence in the face of environmental diversity.

Heterogeneity in the biosphere

Although a simplification, the biosphere is divided into 4 major compartments for the purpose of modeling the environmental fate of chemicals: air, soil, water, and sediment. This is done because these compartments differ widely in terms of key parameters that affect the degradation of organic compounds. Furthermore, those parameters often vary significantly among different regions within a single compartment. For example, there are significant differences in the physical and chemical properties of soils found in various regions of the biosphere, including differences in moisture, organic content, pH, hydraulic conductivity, and so on. It follows, therefore, that the rates of organic compound degradation in these regions may well differ. The situation is further complicated by the fact that conditions found at a specific location within an environmental compartment can also vary on a temporal scale. For example, the degradation of organic compounds in soils from temperate regions can be influenced by seasonal variation in temperature and moisture, while the degradation of organic compounds in ground water can be influenced by fluctuations in the infiltration of surface water, which alter the water chemistry.

The degree to which the rate of degradation of an organic compound varies on spatial and temporal scales depends on the number and magnitude of differences that exist between locations or over time and their impacts on the key parameters. However, with current understanding of degradation processes, it is usually possible to predict the effect on the degradation rate caused by a change in any single parameter. If the difference in the values of a parameter is small, the difference in the rates of organic compound degradation in two regions will be similar and the confidence in the predicted effect will be high. Conversely, as the difference in the two values of the parameter increases, the confidence in the prediction will be proportionately less. Obviously, the confidence in a predicted rate will be less as the number of parameters that vary is increased, since there are often complex interactions among parameters. Consequently, if the regions differ in only a few parameters, and the magnitude of the differences is small, it is possible to use the rate of organic compound degradation observed in one region to confidently predict the rate that can be expected in a second region. To generalize, it is possible to extrapolate, with confidence, data on the rates of organic compound degradation to other regions of an environmental compartment if, and only if, the regions are similar to one another in terms of the key parameters influencing the rates of the dominant degradative processes. It is important to recognize that the degree of certainty in estimates is low for regions of an environmental compartment that differ significantly from the regions from which empirical data have been obtained.

Matters of scale

Heterogeneity exists on a variety of scales. The examples in the preceding section have all alluded to large scales. However, heterogeneity exists on a very small scale as well. Microbial processes in soils can serve to illustrate how "micro-scale" heterogeneity is accommodated into measures of process rates. Numerous studies have shown that microbial processes in soils can vary dramatically on a scale of less than one cubic centimeter. Nonetheless, microbial processes are routinely measured using samples comprised of several hundred grams. It must be at least tacitly recognized that the measured "net rate" for any given process is in fact the average of process rates that vary widely throughout the sample that is used. Other studies have shown that such processes also vary along transects and across landscapes. While this heterogeneity is often not explicitly acknowledged in various studies, it nonetheless exists. Statistically rigorous sampling schemes can be used to account for such variability, and these schemes can be used to interpolate between sites that are not actually sampled.

Coping with uncertainty

For practical reasons, one must devise a reasonable means for predicting the rate of organic compound degradation in the biosphere as a whole, in spite of its heterogeneity on both the macro-and the micro-scale. One approach would be to collect data from as many examples of a given degradation mechanism in a given environmental compartment as possible and then to average them. The result would be the average rate of organic compound biodegradation for "n" soils. If n is sufficiently large, this approach will yield a conservative estimate for the rate at which the organic compound is biologically degraded in soil on a global scale. It should be explicitly recognized that the "global average rate" cannot be considered to be that which would be found at any particular soil. The existence of a high global average rate does not preclude the possibility that in some soils the rate of biodegradation could be quite slow. Conversely, the existence of a low global average rate does not preclude the possibility that the organic compound is rapidly biodegraded in some soils. The practical implications of this are that a given organic compound can appear recalcitrant to biological degradation in certain soils, while being readily metabolized in others. This outcome is not unexpected given the range of environmental conditions that exist in the biosphere and is not by itself cause for undue concern.

Detailed Evaluation of Persistence

The purpose of this section is to describe more fully the steps involved in the detailed evaluation of persistence as outlined in Figure 3-1. The process involves refining pseudo–first-order rate constants for both biotic and abiotic reactions in relevant soil, water, and sediment environments, using data developed in either screening tests or actual environmental samples. These rate constants are then weighted for individual compartments to give an overall persistence value, which is then used to determine if further evaluation (i.e., risk assessment) is needed or the chemical can be classified as nonpersistent and set aside from further evaluation.

Estimation of pseudo–first-order rate coefficients for specific transformations in a single compartment

The detailed evaluation of the environmental persistence of a chemical will continue to require the use of multimedia models as discussed elsewhere in these proceedings. Given the heterogeneity and complexity of the environment and the current state of knowledge, the transformation reactions within a specific environmental compartment c are treated as if they were first-order with respect to the concentration of the compound of interest in that compartment, S_c:

$$(dS/dt)_c = -k_{1,c}\,[S_c] \qquad \text{(Equation 3-28)},$$

where $k_{1,c}$ is a lumped pseudo–first-order rate coefficient. Because many reactions can occur in the compartment, $k_{1,c}$ is given by

$$k_{1,c} = k_{1,c,r1} + k_{1,c,r2} + \cdots k_{1,c,rn} \qquad \text{(Equation 3-29)},$$

where r1 through rn refer to the particular reactions occurring. Thus, the question arises as to how each of the component rate coefficients, $k_{1,c,rn}$, can be estimated from new or historical experimental data. This section addresses that question for the water, soil, and sediment compartments.

General approach

The component pseudo–first-order rate coefficients are influenced by many factors, as discussed previously. Consequently, estimation of values of the pseudo–first-order coefficients for a particular environmental compartment from experimental or empirical data involves correction for those factors. For example, if a k_1 value was developed for 20 °C, but the temperature in the environmental compartment being considered is 10 °C, the numerical value of k_1 must be adjusted using the appropriate temperature coefficient. A similar approach can be used for each factor of importance, with the effect of each factor being multiplicative:

$$k_{1,c,rn} = k_{1,exper,rn} \times F_1 \times F_2 \times F_3 \times \times \times F_n \qquad \text{(Equation 3-30)},$$

where F_1 through F_n are the relevant environmental factors. The importance of the various factors depends on the type of reaction under consideration, as discussed in the material that follows.

Biological degradation processes

The two biotic transformations considered are microbial biodegradation and phytodegradation. Currently, much more knowledge exists for the former than for the latter. Consequently, major emphasis will be given here to microbial biodegradation, although the potential importance of phytodegradation should be recognized.

General principles of microbial biodegradation

Microbial communities in various ecosystems are complex assemblages of multiple species of organisms. Each population occupies a specific ecological niche, which by definition means that their activities are sufficiently different from one another that they coexist within the ecosystem and exclusion of species does not occur. This implies that not all species carry out the same function and that specialization occurs. The relative abundance of the various populations is determined by the resource inputs to the system and the metabolic interactions that arise between populations. As a consequence, some populations are numerically dominant, while most occur in much lower abundance. Under circumstances in which an organic compound is introduced to an environment, the numerical abundance of populations that can utilize the compound as a resource will be determined by 1) the benefit (carbon and energy) that can be derived through its metabolism, and 2) the partitioning of the resource among populations that are functionally redundant, insofar as their ability to utilize the compound. If the amount of resource (organic compound) put into the system is large, and the benefit is large, organisms that metabolize the compound will increase in relative abundance. Conversely, if the amount put into the system is low, or the benefit from its metabolism is low, populations that metabolize the compound will remain relatively rare.

The transformation of organic compounds in the environment is catalyzed through the action of specific enzymes. As described above, these enzymes are synthesized only by a fraction of the total populations present in a microbial community. If one assumes that the relative amount of a specific enzyme is a constant fraction of a cell, the total amount of enzyme present in an ecosystem that can act to degrade a compound is proportional to number of cells within each population that can degrade the compound. This is referred to as the "competent biomass." In most cases the amount of competent biomass is unknown and is difficult to determine with a high degree of certainty; therefore, it must be estimated.

Previous studies done to enumerate populations in various environments in which there has been limited prior exposure to a given organic compound have shown that the numbers of organisms with the potential to degrade the compound are comparatively low. For example, surveys done to enumerate the numbers of bacteria in soils able to degrade biphenyl have shown that in situations in which there are approximately 10^8 total bacteria per g soil, there were only about 10^3 bacteria that were able to mineralize biphenyl. This is only 0.001% of the total microbial community. Other populations might transform biphenyl by some means without mineralizing it (e.g., through hydroxylation of an aromatic ring), but their numbers are unknown. Similar data have been obtained for other organic compounds that are commonly considered to be "xenobiotics." It should be noted that these numbers increase dramatically if acclimation of the community occurs as a result of prior exposure to sufficient amounts of the organic compound.

To estimate the rate of compound degradation, knowledge of the amount of competent biomass is required. In the absence of specific empirical data to indicate otherwise, we propose that one assume that the competent biomass is a small fraction of the total community, and, based on the previous studies described above, we believe that 0.01% to 0.001% of the total microbial number (direct microscopic counts) is a reasonable value (Vashon et al. 1982: Ventullo and Larson 1986). Clearly, this percentage is an average value, with a large distribution on either side depending on the chemical and environment. In the absence of information on the number of specific degraders, however, it represents a reasonable starting point for the extrapolation of pseudo–first-order rate constants to different environmental compartments.

Determination of rate constants for microbial biodegradation

The most important factor influencing $k_{1,Bio}$ is the competent biomass concentration in the compartment under consideration. It is also the most difficult to estimate. However, consideration of fundamental microbiological principles allows reasonable estimates to be made of the biomass concentration present in standard tests to determine biodegradation rate coefficients. However, in making this estimation, consideration must be given to the nature of the test. Although discussion of every possible biodegradation test is beyond the scope of this chapter, the important factors to be considered will be illustrated with two examples.

Ready biodegradability tests have been in use for a long time, and much of the available data on biodegradation was obtained with them. The tests measure oxygen consumption, carbon dioxide production, and/or DOC removal over time and therefore represent complete mineralization, which is a conservative estimate of biotransformation, as discussed earlier. In addition, the data may be analyzed to give a first-order rate coefficient for the test environment, $k_{1,Bio}$, provided that sufficient data are collected to describe the degradation curve. That coefficient must then be divided by the competent biomass concentration in the biodegradability test to give the biodegradation coefficient, k_b, which can be used to estimate the pseudo–first-order coefficient for modeling biodegradation of the compound in the various other compartments. If sufficient data to calculate a rate coefficient are not available from the biodegradation test, it may be necessary to assign a pseudo–first-order rate constant using methodologies such as those described in Tables 3-6 through 3-8.

What, then, should that biomass concentration be? With ready biodegradability tests, the inoculum typically varies over the range of 10^4 to 10^6 cells/ml. Furthermore, the lag period before measurable evidence of biodegradation is obtained varies widely depending on the competent fraction of the biomass and its degree of enzyme expression. Thus, at first glance, it appears difficult to estimate what the competent biomass concentration might be. In reality, however, the nature of the test is such that there is little variability in the competent biomass concentration over the range of data used in parameter estimation. This follows from several facts, which are discussed below and illustrated in Figure 3-3.

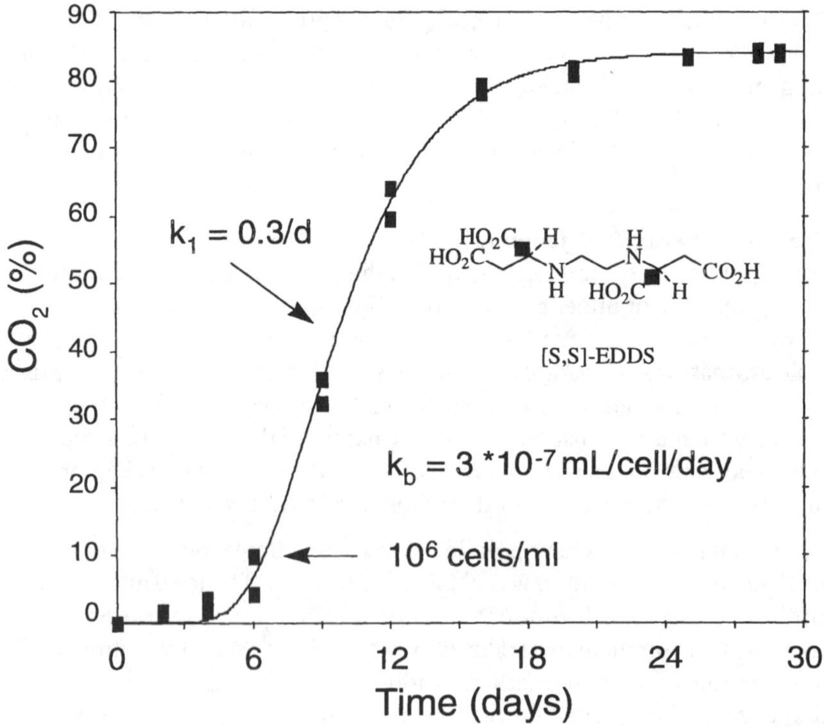

Figure 3-3 Biodegradation of SS-ethylene diamine disuccinic acid in ready biodegradation test

A ready biodegradability test is a growth test in that the substrate concentration added is much larger than the initial biomass concentration in mass units. A typical initial substrate concentration is 20 mg/L. Furthermore, 10^{12} bacterial cells typically have a dry weight of about 1 g (Neidhardt et al. 1990). Thus, an inoculum of 10^5 cells/ml represents approximately 0.1 mg/L, which is small compared with the initial substrate concentration. In addition, the nature of bacterial growth and substrate removal is such that a significant amount of growth must have occurred before a biodegradation curve approximates a first order reaction. Finally, the sensitivity of biodegradation measurements used in screening tests requires that a significant amount of the substrate must be used before biodegradation can be measured, irrespective of the endpoint tested. Typically, at least 10% of the substrate must be consumed before the biodegradation curve conforms to first-order kinetics, thereby allowing the first-order rate coefficient to be estimated.

If the initial substrate concentration is 20 mg/L, this means that 2 mg/L will have been consumed. Because the biomass yield is typically 0.5 mg biomass formed/mg substrate consumed, this means that 1 mg/L of competent biomass will have been formed, which corresponds to approximately 10^6 cells/ml. Consequently, for most ready biodegradability tests, the competent biomass concentration at the point from which the first-order rate coefficient can be obtained is relatively constant and independent of the inoculum size and is equal to 1 mg/L or 10^6 cells/ml.

The value of k_b can be estimated by dividing the first-order rate coefficient from a ready biodegradability test (with typical units of d^{-1}) by 10^6 cells/ml, giving k_b with units of ml-cell^{-1}day^{-1}, or by 1 mg biomass/L, giving k_b with units of L (mg biomass)$^{-1}$d^{-1}. Either of these values of k_b can then be converted into a pseudo–first-order rate constant ($k_{1,Bio}$) for a given compartment by multiplying the k_b value by the competent biomass concentration in that compartment expressed in the appropriate units. It should be emphasized that although this represents a plausible potential approach to estimating second-order (i.e., site-independent, extrapolatable) rate coefficients from standard biodegradation test data, it has not been adequately validated. Such validation would involve comparison of the calculated rate coefficients to values experimentally determined in grab-sample field tests for a reasonable number and diversity of test substances. It would also require that the competent biomass in the compartment or site of interest (i.e., the one for which the half-life must be predicted) be known or measured specifically for purposes of the validation exercise.

River die-away studies constitute another type of biodegradation test for which considerable data exist and therefore that could serve as a source of data for deriving second-order rate coefficients. One major difference between river–die-away and ready biodegradability tests is the level of substrate employed. In the river–die-away test, substrate concentrations are typically in the range of 10 to 100 μg/L, although higher concentrations are sometimes used. Because of the low substrate concentrations, either loss of parent compound or labeled CO_2 production is used as the measured variable. Another difference is in the inoculum, which is often limited to the indigenous bacterial population present in the river water to be tested. In the absence of actual data on competent cell counts, the competent biomass concentration must be deduced from the nature of the response curve.

If evidence of biodegradation starts within a few hours of the beginning of the test, it can be assumed that a significant fraction of the biomass is competent. Thus, the biomass concentration can be estimated from total heterotrophic plate counts. On the other hand, if there is a significant lag before evidence of biodegradation appears, only a small fraction of the inoculum is likely to be competent. In this case, just as in the ready biodegradability test, significant growth will be required before the biodegradation curve conforms to the first-order kinetics typically assumed for data analysis. Consequently, a similar approach can be used to estimate the competent biomass concentration for estimating k_b from the measured first-order rate coefficient. At least 10% of the substrate must be used before the collected data will conform to first-order kinetics. If we again assume a yield of 0.5 mg biomass per mg substrate used, the concentration of biomass present at the beginning of data analysis will be about 5% of the added substrate concentration. Consequently, if 100 μg/L of test substrate is used, the competent biomass at the point that first-order kinetics begins can be assumed to be 5 μg/L. Because 1 μg/L of biomass corresponds roughly to 10^3 cells/ml, the competent biomass concentration associated with biodegradation of 100 μg/L of test substrate would be approximately equal to 5×10^3 cells/ml. Similar estimates could be made for other initial substrate concentrations in the absence of experimental data on the competent biomass concentration.

Data may be available from other tests, as well, including soil or sediment microcosms. Such datasets often are presented in terms of labeled CO_2 evolved and thus could be handled in the same manner to obtain an approximation of the competent biomass fraction. These estimates of competent biomass would then be used to derive second-order rate coefficients from the river, soil, or other die-away pseudo–first-order rate coefficients for the purpose of extrapolating to other compartments and/or sites. However, as with the suggested methodology for deriving extrapolatable rate coefficients from screening data, this proposed scheme for using grab-sample data has not been adequately validated. Further, extrapolation of the rate coefficients so derived to other specific sites requires information (i.e., the competent biomass at the target site) that usually will not be available and must be estimated. To the extent that this necessitates use of generic or default values for competent biomass, an alternative approach that uses results from ready or inherent biodegradation tests to derive default values for first-order rate constants could be employed (Tables 3-6 through 3-8). This approach is fundamentally less elegant than the second-order approach, but it is simpler to apply and, with the appropriate scaling factors, it could be used with multimedia models to estimate persistence.

Determination of rate constants for phytodegradation

Pseudo–first-order rate coefficients for phytodegradation are obtained from measured data only. Kinetic data are generally reported as second-order rate coefficients and can be reduced to pseudo–first-order rate coefficients when substrate concentration is low by assuming a constant biomass. While second-order rate coefficients are available for a few plants and even fewer compounds at the local ecosystem level, they have not been developed at the larger scale. Consequently, at this time, the phytodegradation rate coefficient k_{phyto} will be assumed to be zero until more data become available.

Abiotic degradation processes

Determination of rate constants for hydrolysis in water

Measured values of first- and second-order hydrolysis rate coefficients are preferred. In the absence of measured values, calculated coefficients from QSARs and expert estimates can be used if their reliability can be assessed. A pH that is representative of the aquatic system being modeled is assumed, and this pH remains constant (buffered) throughout the time frame under consideration.

The corresponding second-order rate coefficients are multiplied by the appropriate $[H^+]$ and $[OH^-]$ values. The neutral hydrolysis rate coefficient is pseudo–first-order and is independent of pH. The average pH of the oceans is reported to be pH 7.6 (Snoeyink and Jenkins 1980) therefore, for a global assessment this would be a realistic overall pH. Consequently,

$$k_{hydrol} = k_H [H^+] + k_{H_2O} + k_{OH} [^-OH] \text{ (surface and ground waters)} \qquad \text{(Equation 3-31)}$$

Determination of rate constants for hydrolysis in sediments

Measured hydrolysis rate coefficients are preferred for sediment hydrolysis. However, these values are not generally available. Based on the studies that show that neutral hydrolysis coefficients in the sediment-sorbed phase and the bulk-aqueous phase are similar, it is reasonable to use the same neutral hydrolysis rate coefficient as measured above. This is further justified based on the global pH value of 7.6 because any attenuation or retardation of acid or alkaline hydrolysis would be minimal at the low proton and hydroxide concentrations. Thus, the sediment rate coefficient is given by

$$k_{sediment} = k_{H_2O} \text{ (sediments)} \qquad \text{(Equation 3-32)}$$

Determination of rate constants for direct photolysis reactions near the surface

Measured values for absorption spectra and quantum yields are required. At this time, there is no reliable way to calculate either one of these parameters for organic compounds. Thus, measured values should be used. The first-order rate coefficient is given below:

$$k_{photo} = k_a F \qquad \text{(Equation 3-33)},$$

where k_a is the specific sunlight absorption rate and F is a measure of the spectral overlap between sunlight and absorption spectra of the compound. An average yearly global sunlight irradiance can be calculated using GC-SOLAR (Zepp and Cline 1977).

Determination of rate constants for oxidation in water

Measured reaction rate coefficients should be used for this oxidative process. The first-order rate coefficient for oxidative reaction of certain organic compounds is given by

$$k_{hydrox} = k_{OH} [\cdot OH] \qquad \text{(Equation 3-34)},$$

where the hydroxy radical concentration [·OH] is for clear, near-surface waters.

For fresh surface waters, the hydroxy radical steady-state concentration has been shown to be 10^{-16} mol/L, give or take a factor of 10 (Mill et al. 1980). In seawater, the steady-state concentration is 1 to 2 orders of magnitude lower because of lower concentrations of nitrite, and nitrate, and bromide, which is a quencher (Mill 1999).

Determination of rate constants for reduction in sediments

Pseudo–first-order rate coefficients for abiotic reductions are obtained from measured data only. Kinetic data are generally reported as second-order rate coefficients (k_{red}) and can be reduced to pseudo–first-order rate coefficients when the substrate concentration is low by assuming a constant biomass concentration and thus a constant reductant concentration [R]:

$$k_{1,red} = k_{red} [R] \qquad \text{(Equation 3-35)}.$$

Estimating overall pseudo–first-order rate coefficients for a single compartment

Once pseudo–first-order rate coefficients have been determined for specific biotic and abiotic reactions in a given compartment, they must be summed together to get an overall rate coefficient for that compartment. If a particular reaction is dominant kinetically (i.e., 10-fold faster than the other reactions), that rate coefficient can be used as a surrogate measure for the rest of the group. It is important to note that given the constraints of current multimedia models, bulk substrate concentrations will be used to estimate persistence in a given environmental compartment. This means that the overall pseudo–first-order rate coefficient provided for a given compound in a given compartment must be independent of sorption and other partitioning effects and must reflect the value for the dissolved compound only.

Estimating average pseudo–first-order rate coefficients for multiple compartments

Given the large amount of spatial and temporal heterogeneity in environmental systems, it is difficult to judge a priori when soil, water or sediment systems are sufficiently related to lump pseudo–first-order rate coefficients together to provide an overall weighted average. When characterization information for these compartments is available, weighted-average rate coefficients that span an appropriate range of environmental conditions can be developed. When empirical information is lacking, however, default assumptions can be used. When multimedia modeling is likely to be performed in the detailed analysis of persistence, weighted-average rate constants may not be necessary, since is it likely that rate constants for each compartment will be needed and used as input parameters for the modeling. Under these conditions, individual rate constants for each compartment may be used to estimate persistence, as long as a particular reaction is not kinetically dominant (i.e., 10-fold faster than the other reactions).

Conclusions

In this chapter, we have described both screening-level and detailed approaches for quantifying persistence in soil, surface waters, and sediments. Given the important role that microorganisms play in recycling carbon and other elements in the environment, much of the discussion and kinetic information given has focused on microbial degradation. Clearly, however, abiotic reactions are also important for certain chemicals in some environments. Therefore, major abiotic processes (hydrolysis, photolysis, oxidation/reduction) that can lead to alterations in chemical structure or changes in the distribution of a chemical between the different environmental compartments also have been extensively covered. The sections below summarize the major processes and kinetic equations used to estimate persistence and the key factors that control abiotic and biotic degradation processes.

Estimating rate coefficients

- The transformation reactions within a specific environmental compartment c are treated as if they were first-order with respect to the concentration of the compound of interest in that compartment, S_c:

$$(dS/dt)_c = -k_{1,c}[S_c] \qquad \text{(Equation 3-36)},$$

where $k_{1,c}$ is a lumped pseudo–first-order rate coefficient:

$$k_{1,c} = k_{1,c,r1} + k_{1,c,r2} + \dots k_{1,c,rn} \qquad \text{(Equation 3-37)},$$

where r_1 through r_n refer to the particular reactions occurring.

- If a particular reaction is dominant kinetically (i.e., 10-fold faster than the other reactions), that rate coefficient can be used as a surrogate measure for the sum of the reactions.

Microbial biodegradation

- The reaction rate expression for loss of target compound by biodegradation may be expressed as

$$(dS/dt)_b = -k_b[B][S] \quad \text{(Equation 3-38)},$$

where k_b is the second-order biodegradation rate coefficient,

[B] is the competent biomass concentration, and

[S] is the concentration of the target compound being acted upon. The units of k_b depend on the units of B.

- The value of k_b is specific for the chemical undergoing biodegradation but is influenced by environmental factors such as temperature and pH. Because k_b is normalized for B, it can be used to extrapolate biodegradation data from different tests (primarily screening-level tests) and different systems. It should be noted, however, that while the use of k_b values to estimate persistence in different environmental compartments is conceptually sound, further experimental validation is required to establish the accuracy and utility of this approach.
- A variety of environmental factors are known to influence k_b, B and S, as summarized in Table 3-1. Key factors known to affect B include microbial adaptation, the number of competent bacteria, and the water activity or moisture content. Water solubility and partitioning (K_d and K_{OC}) are important factors that affect S.
- Temperature effects on biotic reactions in natural systems are relatively small and predictable, particularly if temperature changes are gradual and sufficient time is available for the selection of population groups with the appropriate temperature optima.
- Biotic reactions typically are restricted to a narrow range of pH values (i.e., 5 to 9). The effect of pH on biotic rate coefficients tends to be small and comparable to values observed for temperature effects, i.e., a factor of 2 to 3 over a 5 to 9 pH range.
- Although anaerobic environments clearly have some unique biotransformation capabilities (e.g., reductive dehalogenation), biodegradation reactions are generally slower in environments with a low redox potential, and the overall metabolic diversity and activity of the microbial populations may be reduced.

Phytotransformations

- Phytotransformations in soil and water environments can be important pathways for the attenuation of some classes of organic compounds. For example, halogenated hydrocarbons have been shown to degraded in aqueous systems by plants.
- Because of the lack of available data, the phytodegradation rate coefficient may be assumed to be zero until more information becomes available.

Hydrolysis

- Hydrolysis processes in surface water and ground waters can be important loss sinks for certain chemicals.
- Although "pure" hydrolysis should involve only the reactant species provided by water (i.e., H^+, OH^-, and H_2O), the analogous reactions of other chemical species present in the local environment (such as SH^- in anaerobic bogs, Cl^- in seawater, and various ions in laboratory buffer solutions) must be considered in order to obtain a complete picture. These factors complicate an assessment of loss by "hydrolysis" because of the subtle variations in mechanisms for many compounds.
- Hydrolysis in sediments generally is modeled based on aqueous-phase rate coefficients.
- Hydrolysis in soils is not well-defined. Unless there are measured values in soils, this process should not be included.
- Measured values of first- and second-order hydrolysis rate coefficients are preferred. In the absence of measured values, calculated coefficients from QSRRs and expert estimates can be used, if their reliability can be assessed.

Photolytic degradation

- Photolytic processes in aquatic environments can play a significant role in the attenuation of organic contaminants; however, photolysis of organic compounds on soils and in sediments generally is not considered to be significant.
- Direct photolysis is defined by a first-order process wherein the rate of reaction for a given compound is governed by quantum yield, the ultraviolet-visible sorption spectrum, and sunlight intensity.
- Indirect photolysis involves the formation of free radicals and subsequent reaction of organic compounds with the reactive intermediate wherein the rate of reaction for a given compound is governed by the concentration of organic matter and sunlight intensity.

Abiotic redox reactions

- Abiotic redox reactions of organic compounds have been described for relatively few compounds in aquatic and sediment systems. The reactions that have been studied can be very fast and occur across a diverse set of soil/surface water/sediment systems.
- Redox transformations are divided into reductive and oxidative reactions. Reductive reactions generally are more important in sediments and oxidative reactions are more likely in surface waters.

Data sources

- The assessment of degradation typically should rely on empirical data, with known reliability and validity.
- Transformation data may be obtained from either field studies (monitoring data), grab-sample tests, or standardized simulation tests, typically carried out under laboratory conditions. When available, field data are preferred. For most chemicals, however, this type of data will be scarce, necessitating the use of data obtained from grab-sample tests and screening tests (in this order of preference).
- In addition, thoroughly evaluated QSARs may be used to provide additional information on the degradability of a specific chemical. Prediction of biodegradation is hampered because of the lack of reproducibility of biodegradation data and because of the numerous protocols that have been used for biodegradation tests. To date, all QSAR models are better at predicting not-ready biodegradability.

References

Alexander M. 1994. Biodegradation and bioremediation. New York: Academic. 302 p.

Aronson D, Howard PH. 1997. Anaerobic biodegradation of organic chemicals in groundwater: a summary of field and laboratory studies. Prepared for American Petroleum Institute, Washington, DC. http://esc.syrres.com/~esc/anabiodeg.htm.

Aronson D, Citra M, Shuler K, Printup H, Howard PH. 1999. Aerobic biodegradation of organic chemicals in environmental media: a summary of field and laboratory studies. Prepared for U.S. Environmental Protection Agency, Athens, GA.

Aronson D, Howard PH. 1999. Evaluating potential POP/PBT compounds for environmental persistence. Prepared for Chlorine Chemistry Council, Chemical Manufacturers Association.

[ARS] Agricultural Research Service. 1995. Pesticide properties database (ftp://asrr.arsusda.gov/pub/ppdb.common and http://ncsr.arsusda.gov/ ppdb3/).

[ASTER] Assessment Tools for the Evaluation of Risk. 1994. Duluth MN: Environmental Research Laboratory, U.S. Environmental Protection Agency.

Baker JR, Mihelcic JR, Luehrs DC, Hickey JP. 1997. Evaluation of estimation methods for organic carbon normalized sorption coefficients. *Water Environ Res* 69:136–145.

Banks MK, Schwab AP, Govindaraju RS, Kulakow P. 1999. Phytoremediation of hydrocarbon contaminated soils. In: Fiorenza S, Oubre LC, Ward CH, editors. Phytoremediation. New York: CRC.

Barker JF, Patrick GC, Major D. 1987. Natural attenuation of aromatic hydrocarbons in a shallow sand aquifer. *Ground Water Monit Rev* 7:64–72.

Bartholomew GW, Pfaender FK. 1983. Influence of spatial and temporal variations on organic pollutant biodegradation rates in an estuarine environment. *Appl Environ Microbiol* 45: 103–109.

Boethling RS, Howard PM, Meylan WM, Stitler WM, Beauman JA, Tirado N. 1994. Group contribution method for predicting probability and rate of aerobic biodegradation. *Environ Sci Technol* 28:459–465.

Boethling RS, Howard PH, Beauman JA, Larosche ME. 1995. Factors for inter-media extrapolation in biodegradability assessment. *Chemosphere* 30:741–752.

Boethling RS, Howard PH, Stiteler W, Hueber A 1997. Does the semi-continuous activated sludge (SCAS) test predict removal in secondary treatment? *Chemosphere* 35:2119–2130.

Bourquin AW, Przybyszewski VA. 1977. Distribution of bacteria with nitrilotriacetate-degrading potential in an estuarine environment. *Appl Environ Microbiol* 34:411–418.

Brown SC, Grady Jr CPL, Tabak HH. 1990. Biodegradation kinetics of substituted phenolics: demonstration of a protocol based on electrolytic respirometry. *Water Res* 24:853–861.

[CITI] Chemicals Inspection and Testing Institute. 1992. Biodegradation and bioaccumulation data of existing chemicals based on the CSCL Japan. Japan Chemical Industry Ecology-Toxicology and Information Center.

Chiou CT, Porter PE, Schmedding DW. 1983. Partition equilibria of nonionic organic compounds between soil organic matter and water. *Environ Sci Technol* 17:227–234.

Cowan CE, Federle TW, Larson RJ, Feijtel TCJ. 1996. Impact of biodegradation test methods on the development and applicability of biodegradation QSARs. *SAR QSAR Environ Res* 5:37–49.

Cowan CE, Versteeg DJ, Larson RJ, Kloepper-Sams PJ. 1995. Integrated approach for environmental assessment of new and existing chemicals. *Regul Toxicol Pharmacol* 21(1):3–31.

Crosby DG, Wong AS. 1977. Environmental degradation of 2,3,7,8-tetrachlorodibenzo-*p*-dioxin (TCDD). *Science* 195:1337–1338.

[CS] CambridgeSoft Corporation. 1996. Chemical Searching and Information Integration. CS ChemFinder.

Cunningham SD, Berti WR. 1993. The remediation of contaminated soils with green plants: an overview. *In Vitro Cell Dev Biol Plant* 29:207–212.

Degner P, Müller M, Nendza M, Klein W. 1993. Structure-activity relationships for biodegradation. Paris, France: Organization for Economic Cooperation and Development. OECD Environment Monographs No. 68.

Devillers J, Domine D, Boethling RS. 1996. Use of backpropagation neural networks and autocorrelation descriptors for predicting the biodegradation of organic chemicals. In: Devillers J, editor. Neural networks in QSAR and drug design. New York: Academic.

Drossman H, Johnson H, Mill T 1988. Structure-activity relationships for environmental processes. 1: Hydrolysis of esters and carbamates. *Chemosphere* 17:1509–1530.

[EC] European Commission. 1994. Risk assessment of existing substances. Brussels, Belgium: European Commission. Technical Guidance Document XI/919/94-EN

[EC] European Commission. 1996. Technical guidance documents in support of the commission directive 93/67/EEC on risk assessment for new notified substances and commission regulation. Brussels, Belguim: European Commission. EC 1488/94.

Ellenrieder W, Reinhard M. 1988. Athias: an information system for abiotic transformations of halogenated hydrocarbons in aqueous solution. *Chemosphere* 17:331–44.

Ellington JJ, Stancil Jr FE, Payne WD. 1987. Measurement of hydrolysis rate constants for evaluation of hazardous waste land disposal. Volume 1. Data on 32 chemicals. Athens GA: U.S. Environmental Protection Agency. EPA/600/3-86/043, PB87-140349.

Ellington JJ. 1989. Hydrolysis rate constants for enhancing property-reactivity relationships. Athens GA: U.S. Environmental Protection Agency. EPA/600/3-89/063, PB89-220479.

Enviro-Net. 1996. Enviro-Net Material Safety Data Sheet.

EurEco. 1990. Technical data sheets on substances candidates for List 1, Final Report. Luxembourg. Directive 76/464/EEC.

Federle TW, Gasior SD, Nuck BA. 1997. Extrapolating mineralization rates from the ready CO_2 screening test to activated sludge, river water and soil. *Environ Toxicol Chem* 16:127–134.

Fu QS, Barkovskii AL, Adriaens P. 2000. Dioxin cycling in aquatic sediments: Passaic River Estuary. *Chemosphere* (in press).

Gao, J, Garrison AW, Hoehamer C, Mazur C, and Wolfe NL. 1999. Phytotransformations of organophosphate pesticides using axenic plant tissue cultures and tissue enzyme extracts. In Situ and On-site Bioremediation. The Fifth International Symposium; 19–22 April 1999; San Diego, CA.

Gombar VK, Enslein K. 1991. A structure-biodegradability relationship model by discriminant analysis. In: Devillers J, Karcher W, editors. Applied multivariate analysis in SAR and environmental studies. Dordrecht, The Netherlands: Kluwer. p 377–414.

Grady Jr CPL, Smets BF, Barbeau DS. 1996. Variability in kinetic parameter estimates: a review of possible causes and a proposed terminology. *Water Res* 30:742–748.

Haag WR, Yao CCD. 1992. Rate constants for reaction of hydroxyl radicals with several drinking water contaminants. *Environ Sci Technol* 26:005–1013.

Hales SG, Feijtel T, King H, Fox K, and Verstraete W, editors. 1997, Biodegradation kinetics: generation and use of data for regulatory decision making. Brussels: Belgium: Society of Environmental Toxicology and Chemistry (SETAC) Europe. 166 p.

Hermens J, Balaz S, Damborsky J, Karcher W, Müller M, Peijnenburg W, Sabljic A, Sjöström M. 1995. Assessment of QSARS for predicting fate and effects of chemicals in the environment: an international European project. *SAR QSAR Environ Res* 3:223–236.

Howard PH. 2000. Biodegradation. In: Boethling RS, Mackay D, editors., Handbook of property estimation methods for chemicals: environmental and health science. Boca Raton FL: CRC Press. p 281–310.

Howard PH, Banerjee S. 1984. Interpreting results from biodegradability tests of chemicals in water and soil. *Environ Toxicol Chem* 3:551–562.

Howard PH, Hueber AE, Mulesky BC, Crisman JC, Meylan WM, Crosbie E, Gray DA, Sage GW, Howard K, LaMacchia A, Boethling RS, Troast R. 1986. BIOLOG, BIODEG, and fate/exposure: new files on microbial degradation and toxicity as well as environmental fate/exposure of chemicals. *Environ Toxicol Chem* 5:977–80.

Howard PH, Hueber AE, Boethling RS. 1987. Biodegradation evaluation for structure/ biodegradability relations. *Environ Toxicol Chem* 6:1–10.

Howard PH. 1989. Handbook of environmental fate and exposure data for organic chemicals, Volume 1. Chelsea MI: Lewis.

Howard PH, Boethling RS, Jarvis WF, Meylan WM, Michalenko EM. 1991. Handbook of environmental degradation rates. Chelsea MI: Lewis. 725 p.

Howard PH, Meylan WM. 1997. Handbook of physical properties of organic compounds. Boca Raton FL: CRC.

Howard PH, Boethling RS, Stiteler WM, Meylan WM, Hueber AE, Beauman JA, Larosche ME. 1992. Predictive model for aerobic biodegradability developed from a file of evaluated biodegradation data. *Environ Toxicol Chem* 11:593–603.

Hughes JB, Shanks J, Vanderford M, Lauritzen J, Bhadra R. 1997. Transformation of TNT by aquatic plants and plant tissue cultures. *Environ Sci Technol.* 31: 266–271.

Hwang H-M, Hodson RE, Lee RF. 1987. Photolysis of phenol and chlorophenols in estuarine water. In: Zika RG, Cooper WJ, editors. Photochemistry of environmental aquatic systems. Washington DC: American Chemical Society. p 27–43.

[ISO] International Organization for Standardization. 1994. Soil quality: guidance on laboratory testing for biodegradation of organic chemicals in soil under aerobic conditions. Geneva, Switzerland: International Organization for Standardization. ISO Standard 11266.

[ISO] International Organization for Standardization. 1998. Soil quality: guidance on laboratory testing for biodegradation of organic chemicals in soil under anaerobic conditions. Geneva, Switzerland: International Organization for Standardization. ISO Standard 15473.

Jannasch HW. 1967. Growth of marine bacteria at limiting concentrations of organic carbon in seawater. *Limnol Oceanogr* 12:264–271.

Jeffers PM, Wolfe NL. 1997. Degradation of methyl bromide by green plants. In: Seiber JN, editor. Fumigants: environmental fate, exposure, and analysis. Washington DC: American Chemical Society. p 53–59.

Jeffers PM, Wolfe NL. 1998. Green plants: a terrestrial sink for atmospheric methyl bromide. *Geophys Res Lett* 25:43–46.

[JRC] Joint Research Center of the Commission of European Communities. 1996a. Environmental Chemicals Data and Information Network (online database: http://ulisse.ei.jrc.it/img/ ecdin_help.conf?148,100).

[JRC] Joint Research Center of the Commission of European Communities. 1996b. International Uniform Chemical Information Database (IUCLID; CD-ROM). Markdorf: TechniDATA.

Kim H, Hemond HF, Krumholz LR, Cohen BA. 1995. In-situ biodegradation of toluene in a contaminated stream. 1. Field studies. *Environ Sci Technol* 29:108–116.

Klečka GM. 1985. Biodegradation. In: Neely WB, Blau GE, editors. Environmental exposure from chemicals, Volume 1. Boca Raton FL: CRC. p 109–156.

Klečka GM, Maier WJ. 1985. Kinetics of microbial growth on pentachlorophenol. *Appl Environ Microbiol* 49:46–53.

Klopman G, Balthasar DM, Rosenkranz HS. 1993. Application of the computer automated structure evaluation (CASE) program to the study of structure biodegradation relationships of miscellaneous chemicals. *Environ Toxicol Chem* 12:231–240.

Klopman G, Zhang Z, Balthasar DM, Rosenkranz HS. 1995. Computer-automated predictions of aerobic biodegradation of chemicals. *Environ Toxicol Chem* 14:395–403.

Kovárová-Kovar K, Egli T. 1998. Growth kinetics of suspended microbial cells: from single-substrate-controlled growth to mixed-substrate kinetics. *Microbiol Mol Biol Rev* 62:646–666.

Langenberg H, Rorije E, Peijnenburg W. 1996. On the reliability and usefulness of existing QSBRs for risk assessment and priority setting. *SAR QSAR Environ Res* 5:1–16.

Larson RA, Weber EJ. 1994. Environmental photochemistry. In:Reaction mechanisms in environmental organic chemistry. Boca Raton FL: Lewis. p 359–413.

Lichtenstein EP, Schultz KR. 1959. Persistence of some chlorinated hydrocarbon insecticides influenced by soil types, rates of application and temperature. *J Econ Entomol* 52:124–131.

Loonen H, Lindgren F, Hansen BG, Karcher W. 1996. Prediction of biodegradation from chemical structure: use of MITI data, structural fragments and multivariate analysis for the estimation of ready and not-ready biodegradability. In: Peijnenburg WJGM, Damborsky J, editors. Biodegradability prediction. Dordrecht, the Netherlands: Kluwer. p 105–114.

Loonen H, Lindgren F, Hansen B, Karcher W, Niemela J, Hiromatsu K, Takatsuki M, Peijnenburg W, Rorije E, Struijs J. 1999. Prediction of biodegradability from chemical structure: modeling of ready biodegradation test data. *Environ Toxicol Chem* 18:1763–1768.

Loveland TR, Belward AS. 1997. The IGBP-DIS Global 1 km Land Cover Data Set, DISCove First Results. *Int J Remote Sens* 18:3289–3295.

Lyman WJ, Reehl WF, Rosenblatt DH. 1990. Handbook of chemical property estimation methods. Washington DC: American Chemical Society.

Lynch MR, editor. 1995. Procedures for assessing the environmental fate and ecotoxicity of pesticides. Brussels, Belgium: Society of Environmental Toxicology and Chemistry (SETAC) Europe. 54 p.

Mabey W, Mill T. 1978. Critical review of hydrolysis of organic compounds in water under environmental conditions. *J Phys Chem Ref Data* 7:383–415.

Macalady DL, Wolfe NL. 1985. Effects of sediment sorption on abiotic hydrolyses. I. Organophosphorothioate esters. *J Agric Food Chem* 33:167–173.

Mackay D. 1991. Multimedia environmental models: the fugacity approach. Boca Raton FL: Lewis/CRC. 257 p.

Mackay D, Shiu WY, Ma KC. 1992a. Illustrated handbook of physical-chemical properties and environmental fate for organic chemicals, Volume 1. Monoaromatic hydrocarbons, chlorobenzenes and PCBs. Boca Raton FL: Lewis.

Mackay D, Shiu WY, Ma KC. 1992b. Illustrated handbook of physical-chemical properties and environmental fate for organic chemicals, Volume 2. Polynuclear aromatic hydrocarbons, polychlorinated dioxins, and dibenzofurans. Boca Raton FL: Lewis.

Mackay D, Shiu WY, Ma KC. 1992c. Illustrated handbook of physical-chemical properties and environmental fate for organic chemicals, Volume 3. Volatile organic chemicals. Boca Raton FL: Lewis.

Mackay D, Shiu WY, Ma KC. 1995. Illustrated handbook of physical-chemical properties and environmental fate for organic chemicals, Volume 4. Oxygen, nitrogen, and sulfur containing compounds. Boca Raton FL: Lewis.

Magbanua Jr BS, Poole LJ, Grady Jr CPL. 1998. Estimation of the competent biomass concentration for the degradation of synthetic organic compounds in an activated sludge culture receiving a multicomponent feed. *Water Sci Technol* 38(8–9):55–62.

Masunaga S, Wolfe NL, Carreira K. 1993. Transformation of para-substituted benzonitriles in sediment and in sediment extract. *Water Sci Technol* 28:123–132.

Masunaga S, Wolfe NL, Hayase K. 1995. Hydrolysis of para-substituted benzonitriles in water. *Environ Toxicol Chem* 14:1457–1465.

[MedChem] Medicinal Chemistry Project. 1996. Medicinal Chemistry Master database (MedChem; online database http://fox.pomona.claremont.edu/chem/master/mf/ ghindex.html).

Meylan WM, Howard PH. 1995. HYDROWIN: Users Guide. HYDROWIN Software available from Syracuse Research Corporation, Environmental Science Center, North Syracuse, NY 13212.

Mill T. 1980. Chemical and photo oxidation. In: O. Hutzinger, editor. The handbook of environmental chemistry, Volume 2/Part A. Reactions and processes. Berlin, Germany: Springer-Verlag. p 77–105.

Mill T, Hendry D, Richardson D. 1980. Free-radical oxidants in natural waters. *Science* 207:886.

Mill T, Mabey W. 1985. Photochemical transformations. In: Neely WB, Blau GE, editors. Environmental exposure from chemicals, Volume 1. Boca Raton FL: CRC. p 175–216.

Mill T, Haag W, Penwall P, Pettit T, Johnson H. 1987. Environmental fate and exposure studies development of a PC-SAR for hydrolysis: esters, alkyl halides and epoxides. Menlo Park CA: SRI International.

Mill T. 1989. Structure-activity relationships for photooxidation processes in the environment. *Environ Toxicol Chem* 8:31–43.

Mill T. 1999. Predicting photoreaction rates in surface waters. *Chemosphere* 38:1379–1390.

Mopper K, Zhou X. 1990. Hydroxy radical photoproduction in the sea and its potential impact on marine processes. *Science* 250:265–274.

Mulder W, Verhaar H, Hermens J, Peijnenburg WJGM, Rorije E, Sabljic A. 1993. An overview of QSAR for several important environmental parameters. Report for DG-XII. RITOX, University of Utrecht, Utrecht, The Netherlands.

Naziruddin M, Grady Jr CPL, Tabak HH. 1995. Determination of biodegradation kinetics through the use of electrolytic respirometry. *Water Environ Res* 67:151–158.

Neidhardt FC, Ingraham JL, Scgaechter M. 1990. Physiology of the bacterial cell: a molecular approach. Sunderland, MA: Sinauer Associates.

Niemi GJ, Veith GD, Regal RR, Vaishnav DD. 1987. Structural features associated with degradable and persistent chemicals. *Environ Toxicol Chem* 6:515–527.

[OECD] Organization for Economic Cooperation and Development. 1996. OECD Guidelines for testing of chemicals. Paris, France: Organization for Economic Cooperation and Development.

[OECD] Organization for Economic Cooperation and Development. 1997. Guidance document on direct phototransformation of chemicals in water. Paris, France: Organization for Economic Cooperation and Development. OCDE/GD(97)21.

O'Niell W, Nzengung V, Noakes J, Bender J, Phillips P. 1998. Biodegradation of tetrachloroethylene and trichloroethylene using mixed-species microbial mats. In: Wickramanayake GB, Hinchee RE, editors. Bioremediation and phytoremediation. Columbus WA: Batelle. p 233–237.

Paris DF, Steen WC, Baughman GL, Barnett Jr JR. 1981. Second-order model to predict microbial degradation of organic compounds in natural waters. *Appl Environ Microbiol* 41:603–609.

Peijnenburg WJGM, de Beer K, de Haan M, den Hollander H, Stegeman M, Verboom H. 1992. The development of a structure reactivity relationship for the photohydrolysis of substituted aromatic halides. *Environ Sci Technol* 26:2116–2121.

Peijnenburg WJGM, 't Hart M, den Hollander HA, van de Meent D, Verboom HH, Wolfe NL. 1992. QSARs for predicting reductive transformation rate constants of halogenated aromatic hydrocarbons in anoxic sediment systems. *Environ Toxicol Chem* 11:301–314.

Peijnenburg WJGM, de Beer KGM, den Hollander HA, Stegeman MHL, Verboom HH. 1993. Kinetics, products, mechanisms and QSARs for the hydrolytic transformation of aromatic nitriles in anaerobic sediment systems. *Environ Toxicol Chem* 12:1149–1162.

Peijnenburg WJGM. 1994. Structure-activity relationships for biodegradation, a critical review. *Pure Appl Chem* 66:1931–1941.

Perdue EM. 1983. Association of organic pollutants with humic substances: partitioning equilibria and hydrolysis kinetics. In: Christmon RF, Gjessing ET, editors. Aquatic and terrestrial humic materials. Ann Arbor MI: Ann Arbor Science. p 441–460.

Pitter P, Chudoba J. 1990. Biodegradability of organic substances in the aquatic environment. Boca Raton FL: CRC. 306 p.

Punch WF, Forney LJ, Patton A, Wight K, Masscheleyn P, Larson RJ. 1997. BESS, a computerized system for predicting biodegradation potential of new and existing chemicals. In: Chen F, Schuurmann G, editors. Quantitative structure-activity relationships in environmental sciences–VII. Pensacola FL: Society of Environmental Toxicology and Chemistry (SETAC). p 81–92.

Rorije E, Eriksson L, Verboom H, Verhaar H, Hermens J, Peijnenburg W. 1997. Predicting reductive transformation rates of halogenated aliphatic compounds using different QSAR approaches. *Environ Sci Pollut Res* 4:47–54.

Rorije E, Loonen H, Müller M, Klopman G, Peijnenburg W. 1999. Evaluation and application of models for the prediction of ready biodegradability in the MITI-I test. *Chemosphere* 38:1409–1417.

Sabljic AH, Güsten H, Verhaar H, Hermens J. 1995. QSAR modeling of soil sorption: Improvements and systematics of log K_{oc} vs. log K_{ow} correlations. *Chemosphere* 31:4489–4514

Schnoor JL, Litcht LA, McCutcheon SC, Wolfe NL, Carreira LH. 1995. Phytoremediation of organic and nutrient contaminants. *Environ Sci Technol* 29:318A–323A.

Schwarzenbach RP, Stierli R, Lanz K, Zeyer J. 1990. Quinone and iron porphyrin mediated reduction of nitroaromatic compounds in homogeneous aqueous solution. *Environ Sci Technol* 24:1566–1574.

Scow KM, Simkins S, Alexander M. 1986. Kinetics of mineralization of organic compounds at low concentrations in soil. *Appl Environ Microbiol* 51:1028–1035.

Simkins S, Alexander M. 1984. Models for mineralization kinetics with the variables of substrate concentration and population density. *Appl Environ Microbiol* 47:1299–1306.

Snoeyink VL, Jenkins D. 1980. Water chemistry. New York: Wiley. 463 p.

Spain JC, Pritchard PH, Bourquin AW. 1980. Effects of adaptation on biodegradation rates in sediment/water cores from estuarine and freshwater environments. *Appl Environ Microbiol* 40:726–734.

Spain JS, Van Veld PA. 1983. Adaptation of natural microbial communities to degradation of xenobiotic compounds: effects of concentration, exposure time, inoculum and chemical structure. *Appl Environ Microbiol* 45:428–435.

Stewart DKR, Chisholm D. 1971. Long-term persistence of BHC, DDT and chlordane in a sandy loam clay. *Can J Soil Sci* 61:379–383.

Struijs J, van den Berg R. 1995. Standardized biodegradability tests: extrapolation to aerobic environments. *Water Res* 29:255–262.

Tomlin CDS, editor. 1997. The pesticide manual. Eleventh edition. Farnham, Surrey, UK: British Crop Protection Council.

Tratnyek PG, Hoigné J. 1991. Oxidation of substituted phenols in the environment: a QSAR analysis of rate constants for reaction with singlet oxygen. *Environ Sci Technol* 25:1596–1604.

Tunkel J, Howard PH, Boethling RS, Stiteler W, Loonen H. 2000. Predicting ready biodegradability in the MITI test. *Environ Toxicol Chem* (in press).

Tyle H, Niemela J. 1999. Use of QSARS for selection of POPs. Draft Report. Danish EPA.

University of Akron. 1996. Hazardous Chemical Database. (http://odin. chemistry.uakron.edu/cgi-bin/wwwwais).

[USEPA] U.S. Environmental Protection Agency. 1994. USEPA/EC Joint project on evaluation of quantitative structure activity relationships (QSARs). Washington DC: U.S. Environmental Protection Agency. EPA 743-R-94-001.

[USEPA] U.S. Environmental Protection Agency. 1996. Chemical substance factsheets (online database: gopher://ecosys.drdr. Virginia.EDU: 70/11/library/gen/ toxics).

[USEPA] U.S. Environmental Protection Agency. 1998. Fate, transport, and transformation test guidelines. Washington DC: U.S. Environmental Protection Agency. EPA 712-C-98-090.

Vanderford M, Shanks JV, Hughes JB. 1997. Phytotransformation of trinitrotoluene (TNT) and distribution of metabolic products in Myriphyllum aquaticum, *Biotechnol Lett* 19:277–280.

Vashon RD, Jones WJ, Payne AG. 1982. The effect of water hardness on nitrilotriacetate removal and microbial acclimation in activated sludge. *Water Res* 16:1429–1423.

Ventullo RM, Larson RJ. 1986. Adaptation of aquatic microbial communities to quaternary ammonium compounds. *Appl Environ Microbiol* 51:356–361.

Verschueren K. 1983. Handbook of environmental data on organic chemicals. Second edition. New York: Van Nostrand Reinhold. 1310 p.

Vogel TM, Reinhard M. 1986. Reaction products and rates of disappearance of simple bromoalkanes, 1,2-dibromopropane and 1,2-dibromoethane. *Environ Sci Technol* 20:922–997.

Vollner L, Klotz D. 1994. Behavior of DDT under laboratory and outdoor conditions in Germany. *J Environ Sci Health*. B29:161–167.

Ward DM, Brock TD. 1978. Hydrocarbon biodegradation in hypersaline environments. *Appl Environ Microbiol* 35:353–359.

Webster E, Mackay D, Wania F. 1998. Evaluating environmental persistence. *Environ Toxicol Chem* 17:2148–2158.

Wolfe NL, Metwally ME, Moftah AE. 1989. Hydrolytic transformations of organic chemicals in the environment. In: Sawhney BL, Brown K, editors. Reactions and movement of organic chemicals in soils. Madison WI: Soil Science Society of America and American Society of Agronomy. p 229–242.

Wolfe NL, Jeffers PM. 2000. Hydrolysis. In: Boethling RS, Mackay D, editors. Handbook of property estimation methods for chemicals: environmental and health science. Boca Raton FL: CRC Press. p 311–334.

Zepp RG, Cline DM. 1977. Rates of direct photolysis in the aqueous environment. *Environ Sci Technol* 11:359.

Zepp RG, Schlotzhauer PF. 1979. Photoreactivity of selected aromatic hydrocarbons in water. In: Jones PW, Leber P, editors. Polynuclear aromatic hydrocarbons. Ann Arbor MI: Ann Arbor Science Publishers. p 141–158.

Zepp RG. 1982. Experimental approaches to environmental photochemistry. In: Hutzinger O, editor. The handbook of environmental chemistry, Volume 2/Part B. Berlin, Germany: Springer-Verlag. p 19–41.

Zepp RG, Braun AM, Hoigne J, Leenheer JA. 1987. Photoproduction of hydrated electrons for natural organic solutes. *Environ Sci Technol* 21:485–490.

Zepp RG, Wolfe NL. 1987. Abiotic transformation of organic chemicals at the particle-water interface. In: Stumm W, editor. Aquatic surface chemistry. New York: Wiley. p 423–454.

Zepp RG. 1991. Photochemical fate of agrochemicals in natural waters. In: Frehse H, editor. Pesticide chemistry, advances in international research, development, and legislation. Proceedings of the Seventh International Congress of Pesticide Chemistry. New York: VCH Publishers. p 329–345.

CHAPTER 4

Inter-media Partitioning and Transport

Anders W. Andren, Don Mackay, Joseph V. DePinto, Katharine Fox, Louis J. Thibodeaux, Michael McLachlan, Stefan Haderlein

It is evident that the overall persistence of a chemical in the environment is affected by how the chemical partitions between different media such as air, water, and soils. For example, a substance may have a very short degradation half-life in air, but this is of little consequence if the substance is nonvolatile and remains attached to solid particles. Furthermore, the rate of inter-media transport can be a significant determinant of overall half-life. For example, the effective half-life of a substance that degrades slowly in water but rapidly in air may be controlled by how fast it evaporates. The concepts described in this chapter lead to a set of descriptive equations that contain parameter values determined by the properties of the chemical and those of the environment. These equations are incorporated into the mass-balance models (see Chapter 5), which describe how the chemical behaves in the environment. This behavior results in persistence and potential for long-range transport, 2 attributes that can be estimated using a validated model as a contributor to the regulatory process.

The aim of this chapter is to review the present state of the science of equilibrium partitioning and inter-media transport.

Environmental Media

The first and most fundamental step is to define the relevant media. It is relatively easy to list a large number of compartments, media, or phases into which chemicals partition. It is less easy to aggregate them into a smaller number of manageable compartments. In simplistic terms, if the number of compartments is large, so too are the number of potential partition coefficients and the number of inter-media

CHAPTER PREVIEW

Evaluation of Persistence and Long-Range Transport of Organic Chemicals in the Environment. Gary Klečka et al., editors.
 ISBN 1-880611-22-8

transfer processes that must be considered. This produces a more-complex model that can provide a realistic description of the system and more information and greater accuracy in the analysis. This aggregation-decision process may be reflected in a multi-tier system in which the early tiers treat simpler systems with fewer compartments.

The simplest conceivable system presented in Table 4-1, Option 1 and Figure 4-1, is a 3-compartment system of air, water, and soils in which bottom sediments are included in the soil compartment. This treatment is consistent with the suggested availability of half-life estimates. Knowledge of the air–water and organic carbon–water partition coefficients may be sufficient to characterize the equilibrium partitioning of a chemical in this simple system. A key issue is whether or not sediments and terrestrial soils should be separated. Factors in favor of this separation include the observation that there is little or no photolysis in sediments, conditions may be anaerobic rather than aerobic, there is no direct sediment–air contact, and sediments may significantly affect the residence time of chemicals in the water compartment. On the other hand, volumes of sediments are relatively small, at least in freshwater systems. This system could be defined as being at equilibrium (Level I or II) or at non-equilibrium with defined inter-media transport expressions.

Table 4-1 Options for environmental compartmentalization

Option 1	Option 2	Option 3
Air	Air	Air (possibly stratified)
	Aerosols	Aerosols (possibly segmented into size classes)
Water	Fresh water	Fresh water (possibly stratified)
	Suspended particles	Freshwater particles
	Ocean water	Freshwater biota
	Suspended particles	Ocean water
		Ocean water particles
		Ocean water biota
Soil plus sediment	Soils, including vegetation and ground water	Soils (possibly stratified)
		Vegetation
		Ground water
	Freshwater sediments	Freshwater sediments (possibly stratified)
		Freshwater sediment biota

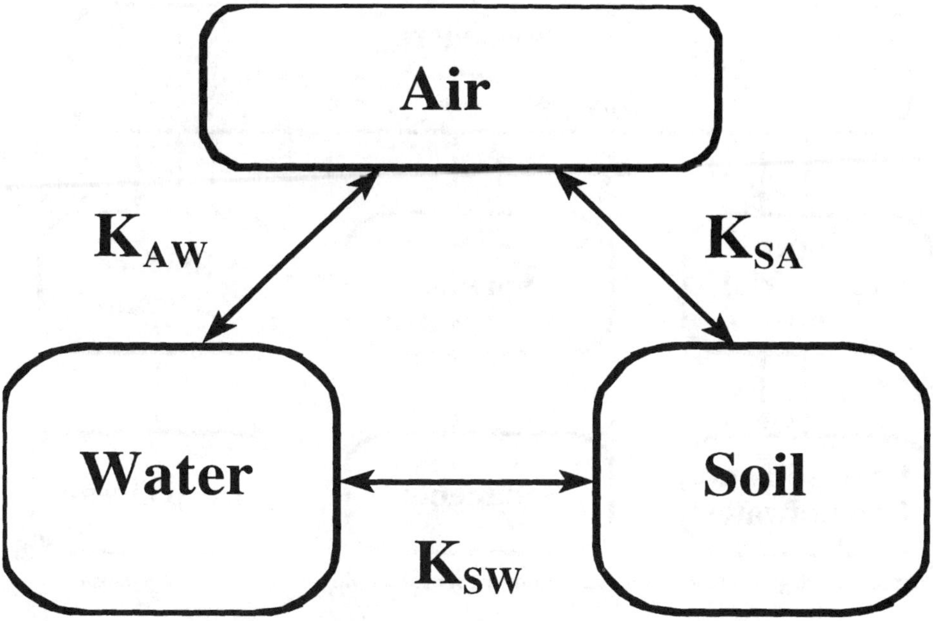

Figure 4-1 Diagram of simple 3-compartment system with equilibrium partitioning between compartments

Proceeding to a higher level of complexity and flexibility, it is suggested that the media in Table 4-1, Option 2 be included, the resulting system being depicted in Figure 4-2. As illustrated in Figure 4-2, complexity has been added to take into account the major inter-media transfer processes and additional compartments that are necessary to overcome the major sources of uncertainty inherent in the spatial aggregation shown in Figure 4-1. Using inter-compartment transfer rates, as opposed to inter-compartment equilibrium partitioning, permits an analysis of persistence as a function of which compartment is receiving the chemical discharge. This increase in robustness and accuracy in estimating persistence comes at a price of greater model resolution and data demands. The significant changes, therefore, are of 2 types:

- A compartment may be separated or disaggregated spatially. For example, water can be divided into fresh and marine systems.
- A compartment may be redefined as consisting of a combination of different phases. For example atmosphere can be treated as a gas plus an aerosol phase.

The system depicted in Figure 4-2 is similar to that used in the Equilibrium Criterion (EQC) series of models of Mackay et al. (1996) except that it specifically treats ocean water. Air is now treated as gaseous air plus aerosols, the latter being important as a deposition vector for substances such as benzo(a)pyrene and

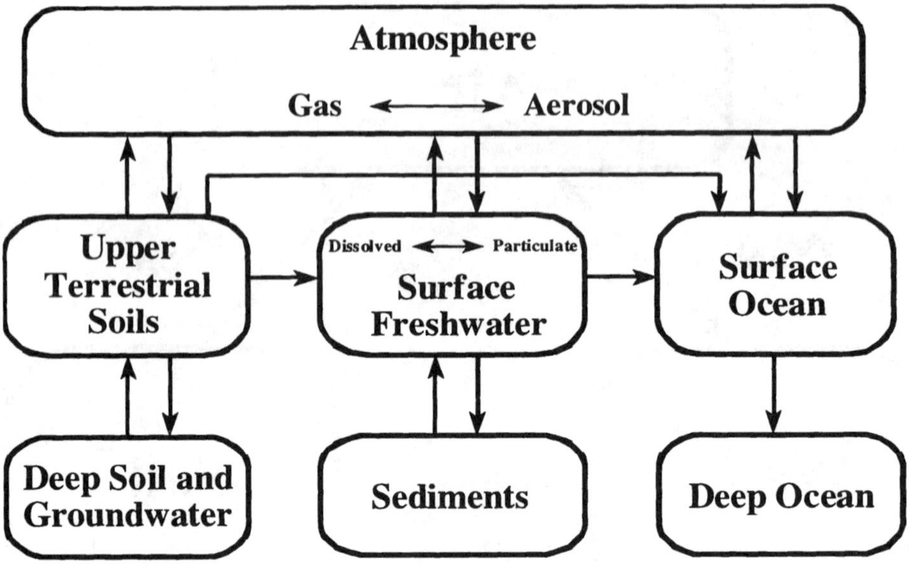

Figure 4-2 Diagram of more-complex multimedia system showing kinetic inter-media transfers

octachlordibenzodioxin, both of which are strongly associated with the particulate phase. Terrestrial soils consist of mineral matter, organic carbon (including humic material, vegetation, and terrestrial biota), air, and water (including ground water). There is uncertainty about the optimum method of including vegetation in such models. The importance of vegetation as a modulator of the environmental fate of chemical is currently a topic of debate and research. It is likely that, for at least some chemicals, forests and grasses play an important role in scavenging chemicals from the atmosphere and providing a substrate for degrading reactions.

The water compartment of Option 2 is also treated in more detail by disaggregating it from its single compartment into three interacting compartments: fresh water, freshwater sediments, and surface ocean water. As discussed earlier, this increased spatial resolution is deemed necessary because sediments significantly affect the residence time of chemicals in freshwater systems. It is possible for the surface sediment layer of lakes and rivers to contain a large fraction of the mass of a chemical residing in these systems; the chemical transformation half-life and overall residence time will be very different in sediments versus overlying water. Bottom sediments interact with overlying water through reversible inter-media mass-transfer processes. Particle deposition can move chemical mass from the water column to the sediments, particle resuspension can move chemical mass out of the sediments into the water column, and sediment–water diffusion can generate a net movement of mass across the interface. In lakes and the lower portions of major rivers and estuaries, there is a net downward flux of solids and associated chemical mass out of the surface sediments and into deep, less-accessible sediments. This sediment and associated chemical burial represents a permanent loss in this model framework, which is viewed as a justifiable assumption based on many studies of these processes in surface freshwater systems (e.g., DePinto et al. 1993). Through this combination of sediment–water processes, bottom sediments can significantly buffer the response time of a water column to changes in chemical inputs and thus profoundly affect overall persistence.

The surface ocean compartment consists of the upper, mixed layer (top 100 to 200 m) of the ocean, which receives the runoff from the fresh water compartment and is accessible to the atmosphere in relatively short times of weeks to months. It is necessary at this level to distinguish oceans from freshwater because of the potentially significant differences in volumes, residence times, transformation rates, air–water exchange fluxes, and potential for long-range transport. It is generally assumed that the transfer of mass from this ocean surface layer to the deep ocean by particle deposition and hydrodynamic transport represents a permanent loss of chemical from the system.

In addition to the increased spatial/intra-compartment degradation resolution, it is also desirable to redefine the chemical present in the fresh water as being in dissolved and particulate phases. This is necessary because many of the chemicals of concern are hydrophobic. As a result, a substantial fraction of the chemical in the freshwater compartment is in the particulate phase. For example, for a chemical with a partition coefficient, log K_P, of 6 in a surface water with a suspended solids concentration of 1 mg/L, approximately half of the chemical will be in the particulate phase at sorption equilibrium. This partitioning of the freshwater column compartment into 2 phases permits a more accurate description of inter-media transfer rates as well as intra-media transformation processes. For example, air–water exchange is dependent on the gradient between the dissolved phase of the chemical in the water and the partial pressure of the chemical in the atmosphere. Hence, estimates of partitioning of the chemical in both the water column and the air are necessary in order to use typical mass-transfer rate coefficients that have been calculated or measured for surface waters. The same argument can be applied to transfer across the sediment–water interface. Settling and resuspension apply only to the particulate phase of the chemical, while sediment–water diffusion acts only on the dissolved phase of the chemical. Further, the transformation half-lives (photolysis, hydrolysis, biotransformation) of most chemicals are dependent on the phase distribution of that chemical.

It is possible to proceed to higher levels of disaggregation. Some options are as follows:

- The atmosphere can be segmented vertically into several layers, which differ in temperature and accessibility to the surface. Aerosols can be segmented into different size ranges with different deposition characteristics. Whereas rain is introduced as a driving factor for wet deposition, and is, of course, variable with climate, it may be desirable to include snow as a deposition vector that can be important in cold climates.

- Soils can be segmented vertically to account for the differences in behavior of surface soils that are subject to solar radiation and more extreme temperature variation. There are also differences in organic matter contents with depth. A more rigorous treatment of leaching to ground water may be included.
- Similarly, freshwater systems may be differentiated on the basis of whether they are lentic (impounded) or lotic (fluvial). This is done to capture differences in compartment volumes, advective residence times, inter-media transfer rates, and potential for long-range transport. Impounded waters may also be stratified vertically because of thermal gradients, again permitting a more realistic representation of cross-media interacting compartments. As in air and soil systems, phase fractionation in surface waters can be highly resolved. For example, sorbent material in surface waters can be differentiated on the basis of whether it is biotic (plankton, bacteria), abiotic (soil-derived particulate matter) or non-settleable organic matter (dissolved organic carbon). This classification permits a more-detailed description of organic carbon content and character (and thus binding properties), sorbent sources, and sorbent transport and transformation rates. This approach was taken in conducting a mass-balance model development and application to address research and management questions regarding polychlorinated biphenyl (PCB) congeners in Green Bay, Lake Michigan (Bierman et al. 1992; DePinto et al. 1994).

In the judgment of the group of experts, the 3-compartment system (Option 1) may be adequate for simple evaluation of chemical fate at a very large scale in which volumes of oceanic water greatly exceed volumes of fresh water. This option may be preferred for global evaluation purposes.

For evaluating the fate of chemicals on a regional or continental scale, Option 2, which is similar to the EQC model structure, is preferred because it explicitly separates fresh and oceanic water and it can better reflect the conditions that apply in regions of human habitation. It is likely that the global ocean:land area ratio of 7:3 in Option 1 is best reduced to perhaps 1:3, reflecting the reduced human access to oceans. Option 3 is more appropriate for use in full risk assessments or in site-specific research studies because of its complexity.

Phase Partitioning

Background

Any compound released into the environment will become distributed between the various environmental phases according to its physicochemical properties. In this context the term "phase" refers to any of the various gaseous, liquid, or solid media that make up the different compartments of the environment. For instance, the environmental compartment "atmosphere" consists primarily of the gaseous or air phase but also of various condensed phases that are suspended in air, including solid phases (aerosols, e.g., desert dust, ash, soot, snow, hail) and liquid phases (e.g., rain, fog, sea spray). Other environmental compartments such as soils or sediments may consist of an even more complex set of different natural or anthropogenic phases.

In order to arrive at a degree of complexity that is manageable in terms of the availability of data (i.e., sufficiently transparent for regulatory purposes but still comprised of the environmental phases that are most relevant for distribution processes of the compounds under consideration), we confine discussion here to a limited set of phases. These phases generally have well-defined, specific storage capacities and mechanisms of interaction for persistent organic pollutants (POPs) and may consist either of a single natural phase (e.g., air) or represent a common property of various different environmental phases (e.g., organic matter present in various natural phases). Five phases are considered, as shown in Figure 4-3.

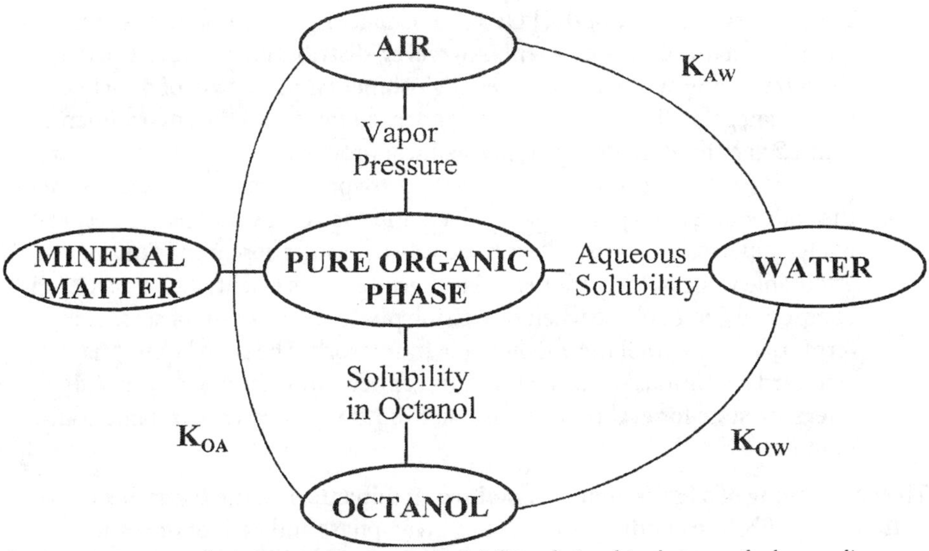

Figure 4-3 Diagram depicting partitioning and solubility relationships between the key media involved in laboratory determinations of physicochemical properties

- The *air phase*, although having a rather small specific storage capacity for most POPs, is by far the most important phase by volume on the global scale. It is also the primary vector for long-range transport of most chemicals.

- The *water phase* is the second most abundant phase on a global scale and is present in significant quantities in virtually all environmental compartments. However, similar to the air phase, its affinity and specific storage capacity for most POPs is rather low.
- The *octanol phase* is considered a surrogate for various condensed lipophilic materials present in natural phases such as non-living natural organic matter present in soils, sediments, or aerosols and certain lipid-like constituents of plants, animals, and microorganisms. The fraction of such organic phases on a global scale is quite low, but they exhibit both a high affinity and a high specific storage capacity for hydrophobic organic contaminants. Concentrations of POPs are usually highest in these phases; thus, they are key vectors for exposure (e.g., in foods).
- The *mineral phase* is a combination of various natural solid hydrophilic phases present in the matrix of soils and sediments or in aerosols. Mineral phases make up a large fraction of soils and sediments, but their affinity and specific storage capacity for hydrophobic organic contaminants in aqueous environments are rather low. However, the distribution of POPs from the gaseous phase to mineral phases may be significant, and certain polar chemicals may interact with specific minerals such as clays.
- The *pure organic phase* of the POP is not usually significant in terms of mass or volume of natural compartments. However, distribution processes between the pure chemical phase and other environmental phases are of direct significance if POPs are released in condensed form into the environment. Thus, 2 very fundamental physicochemical properties of POPs, i.e., vapor pressure (P^S) and aqueous solubility (S_{aq}) are partitioning processes between the compound in its pure organic phase and the air or water phase, respectively. Superscript s is used here to denote the saturation or maximum attainable value of the parameter. Furthermore, it is essential to consider the compounds' pure phase when discussing phase-distribution processes in terms of a conceptual thermodynamic framework. The pure liquid phase at standard conditions of 25 °C and 1 atmosphere pressure is a very useful reference state for evaluating phase-distribution processes of organic compounds.

The partitioning of a hydrophobic organic contaminants, i.e., the reversible distribution at equilibrium conditions between a given phase and each of the other phases discussed above, can be described by the respective distribution coefficient, K_{xy}, which is a compound specific property:

$$K_{xy} = C_x/C_y \qquad \text{(Equation 4-1)},$$

where C_x and C_y are the concentrations of the compound in phases x and y, respectively.

It is noteworthy that the equilibrium partitioning coefficient for a given phase transfer process and at given conditions, K_{xy}, may be reported in a variety of units and thus may have very different numerical values, depending on the units used for the concentration of compounds in the respective phases. For most practical purposes, i.e., for the type of compounds and the range of (rather low) concentrations considered here, the phase distribution constants, K_{xy}, can be considered to be independent of the compound's concentration. As with other thermodynamic equilibrium constants, values of K_{xy} vary with temperature and pressure. Other environmental factors such as the presence of solutes (e.g., salinity) affect the partitioning constants between the water phase and other phases and have to be taken into account when applying particular coefficients to environmental compartments with different compositions of the aqueous phase, such as fresh water and seawater.

The following section discusses in more detail the individual partition coefficients necessary to quantify the partitioning of POPs between the various environmental phases considered. First, we address the most fundamental partition coefficients, describing the distribution of POPs present in the pure phase, the gas phase (vapor pressure), and the water phase (aqueous solubility). Next, we treat dimensionless partition constants between the air and water phases (K_{AW}, which can also be expressed as the related Henry's Law constant, H), between octanol and water phases (K_{OW}) and between the octanol and air phases (K_{OA}). For all distribution coefficients, we will discuss their significance for quantifying partitioning processes of POPs within and between the various environmental compartments, their thermodynamic foundation, experimental and estimation methods by which these coefficients can be determined, and the ways their values vary at different environmental conditions (pressure, temperature, salinity).

It is important to keep in mind that, when choosing the phases and the related partition coefficients indicated above, we have purposely confined ourselves to environmental processes that are pertinent to neutral hydrophobic organic compounds whose interactions with other solutes or phases are dominated by dispersive forces. Thus, one should not be surprised if the approach chosen falls short if applied to the large group of organic contaminants that are neutral but are able to undergo acid–base or donor–acceptor interactions or to permanently charged organic contaminants that are subject to coulombic forces.

General partitioning

As discussed above, the physicochemical properties that most often are used to determine the persistence and transport of chemicals in the environment include the saturation vapor pressure (P^S), aqueous solubility (S_{aq}), air–water partition coefficient (K_{AW}) or Henry's Law coefficient (H), octanol–water partition coefficient (K_{OW}), and octanol–air partition coefficient (K_{OA}). These partition coefficients control, or can be used to predict, the equilibrium partitioning and several transfer rates for chemicals among various media. Most regulations concerning the introduction of chemicals new to commerce, or requiring assessment of chemical fate and effects routinely require these saturation properties or partition coefficients. As a result, extensive databases exist for most candidate POPs, but, as is discussed later, there are often conflicting reports of the actual numerical values.

Saturation vapor pressure

The saturation vapor pressure describes the equilibrium partitioning of a pure chemical between its pure phase and the vapor phase. More specifically, P^S is the pressure of a pure chemical vapor that is in equilibrium with the pure liquid or solid phase (Reid et al. 1977). The saturation vapor pressure for a solid is also called the sublimation pressure. The vapor pressure, or concentration, of a chemical in a mixture of volatile chemicals is called the partial vapor pressure, P.

An examination of P^S values for organic chemicals indicates an 11 order-of-magnitude range, i.e., from about 10^5 to 10^{-6} Pa (1 Atm = 101325 Pa). The P^S of a compound is strongly dependent on temperature. The relationship between P^S and temperature can be approximated by the Clausius-Clapeyron equation:

$$d \ln P^S/dT = \Delta H_{vap}(T)/RT^2 \qquad \text{(Equation 4-2)},$$

where T is the temperature of interest, R is the gas constant (8.314 Pa m^3/mol-K), and H_{vap} is the enthalpy of vaporization, which changes slightly as a function of temperature. Equation 4-2 is often integrated to give the form

$$\ln P^S = A - \Delta H_{vap}/RT \qquad \text{(Equation 4-3)},$$

where A is a constant.

In the range from ambient temperatures up to the normal boiling point, T_b, the temperature dependence of the heat of vaporization can be expressed by Equation 4-4 (Prausnitz et al. 1986):

$$\Delta H_{vap}(T) = \Delta H_{vap}(T_b) + \Delta C_p(T_b)(T - T_b) \qquad \text{(Equation 4-4)},$$

where $\Delta C_p(T_b)$ is the difference in heat capacity between the organic vapor and liquid at the boiling point. Equations 4-3 and 4-4 may be combined, integrated, and simplified to yield

$$\ln P^S = (-\Delta H_{vap}(T_b)/R)(1/T - 1/T_b) \qquad \text{(Equation 4-5)}.$$

Equation 4-5 describes the relationship between the boiling point and vapor pressure of a liquid and the environmental temperature of interest. The formulation must be modified to describe the temperature dependence for a solid. This is best done by relating the vapor pressure of the solid to that of the supercooled liquid. The supercooled liquid state is a hypothetical state that is obtained by extrapolating the relationship described by Equation 4-5 below the melting-point temperature of the solid. The following approximation results:

$$\ln P^S_S = \ln P^S_l - \Delta S_{fus}(T_m)/R\,(T_m/T - 1) \quad \text{(Equation 4-6)},$$

where $\Delta S_{fus}(T_m)$ is the entropy of fusion at the melting point of the organic solid.

The thermodynamic framework outlined above has served as one of the most important points of departure for semi-empirical methods used to estimate vapor pressures of chemicals (Prausnitz et al. 1986; Schwarzenbach et al. 1993). The temperature effect on vapor pressures is considerable for organic solids as well as for liquids. The vapor pressure of organic compounds typically increases by about a factor of 2 for every 10 °C increase in temperature.

A variety of techniques may be used to measure vapor pressure, ranging from the use of manometers to gas chromatographic analysis of streams of inert gas that are brought into equilibrium with the pure solid or liquid phase. Measuring low vapor pressures is very demanding experimentally and reported data therefore often disagree. For a series of chemicals such as PCBs, plots of log vapor pressure versus molecular mass or another molecular descriptor are often quite linear; therefore, there is a potential to estimate vapor pressure for unmeasured substances. In addition, various quantitative structure-activity relationship (QSAR) predictive methods are available by which information or molecular structure is used to estimate a value. The series of handbooks by Mackay et al. (1992–1997) can be consulted for details.

Aqueous solubility

The aqueous solubility, (S_{aq}) of a compound plays an important role in controlling its partitioning between air, water, soil, and biota. The aqueous solubility of a compound can be defined as the aqueous concentration of a solute that is in equilibrium with the pure chemical phase. The range of solubility of organic chemicals in water at environmental temperatures extends from completely miscible to less than 10^{-12} mol/L for those that are very hydrophobic.

A considerable amount of data on the aqueous solubility of organic compounds is now available, although the temperature dependence, especially for organic chemicals that are solids or gases at environmental temperatures (40 °C to –30 °C) is generally not well documented. Furthermore, large discrepancies may exist in experimental values, especially for chemicals with solubilities less than about 10^{-6} mol/L. Useful data bases have been presented and reviewed by Howard and Meylan (1997), Mackay et al. (1992–1997), Schwarzenbach et al. (1993), and Baum (1998).

The thermodynamic framework for examining aqueous solubilities have been presented by a number of authors, including Prausnitz et al. (1986), Yalkowsky and Banerjee (1992), Schwarzenbach et al. (1993), and Mackay et al. (1995). The mole fraction solubility, x_W, for organic liquids, solids, and gases may be expressed by Equations 4-7 through 4-9, respectively:

$$x_W = 1/(\gamma_W) \quad \text{(Equation 4-7)},$$

$$x_W = P^S(S)/(\gamma_W P^S(L)) \quad \text{(Equation 4-8)},$$

$$x_W = P_G/(\gamma_W P^S(L)) \quad \text{(Equation 4-9)}.$$

Here γ_W is the activity coefficient of organic solute in saturated aqueous solution, $P^S(S)$ is the saturation vapor pressure of the organic solid, $P^S(L)$ is the vapor pressure of pure supercooled liquid, and P_G is the gas-phase partial pressure. The activity coefficient is referenced to a standard state according to Raoult's Law, i.e., when x is zero, it approaches an infinite dilution value, and when x is 1.0, it equals 1.0.

The temperature dependence of the mole fraction solubility of an organic chemical dissolved in water may be described by Equation 4-10:

$$d \ln x_W/dT = (\Delta H_S/RT^2) \quad \text{(Equation 4-10)},$$

where ΔH_S is the enthalpy of solution.

An examination of data for organic liquids indicates that aqueous solubilities show a weak temperature dependence. The temperature dependence for organic liquids is therefore often ignored, except for situations that require the most exact outcomes. On the other hand, the solubility of organic solids and gases is strongly dependent on temperature. Aqueous solubilities for rigid organic molecules that are solids generally increase by a factor of 2 for each 25 °C increase in temperature (Yalkowsky and Banerjee 1992, Schwarzenbach et al. 1993).

The aqueous solubilities of organic solutes are generally reduced in the presence of dissolved salts. The effect of dissolved salts may be expressed by the Setchenow equation, which has the form

$$\text{Log}\,(S_W/S_S) = K_S C_S \quad \text{(Equation 4-11)},$$

where S_W is the molar solubility in pure water,
S_S is the molar solubility in salt solution,
C_S is the molar concentration of salt in solution, and
K_S is the Setchenow or salting out coefficient.

According to data reviewed in Schwarzenbach et al. (1993), neutral organic compounds show a 10% to 50% decrease in solubility when determined in seawater as opposed to distilled water.

Techniques for measuring solubility are similar in principle to those used to measure vapor pressure; the gas is simply replaced by water. Again, the measurement of low solubility is very demanding, and reported data often disagree. Predictive methods exist, however, as reviewed by Yalkowsky and Banerjee (1992).

Air–water partitioning

The air–water partition coefficient can be expressed as a dimensionless ratio of concentration, K_{AW} or C_A/C_W, or as the Henry's Law constant, H, which is the ratio of partial pressure to concentration in aqueous solution P_G/C_W and thus has units such as Pa m^3/mol. K_{AW} and H are related through the gas constant temperature group,

$$K_{AW} = H/RT \quad \text{(Equation 4-12).}$$

H is frequently evaluated as the ratio of vapor pressure and solubility, although this obviously fails for substances that are water-miscible. It is noteworthy that higher-molecular-weight substances tend to experience both a lower vapor pressure and lower solubility; thus, H may not change significantly in a series of chemicals such as PCBs. Therefore, K_{AW} or H, rather than P^S alone, is the true indicator of volatility from aqueous solution.

Experimentally, the obvious approach is to bring a dilute solution of the substance in water into equilibrium with air in a closed system at constant temperature. P and C are then measured. In practice, a variety of related methods are used involving analysis of only one phase. Predictive methods are available that use molecular structure as input as described by Mackay et al. (1992–1997).

Octanol–water partition coefficient

Nonpolar organic compounds exhibit a high tendency to partition from the aqueous phase to natural organic phases, including non-living organic material as well as biofilms, algae, or higher organisms present in soils, sediments, or surface waters. The fraction of such organic phases on a global scale is quite low, but these phases exhibit a high specific storage capacity for hydrophobic organic contaminants.

Partition coefficients of a given nonpolar organic solute between various natural organic phases and the aqueous phase often vary only within a relatively small range when they are normalized to the organic carbon content of the organic phases (Karickhoff et al. 1979). Thus, the characteristics of the various types of natural organic phases are fairly similar with respect to partitioning of nonpolar organic solutes, which enables us to mimic the various natural organic phases by a standardized surrogate. n-Octanol has widely been used as a model phase of living and non-living natural organic material. Partition coefficients between n-octanol and water are defined by

$$K_{OW} = C_O / C_W \quad \text{(Equation 4-13).}$$

K_{OW} is usually reported as dimensionless values, since the concentrations of the compounds in octanol (C_O) and water (C_W) are commonly given in similar units (e.g., in mole/L). Thus, a dimensionless K_{OW} value of, for example, 1000, indicates that the total amount of the contaminant in the octanol phase is 1000 times higher than in an equal volume in the aqueous phase. Measured K_{OW} values are available for a large number of solutes, including most POPs (Mackay et al. 1992–1997), and span a wide range for typical nonpolar organic contaminants, i.e., 10 to 10^7. Standardized methods and guidelines for designing octanol–water partitioning experiments are also available in the literature.

The success of this simple model system for describing hydrophobic partitioning processes can be rationalized by examining the thermodynamic properties that contribute to K_{OW}. It can be shown that

$$K_{OW} = C_O/C_W = v_W\gamma_W/v_O\gamma_O \quad \text{(Equation 4-14)},$$

where γ_W and γ_O are the activity coefficients of the solute in the water and octanol phases, respectively, and v_W and v_O are numerical values of the molar volumes of water and octanol, respectively. The γ_O values of nonpolar organic solutes generally are quite similar and tend to be fairly close to unity (the value for ideal mixtures). It is thus apparent that K_{OW} is primarily determined by the magnitude of γ_W, which is a measure of the "incompatibility" of the organic solute with water, i.e., the excess free energy required to transfer the solute from its own liquid phase to the aqueous phase. Consequently, excellent correlations between K_{OW} values and physicochemical properties of nonpolar solutes such as aqueous solubility and molecular size (e.g., expressed as hydrophobic molecular volume or surface area) have been found. Such correlations can, among various other methods, be used to estimate unknown K_{OW} values from known compound properties such as molecular size and topology. The driving force for the high tendency to partition from an aqueous phase into organic phases is thus dominated primarily by the unfavorable energetics of retaining nonpolar organic solutes in the aqueous phase rather than by their interactions with the organic phases. The text by Yalkowsky and Banerjee (1992) gives a full account of this theory and estimation methods.

K_{OW} values are primarily determined by the magnitude of γ_W, i.e., by the excess free energy required to transfer the solute from its own liquid phase to the aqueous phase. Changes in temperature and salinity will affect K_{OW} values in a manner similar to that discussed for aqueous solubility. Thus, a decrease in temperature and/or an increase in salinity will increase K_{OW} values and thus favor partitioning of POPs into natural organic phases. Pressure changes within the environmentally relevant ranges do not strongly affect K_{OW} values.

There are well-developed methods for determining Kow experimentally, and extensive databases and predictive methods exist, as discussed by Baum (1998). Values in excess of 10^6 are difficult to measure accurately because of the very low concentrations in water and thus reported data often disagree.

Octanol–air partition coefficient

The octanol–air partition coefficient (K_{OA}) has been shown to be a useful predictor for the partitioning of nonpolar organic chemicals between air and condensed organic phases in the environment. Implicit in the use of this partition coefficient is again the assumption that the sorption properties of condensed organic phases for nonpolar organic chemicals are similar to the sorption properties of octanol. This octanol model has proven to be very useful in describing the partitioning behavior of known POPs, and K_{OA} is used to predict air–soil, air–plant, and air–aerosol partition coefficients.

K_{OA} has been measured in the laboratory for a number of chemicals (Harner and Mackay 1995; Harner and Bidleman 1996; Kömp and McLachlan 1997a). The magnitude of this partition coefficient is very large for POPs, with values commonly lying in the range of 10^8 to 10^{10} at 25 °C. The measurement of such large partition coefficients is difficult, and, as a result, measured values are available for only a few compounds. In the absence of measured values, K_{OA} can be estimated by dividing the octanol–water partition coefficient K_{OW} by the air–water partition coefficient K_{AW}. K_{OW} and K_{AW} are generally available, either from laboratory measurements or from calculation using QSARs. Consequently, calculated, not measured, K_{OA} values are commonly used for predicting environmental partition coefficients.

There is evidence indicating that calculated K_{OA} values are often lower than measured ones (Harner and Mackay 1995; Kömp and McLachlan 1997a). It has been suggested that this may result from the fact that K_{OW} is actually a partition coefficient between octanol saturated with water and water saturated with octanol, which will tend to be lower than the partitioning between pure water and pure octanol. This does not disqualify the use of calculated K_{OA} values for predicting environmental partitioning. Indeed, octanol saturated with water may be a better model of condensed organic phases than is pure octanol. Because most correlations between environmental partitioning coefficients and K_{OA} are based on calculated K_{OA} values, it is prudent to continue using calculated values at this time.

The temperature dependence of K_{OA} is very significant. For many substances, K_{OA} increases by a factor of 2.5 to 3.5 for every 10 °C decrease in temperature. Measurements have shown that the temperature dependence of K_{OA} is in most cases similar to the temperature dependence of the vapor pressure of the subcooled liquid chemical (Harner and Bidleman 1996; Kömp and McLachlan 1997a). Hence, as a first approximation, the temperature dependence of K_{OA} can be calculated using the enthalpy of vaporization of the subcooled liquid. It should be noted that if values of K_{AW}, K_{OW}, and K_{OA} are measured or predicted, it is possible to perform a consistency check that K_{OA} is K_{OW}/K_{AW}. This can be valuable for checking the accuracy or consistency of reported values.

A complication arises when calculating diffusion rates between media such as air and water in which temperatures often differ. For example, air is usually warmer than water in the summer and vice versa in the winter. It is not then obvious at which temperature the partition coefficient should be calculated. In equilibrium fugacity, models Z values can be defined for each medium at its specific temperature, or an average temperature can be used. When calculating fluxes under non-equilibrium conditions, it is usual to assume that the temperature that applies is that of the phase in which the most resistance to transfer occurs.

Finally, it should be appreciated that the volatility of a substance, i.e., its tendency to evaporate, is a function of the medium from which it is evaporating. This is important in the context of this study because evaporation affects persistence and can contribute to long-range atmospheric transport.

For a pure chemical, the correct descriptor of volatility is the vapor pressure P^S. For a chemical in solution in water, it is K_{AW}. For a chemical in soils or associated with organic media such as vegetation, it is $K_{AO;}$ the air–octanol partition coefficient is expressed usually as the reciprocal K_{OA}.

Property Estimation Techniques

It is now possible to predict most physicochemical properties that are important to the environmental fate of a chemical, with a reasonable indication of the amount of error expected for a new chemical and without having to select a model based upon similarity of chemical structure. Much of the original predictive techniques for physicochemical properties, transport, and degradation processes were summarized by Lyman et al. (1982). Many of the property databases and estimation techniques were updated by Neely and Blau in 1985. Howard and Meylan (1997) have reviewed new property databases and property estimation techniques. The report by Howard and Meylan (1997) also evaluates a number of computer programs that are now available to compute many of the properties of interest. Table 4-2 lists a number of the most useful programs now available.

Environmental Phase Partitioning

If the key environmental media are taken as indicated in Option 2 of Table 4-1, it becomes essential to quantify partitioning involving aerosols, plants, soils, sediments and biota as well as the air–water system treated earlier. In doing so, it must be appreciated that these coefficients depend strongly on the nature of the medium, e.g., the organic matter content of soils. The key partition coefficients are shown in Figure 4-4.

Air–aerosol partitioning

Despite the fact that concentrations of aerosols are very low (e.g., 5 to 100 μg/m^3), an appreciable fraction of relatively nonvolatile chemical in the atmosphere may be associated with aerosol particles. There is some debate as to whether the partition-

Table 4-2 Computer software for estimating physicochemical properties

Estimated parameter	Estimation method	Software
Log K_{ow}	Meylan and Howard (1997)	LOGKOW KOWWIN
	Hansch and Leo (1979, 1995)	PC-CLOGP,CLOGPTM, MacLogP
	Proprietary fragment constant method (ACD 1995)	ACD/LogP
	Broto et al. (1984); Rekker and de Kort (1979); Rekker (1977)	ProLogP
	Ghose et al. (1988)	ATOMLOGP
	Devillers et al. (1995, 1997)	AUTOLOGP
Water solubility	Syracuse Research Corp (1996)	WS/KOW
	USEPA (1987)	PCCHEM
	Myrdal et al. (1995)	AQUAFAC
	Jensen et al. (1981)	UNIFAC
Vapor pressure	Syracuse Research Corp (1995)	MPBPVP
Henry's law constant	Syracuse Research Corp (1996)	HENRY
Soil–sediment adsorption coefficient	From Log P (USEPA 1987)	PC-KOC

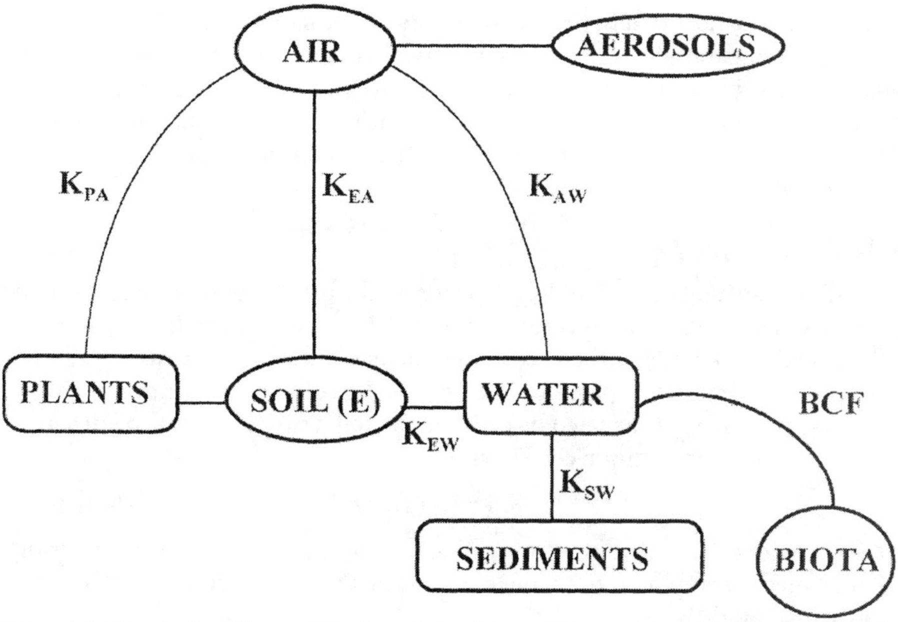

Figure 4-4 Diagram depicting partitioning relationships and transport pathways between key environmental media

ing is absorption into organic matter or adsorption to surfaces that may include organic material (Pankow 1988; Pankow and Bidleman 1991; Bidleman 1988). In any event, it is the total that is measured and correlated.

The traditional method of correlating this phenomenon is the Junge-Pankow equation that regards the process as adsorption to a defined available area. The chemical-specific parameter applied is the liquid or subcooled liquid vapor pressure of the chemical $P^S(L)$. The usual approach is to measure a gas–aerosol partition coefficient, K_P, and correlate it as follows using fitted constants A and B:

$$\log K_P = A - B \log P^S(L) \qquad \text{(Equation 4-15).}$$

Examples are the equations of Bidleman and Pankow cited above and others reviewed by Finizio et al (1997).

More recently it has been suggested that the octanol–air partition coefficient will give a satisfactory correlation with K_P using an equation similar to that above (Finizio et al. 1997):

$$\log K_P = A + B \log K_{OA} \qquad \text{(Equation 4-16).}$$

Examination of the dependence of K_{OA} on vapor pressure shows that K_{OA} is approximately inversely proportional to $P^S(L)$; thus, the success of this equation is not surprising. An advantage of the K_{OA} approach is that data on K_{OW} and K_{AW} are often available and therefore there is no need to estimate $P^S(L)$ from solid state vapor pressures.

It is important to note that both K_{OA} and K_P are highly temperature-dependent; thus, at low temperatures, such as those at high altitudes and high latitudes, there is enhanced partitioning to aerosol particles. Relative humidity is theoretically an important factor (Thibodeaux et al. 1991), but the limited field data suggest the effect to be of secondary importance. The effect of relative humidity on partitioning to soils is more pronounced.

Soil–air partitioning

The soil–air partition coefficient K_{EA} depends on the properties of the chemical, the properties of the soil, the temperature, and the relative humidity (Valsaraj and Thibodeaux 1992). Because there have been many more measurements of soil–water and sediment–water partition coefficients than of soil–air partition coefficients, K_{EA} is normally calculated by dividing the soil–water partition coefficient K_{EW} by the air–water partition coefficient K_{AW}:

$$K_{EA} = K_{EW}/K_{AW} \qquad \text{(Equation 4-17).}$$

In the presence of sufficient moisture, i.e., in wet soils, K_{EW} can be estimated using the same method employed to estimate the sediment–water partition coefficient K_{SW} discussed shortly.

Recently, direct measurements of K_{EA} have been made for a range of POPs, and excellent agreement was found between the measured values and values calculated using Equation 4-17, suggesting that this approach is indeed valid (Hippelein and McLachlan 1998). K_{EA} is a strong function of temperature. Direct measurements of the temperature dependence of K_{EA} for PCBs in soil showed that K_{EA} increased by about a factor of 3 for every 10 °C decrease in temperature. The enthalpies of phase change from soil to air were in some cases lower and in some cases higher than were the enthalpies of phase change from octanol to air (Hippelein 1997). However, the temperature dependence of K_{OA} can be used as a first approximation of the temperature dependence of K_{EA} for nonpolar compounds that partition primarily into the organic carbon fraction of the soil.

K_{EA} is also a function of the relative humidity of the soil. This has been demonstrated theoretically by Valsaraj and Thibodeaux (1992) and in the field for a variety of volatile and semi-volatile POPs (Hippelein 1997). As the water content of a soil decreases from saturation, K_{EA} usually decreases because of the loss of the partitioning capacity of the aqueous phase. When the relative humidity of the soil drops below 100% (which does not occur until the soil is quite dry), K_{EA} begins to increase exponentially until a relative humidity of about 30% is reached. Beyond this point, the rate of increase in K_{EA} accelerates. While the influence of relative humidity is less than the influence of temperature on K_{EA} (e.g., a factor of 13 for PCB 52 decreasing from a relative humidity of 100% to 30%), the combined effect of these variables can be dramatic. Hippelein (1997) measured a K_{EA} for PCB 52 at 5 °C and 30% relative humidity that was 18,000 times higher than that measured at 60 °C and 100% relative humidity. In temperate climates, the relative humidity in surface soil seldom drops below 100%. However, in arid climates or on soil surfaces exposed to direct sunlight or the heat during dry times of the year, this effect is important.

Plant–air partitioning

The plant–air partitioning coefficient K_{PA} depends on the properties of the chemical, the properties of the plant, and the temperature. For known POPs, partitioning from the air occurs primarily to the foliage. Partitioning into the stem/trunk and roots is important only for chemicals that are readily transported out of the leaves in the phloem. Most known POPs are hydrophobic, and their phloem mobility is low.

Plant foliage consists of an aqueous phase and a condensed organic phase, with the plant–air partition coefficient being the sum of the water–air and condensed organic phase–air partition coefficients weighted according to the volume fraction of these phases in the plant.

$$K_{PA} = f_W/K_{AW} + f_{CO}K_{COA} \qquad \text{(Equation 4-18)},$$

where

f_W = volume fraction of water in the plant (typically 50 to 90%),

K_{AW} = air–water partition coefficient (on a volume basis),

f_{CO} = volume fraction of the condensed organic phase, and

K_{COA} = condensed organic phase–air partition coefficient (on a volume basis).

Whereas f_W/K_{AW} can be easily defined, the quantification of the second term is more difficult. There are relatively few measurements of K_{PA} for more hydrophobic chemicals that preferentially partition into the condensed organic phase. Studies of partitioning into azalea leaves using uptake and clearance experiments conducted in chambers have suggested that $f_{CO}K_{COA}$ is approximately linearly related to the octanol–air partition coefficient K_{OA} (Bacci et al. 1990). This suggests that, for partitioning purposes, the condensed organic phase of the plant can be viewed as equivalent to a certain volume of octanol, and this has led to the following modification of Equation 4-18, which has been incorporated into a number of models of accumulation of organic chemicals in plants:

$$K_{PA} = f_W/K_{AW} + f_O K_{OA} \qquad \text{(Equation 4-19)},$$

where f_O is the volume fraction of the plant that behaves as if it were octanol.

However, other research into the partitioning of organic chemicals into isolated plant cuticles (cuticles are one of the major tissues in which organic chemicals are stored in plants) has indicated that the relationship between $f_{CO}K_{COA}$ is exponential and not linear (Kerler and Schönherr 1988):

$$K_{PA} = f_W/K_{AW} + f_X K_{OA}{}^n \qquad \text{(Equation 4-20)},$$

where f_X and n are empirical constants specific to the plant and possibly to the chemical.

Measurement of K_{PA} for PCBs in 5 different grass and herb species yielded results that were fitted well by Equation 4-20. However, the value of n varied widely between the five species, ranging from 0.57 to 1.15. This indicates that the storage properties vary widely from plant to plant. The partition coefficients of the lower chlorinated PCBs varied by up to a factor of 20, despite similar plant-lipid contents and cuticle volumes (Kömp and McLachlan 1997b).

There is as yet no unifying concept to explain the large interspecies variability in plant partitioning. The partitioning of PCBs and several nonpolar aromatics in grass at 25 °C may be reasonably approximated by using Equation 4-19 with f_O of 0.01 (Tolls and McLachlan 1994; Kömp and McLachlan 1997c). A field study using PCBs yielded values in Equation 4-20 of 38 and 14 for f_X and 0.69 and 0.76 for n in a spruce and mixed beech/oak canopy, respectively (Horstmann and McLachlan 1988). However, if these equations are extrapolated to other plant species and substance classes, errors can be expected in excess of an order of magnitude.

K_{PA} has been shown to be very sensitive to temperature. The temperature sensitivity (i.e., the enthalpy of phase change) also varies from plant species to plant species (Kömp and McLachlan 1997c), and there is currently no way of predicting this variability. However, the enthalpies of phase change for plant to air generally lie in the same range as do the enthalpies of phase change from octanol to air, and hence, the latter can be used to calculate a first approximation of the temperature dependence of K_{PA}.

This is clearly a subject that deserves more research because interactions between chemicals in the atmosphere and in vegetation could have a major effect on both persistence and transport. For example, if there is appreciable photolysis on leaf surfaces, this could reduce persistence. Substances that are strongly absorbed by foliage may be effectively removed from the atmosphere, being destined to accumulate in litter, evaporate to air, or enter surface waters as the foliage decays.

Partitioning between water and biota

Living organisms contribute to the pool of natural organic matter in all terrestrial and aquatic systems. In certain environmental compartments, primarily in the surface layers of marine and freshwater systems, organisms including phytoplankton, zooplankton, and higher organisms dominate or even exclusively represent the organic phases present (Thibodeaux et al. 1993). Similar to other natural organic phases, biota exhibit a high affinity and specific capacity for hydrophobic organic contaminants. Thus, under certain conditions, uptake by partitioning into biota may be the dominant removal mechanism for hydrophobic POPs in surface waters which even can drive partitioning of contaminants from the air phase into the water. This can reduce the dissolved concentration and increase air-to-water fluxes. Furthermore, partitioning of contaminants from the aqueous phase to phyto- and zooplankton is one of the underlying mechanisms for biomagnification of POPs through the food web of marine and freshwater systems.

A quantitative measure for the partitioning of solutes between the water and biota phases is the bioconcentration factor (BCF), which is defined in a manner similar to partition coefficients such as K_{OW}:

$$BCF = C_B/C_W \qquad \text{(Equation 4-21)}.$$

C_B in Equation 4-21 is the concentration of the solute in the organism, which can be expressed in various ways, e.g., normalized to the dry weight or the lipid content of the organisms. Note that depending on the units and reference phase chosen for C_B, the numerical values and dimensions of BCF vary accordingly. Dimensionless BCFs can be obtained if C_B is given per volume of lipid content. Dimensionless BCF values can be compared to the respective K_{OW} values of the compounds.

For persistent neutral organic compounds, good linear correlations between log BCF and log K_{OW} values can be expected for a wide range of hydrophobicities of POPs, provided that the BCF values of the dataset were determined for the same, or at least similar, groups of organisms. However, the uptake of extremely hydrophobic compounds (log $K_{OW} > 7$) by biota is generally overestimated by such relationships since there may be appreciable sorption in the water causing reduced bioavailability, the permeability of biological membranes for such large compounds may be reduced as a result of steric effects, and the time for equilibration may be very long.

An invaluable and simple predictive approach is to assume that the lipid in an organism is equivalent in solvent properties to octanol; thus, BCF equals K_{OW} when C_B is expressed as a lipid-normalized concentration.

It is important to note that BCFs in general cannot be considered thermodynamic equilibrium constants since organisms must be viewed as "reactive sorbents" or "reactive solvents." Because of various uptake, transformation, and excretion processes operating in living organisms, BCFs instead represent steady-state partitioning coefficients. Because differences in the general status and the metabolic activity between different species, and even between different sexes or age groups of the same species, of organisms may be significant, BCF values can vary considerably among the various biota. The problem of inter- and intraspecies variability of BCFs is moderated if lipid normalized values are compared. Organic contaminants present in lipid tissues of organisms are less subject to transformation or excretion processes. For such values, significant correlations between the hydrophobicities of POPs (expressed as log K_{OW}) and log BCF values are found for organisms of the same trophic level. An important complication is the presence of biomagnification, a process by which an organism develops a higher lipid-normalized concentration of chemical than that in its food. This occurs when log K_{OW} exceeds about 5 and is a function of the nature of the diet. Considerable and successful efforts have been devoted to developing models of chemical partitioning into single organisms and entire food webs.

Generally, the effects of changes in environmental factors such as temperature and water composition (e.g., salinity) on BCF values ought to be similar to those discussed for K_{OW} values. A quantification or even interspecies generalization of these effects is hampered by the scatter of the available data, e.g., because of concurrent effects of changing environmental factors on the general status and the metabolic activity of the biota. Several models are available to predict bioconcentration and bioaccumulation, e.g., Gobas (1993), Thomann (1989), Clark et al (1990). Food web models have been described by Campfens and Mackay (1997) and Gobas (1993).

Sediment–water partitioning

Partitioning of nonpolar organic chemicals between water and particulate matter in soils and aquatic systems (loosely referred to here as sediment) is generally well-correlated with their partitioning to octanol. The reason is that octanol proves to be a reasonably good surrogate for the type of natural organic matter that is associated with soils and aqueous particles. Hence, the sediment–water partition coefficient (K_{SW} [L/Kg]) for these systems is dependent on the same factors as the octanol–water partition coefficient (K_{OW}).

Because K_{OW} values are widely available for organic chemicals of concern, several investigators have attempted to relate K_{SW} to the specific properties of the chemical, including K_{OW} or its well-correlated solubility (e.g., Chiou et al. 1977; Karickhoff et al 1979; Schwartzenbach and Westall 1981). All of these approaches begin by normalizing to the organic carbon content of the solids in question, namely

$$K_{OC} = f_{OC} \times K_{SW} \qquad \text{(Equation 4-22).}$$

Here f_{OC} is the weight fraction of the total organic carbon in the solid (gC/g(dw)) and K_{OC} is organic carbon partition coefficient (L/Kg). A regression equation is then developed that relates K_{OC} to K_{OW} and/or solubility of the chemical in water. A comprehensive review of such regressions is presented by Schwartzenbach et al. (1993). In general, the regressions improve as they are applied to smaller groupings of chemicals with similar structures and narrower ranges of organic carbon content and character.

An example of a very simple relationship developed by Karickhoff et al. (1979) is given by

$$K_{OC} = 0.69 \times K_{OW} \qquad \text{(Equation 4-23).}$$

In general, the relationship is linear with a slope of the log K_{OC} versus log K_{OW} plot approaching 1.0 for solids having low organic carbon content ($f_{OC} < 0.05$), such as those encountered in soil and bottom sediment systems. However, when applied to organic particles in lakes, which have f_{OC} values in the range of 0.1 to 0.4 (the higher values are for live phytoplankton), the slope of the relationship is closer to 0.2 (Eadie et al. 1992; DePinto et al. 1994). It has been suggested that the organic carbon associated with particulate matter in lakes is on average of a lower molecular weight and has a reduced affinity for hydrophobic chemicals. Having expressed this caveat, it is probably safe to use a relationship such as Equation 4-23 and a sediment organic carbon fraction of 0.05 for the purposes of the screening models described in this chapter.

Summary

The fundamental partitioning properties of most POPs in pure systems (e.g., distilled water) are well-established. There is a growing capability of estimating these properties from molecular structure. When values are extreme, e.g., vapor pressures are very low or K_{OW} is very high, there are often conflicting reported values in the literature. Regrettably, funding agencies have been slow to recognize the importance of obtaining reliable property data, especially as a function of temperature. The result is that many assessments of fate are subject to error. For complex mixtures, such as toxaphene, this problem is particularly acute.

Partitioning between environmental media is only approximated because of the complexities of variable and uncertain composition. However, if reliable physico-chemical data are available, reasonably accurate predictions can be made, provided that those making the prediction are aware of the many pitfalls that exist as a result of potentially inappropriate application of general predictive methods. Phase partitioning is thus one of best-quantified aspects of environmental science.

Inter-media Transport

The purpose of this section is to briefly review some basic concepts concerning transport coefficients, both intra- and inter-media and advective, and to give the reader some guidance in selecting appropriate values for use in POP models. The general field is highly developed. References include Thibodeaux (1996) for inter-media coefficients, Schnoor (1996) for intra-media aquatic coefficients, and Warneck (1988) for atmospheric boundary exchanges.

Transport of a chemical between compartments can be facilitated by the physical movement of a material phase with which the chemical has become associated (generally, by partitioning), as well as by concentration gradients. Level III models (Mackay 1991) incorporate such transfer mechanisms. This section describes the most important mechanisms by which inter-media transport will take place for persistent organic chemicals that may be involved in long-range transport. The lines connecting the media in Figure 4-4 depict these processes.

Transport from air to soil or water

Persistent organic chemicals associated with the air phase either are "dissolved" in the gaseous component of the air or associated with aerosol particles. Transfer to soil or water can occur in dry weather as a result of deposition or condensation of aerosol particles or of equilibrium partitioning from the gas phase. However, if precipitation occurs, this plays a major role in inter-media transport because of "scavenging" of both particulate and gaseous material by the precipitating rain or snow. This process is also referred to as "washout."

The removal during washout is dependent upon the surface area of the scavenging material. The relative amounts of the persistent organic chemical removed from the aerosol and gaseous phases depend upon both the surface area of the scavenger and the physicochemical properties of the chemical. For persistent organic chemicals, the major removal process can be via either the particulate or the gaseous removal mechanism. Recently, it has been shown that snow, when present, is a more effective scavenger than is rain (Franz and Eisenreich 1998).

The deposition process of aerosols onto a terrestrial or water surface has advanced to the point such that velocities of deposition, V_D, can be estimated. Wind velocity in the atmosphere is the primary driving parameter. The following relationship is recommended for aged, atmospheric aerosols with ~ 1 μm radius:

$$V_D = 0.0014\ V_X \qquad \text{(Equation 4-24)},$$

where V_X is the wind speed. For a 5 m/s wind speed, $V_D = 0.7$ cm/s. Warneck (1988) gives data ranges for grass and water surfaces of .01 cm/s for 0.1 μm up to ~ 5.0 cm/s for ~ 1 μm particles.

Dry deposition of chemical species also occurs without particles. Again, atmospheric turbulence and wind speed are used to parameterize this coefficient, k_G. Data on POP-like substances suggest a range of 200 to 1200 cm/h for wind speeds of 2 to 15 km/h (0.3 to 4 m/s). Theory suggests that k_G varies with wind speed to the 0.78 power (Thibodeaux 1996).

The modes of transport of POP by particles and vapor are different; therefore, care in using reported coefficients is advised. As discussed earlier, the deciding factor is the fraction sorbed onto the particles. It is expected that many POPs of low volatility are appreciably particle-bound and that delivery to a terrestrial or water surface will be primarily by particles. In such cases, the deposition velocity is the transport coefficient; otherwise it is preferably expressed as V_{OD}, where

$$V_{OD} = V_D\phi + (1 - \phi)K_G \quad \text{(Equation 4-25).}$$

Here ϕ is the fraction on the aerosol that can be estimated from the partition coefficient and the aerosol concentration (Bidleman 1988). Most data supporting the above models for deposition are for flat or grass surfaces. Recently, measured rates of deposition of POP-like substances onto tree-covered terrain suggest that the effective deposition velocity V_D may be considerably higher, presumably because of the efficient interception by the tree canopy.

Rain and snow scavenging are important process for removing POPs from the atmospheric boundary layer (ABL). The process is tied directly to the precipitation (i.e., rain or snow) rate p in cm/h. The effective transport coefficient, V_W, includes the so-called washout ratio to yield the relationship

$$V_W = pR \quad \text{(Equation 4-26),}$$

where R is the washout ratio. Under ideal gas to water equilibrium conditions, R is equal to K_{WA}, the inverse of Henry's constant in dimensionless units. However, field-measured washout ratios are preferable because they also include the washout of the particle-bound fractions. A rule of thumb is that each volume of rain scavenges aerosols from 200,000 volumes of atmosphere; thus, for particle-bound material, R is 200,000.

A more general form of Equation 4-26 is

$$V_W = p\,[200{,}000\phi + K_{WA}(1 - \phi)] \quad \text{(Equation 4-27).}$$

The flux from air to soil or water is then $V_W C_A$, where C_A is the concentration in air. There may, however, be evaporation of this deposited chemical, as discussed in the next section.

In general, the above summarizes the air-side transport parameters for the atmospheric ABL above both terrestrial and water surfaces. The assumption has been that the air-residing POPs were to be delivered to the soil, vegetation, or water surfaces. The reverse process, that of chemical transport from these surfaces into the ABL, includes vaporization, dust resuspension, and sea-spray generation and entrainment and is discussed below.

Transport from water or soil to air

Transfer of persistent organic chemicals from water or soil to air is influenced primarily by wind velocity. In addition to the evaporation enhancement caused by ventilation of the surface air layer, wind can resuspend soil particulate matter and can also remove aqueous droplets from turbulent water surfaces. Persistent organic chemicals associated with these particles or droplets will thus be transferred to the air compartment. For the purposes of this work, the latter two processes are usually of secondary importance and are mentioned only in passing. The conventional approach is to apply mass-transfer coefficients describing the various resistances that may be experienced by the evaporating chemical. For water surfaces, there are resistances in both water and air "films." For soils there may be resistances attributable to diffusion in pore water, pore air, and a boundary layer, as described in the reports of Jury et al. (1983, 1984a, 1984b, 1984c). To a first approximation, these transfer coefficients are independent of the chemical, and models such as EQC can be formulated that are applicable to a variety of chemicals. The reader is referred to the texts by Thibodeaux (1996) or Mackay (1991) for details of the equations used.

Air–vegetation exchange

Whereas exchange between the atmosphere and soil or water has been well studied, and sound predictive equations exist for these processes, the situation is less satisfactory for air–vegetation exchange. Accordingly, we devote more time to this issue.

Exchange of organic chemicals between the atmosphere and the earth's surface occurs by wet deposition, dry deposition of chemical bound to particles, and diffusive exchange of gaseous chemical. When vegetation is present, it too is subject to these exchange processes and, in some cases, influences them. The following builds on the concepts that have been introduced and explores how air–surface exchange processes are influenced by vegetation.

Wet deposition occurs as snow, rain, hail, and fog and can contain organic chemicals dissolved in the aqueous phase, sorbed to the surface of the precipitation, or sorbed to particulate matter in the precipitation. With the exception of orographic fog, wet deposition is a sedimentation process driven by gravity. Apart from edge effects resulting in increased capture at the windward edge of canopies, sedimentation is not influenced by the presence of vegetation. Therefore, it is reasonable to assume that wet deposition to vegetated surfaces is the same as to non-vegetated surfaces and can be calculated as described previously.

Dry particle-bound deposition occurs by 3 mechanisms depending on the size of the particles:

- very small particles diffuse to surfaces much like gases,
- very large particles are deposited by sedimentation in a manner comparable to wet deposition, or
- particles of intermediate size impact on surfaces as a result of their inertia in the turbulent air surrounding the vegetation.

The deposition velocity of large particles is very high and decreases to a minimum for particles in the impaction range. The deposition velocities then increase somewhat with decreasing particle size as diffusive deposition becomes dominant as a result of the higher diffusiveness of smaller particles. The net deposition velocity of a particle-bound chemical depends on the distribution of the chemical among the particles in the atmosphere. The substance will deposit fastest when it is predominantly on large particles.

Sedimentation is not appreciably influenced by the presence of vegetation. However, deposition via diffusion and impaction is generally higher to a leaf canopy than to a bare soil because of the higher turbulence resulting from the greater surface roughness and the greater surface area available for deposition. Hence, the presence of vegetation will result in a higher dry particle-bound deposition velocity. Just how much higher this velocity is depends on the particle size distribution of the chemical. Chemicals found primarily on large particles deposit nearly exclusively by sedimentation, and their dry particle-bound deposition is largely unaffected by the presence of vegetation. On the other hand, a chemical found primarily on a small particle size class that has a markedly higher deposition to vegetation than to a bare surface has a markedly higher particle-bound deposition velocity. Measurements of the non-sedimenting, dry particle-bound deposition velocities of organic chemicals to vegetation are rare. In a study of deposition of polychlorinated dibenzo-*p*-dioxins/dibenzofurans (PCDD/Fs) and polycyclic aromatic hydrocarbons (PAHs), two compound classes found primarily on small particles, the non-sedimenting, dry particle-bound deposition velocities to a spruce and a mixed deciduous canopy were 7 m/h (Horstmann and McLachlan 1998), a factor of 12 higher than the estimated deposition velocity to a soil without vegetation. However, when one considers that 60% to 75% of the total dry particle-bound deposition of these compounds occurred via sedimentation, the net increase in the overall dry particle-bound deposition velocity was a factor of 1.5. Because forest canopies can be expected to have a more pronounced impact on particle deposition than do other plants (Jonas and Heinemann 1985), this factor can be cautiously employed as a preliminary upper limit for the influence of vegetation on the dry particle-bound deposition velocity.

Diffusive exchange of gaseous chemical is determined by the rate at which the compound is transported from the free atmosphere to the vicinity of the surface, the surface area available for diffusion, and the thickness of the laminar boundary layer surrounding the surface through which the compound must diffuse. Vegetation tends to increase the rate of transport to the vicinity of the surface, increase the surface area available for deposition, and reduce the thickness of the laminar boundary layer—at least in the upper parts of the canopy. All of this tends to increase the deposition velocity. Again, measurements of gaseous deposition velocities of organic chemicals, especially POPs, to vegetation are rare. Studies of PCDD/F deposition to grass canopies in Germany yielded an average gaseous deposition velocity of 8 m/h (Welsch-Pausch 1998). This is just a factor of 4 higher than the value suggested for bare soil above. Measurements to spruce and deciduous forest canopies gave much higher values of 28 m/h and 130 m/h, indicating that forest ecosystems can have an important impact on dry gaseous deposition of POPs in the atmosphere (Horstmann and McLachlan 1998). Until further measurements become available, these values may be taken as initial estimates of dry gaseous deposition velocities to forest and grassland ecosystems.

Transport from soil to water

Persistent organic chemicals can be transported from soil to surface water by runoff processes following a precipitation event. Both dissolved chemicals and chemicals associated with eroded particulate soil matter are included in this process. Surface soils are capable of intercepting, holding, and degrading POPs delivered from the ABL. Once at the soil–air interface, downward movement occurs by combined gas-phase diffusive motion plus advection by water from precipitation. The depth of POP penetration, h, into the soil is determined using the Damkohler number algorithm, which captures the above processes (Cowen et al. 1995). Once determined, the soil-side transport coefficient becomes

$$K_S = D_{AS}/h + V_I \qquad \text{(Equation 4-28)},$$

where D_{AS} is the effective diffusion coefficient for the POP in air-filled pores and V_I is the effective contaminant velocity that results from the water-infiltration rate.

The diffusive-advective transport of POPs between the air–soil, water–sediment, and air–water compartments involves the above transport coefficients in a complex connected fashion. For example, POP deposition onto water involves particles and gaseous deposition as parallel processes. Its continuation through the air–water interface into the surface waters is in series with the air-side processes. These coefficients, therefore, combine approximately to reflect these processes. In addition, the final combined overall coefficient for interphase transport is direction dependent. These formulations are beyond the scope of this chapter and are described elsewhere (Thibodeaux 1996) and are used in multimedia fate models (Brandes et al. 1996; Mackay 1991).

Sediment–water exchange

Hydrophobic organic chemicals in aquatic systems can exist in a dissolved phase or while associated with colloidal or particulate suspended matter. Bottom sediments of these systems interact with overlying water through reversible inter-media mass-transfer processes. Particle deposition provides a mechanism for transport of sorbed organic chemicals from the water column to the sediments. Particle resuspension can move chemical mass out of the sediments into the water column. Sediment–water diffusion will generate a net movement of mass across the interface in a direction dictated by the concentration gradient.

Deposition in association with particles

Deposition of particles in surface waters is dependent on gravitational settling as well as fluid-driven transport. Under laminar conditions, particles settle according to Stokes' Law, which states that the settling velocity is a function of the particle's density, diameter, and shape. Suspended particles in natural waters will vary in density from about 1.03 g/cm^3 for non-diatom phytoplankton to 2.65 g/cm^3 for silt. Typical particle diameters will range from about 1 to 100 mm. Given appropriate combinations of these properties, we might expect to see Stoke's settling velocities in natural waters to range from ~ 0.01 to 30 m/d.

In practice, few models use Stokes Law to determine water-column deposition rates for particulate matter. The reason is that particles in natural water have variable densities and shapes and tend to aggregate over time, making direct application of Stokes Law problematic. But the most important factor in invalidating the use of Stokes settling velocities is that flow and circulation in natural system is turbulent. It is possible for turbulence to move particles vertically at much higher rates than does gravitational settling. This mechanism can either increase or decrease deposition rates depending on the type of system (lake or river) and the turbulent gradient in that system (higher near the surface in lakes and higher in the middle of an open-channel system).

Estimates of particle-deposition velocities in natural waters are therefore usually based on empirical data (e.g., sediment traps or solids model calibration). Based on considerable experience in aquatic systems, particle-deposition velocities for the types of models being applied in this exercise are usually in the range of 0.5 to 2.5 m/d for lakes and 1 to 20 m/d for rivers.

Given a water-column particle-deposition velocity, n_S, the gross deposition rate of a chemical to sediments by this process is deduced as

$$n_S \times A \times C_{TW} \times f_p \text{ [mass/time]} \qquad \text{(Equation 4-29)},$$

where

A is the projected sediment area,

C_{TW} is the total chemical concentration in the water column, and

f_p (the fraction of the chemical in the particulate phase as determined by the particle–water partition coefficient K_p and m, the suspended particle concentration in the water column) is $K_p m/(1 + K_p m)$.

For a given particle-deposition velocity, chemicals with higher partition coefficients in systems with higher suspended solids concentrations will have a higher gross deposition flux.

Deposition in dissolved state

Although the solubility of most hydrophobic POPs is small, the fraction of the substance in solution in natural waters is in the range of 50% to 90%, depending on K_p and m. In this case the rate of the in-solution transport mechanism may be comparable to or higher than that of particle deposition. In-bed processes, particularly those dominated by bioturbation and biodiffusion deliver clean sediment particles below the interface to the surface. Because transport through the benthic boundary layer is rapid, clean surface-residing clay and silt particles readily sorb these substances. Continuing bioturbation delivers this material to depths of up to 10 cm below the interface.

Bottom sediment resuspension

Resuspension of bottom sediments can be caused by physical or biological factors. There is almost always a low level of background resuspension occurring in aquatic systems as a result of bioturbation and/or near-bottom turbulence. However, the major cause of resuspension in lakes and rivers is either wind or flow-induced bottom shear stress that exceeds the critical shear stress for sediment scour. In systems with non-cohesive (e.g., sandy) sediments, scour is a well-established function of bottom shear stress. However, most natural waters have highly cohesive bottom sediments, and it is in these organic, cohesive sediments that most of the organic chemicals reside. Water-current–driven resuspension of cohesive bottom sediments is determined by using an empirical relationship with shear stress.

As a result of flume experiments and observation of resuspension events, a fair predictive ability now exists by which the entrainment flux can be estimated as a function of the sediment properties and the shearing regime at the sediment interface.

As mentioned previously, wind-or flow-driven resuspension events occur on an intermittent basis. It is the large events that are most significant in entrainment and subsequent advective transport of sediments and associated chemicals. When averaged over an annual cycle, lake-wide average resuspension velocities range from about 0.5 to 5 mm/y for lakes, with larger values occurring in shallower lakes with higher wind shears. In rivers, the long-term average resuspension rates are generally such that no net sediment accumulation occurs. However, in the lower reaches of large rivers and estuaries, resuspension rates will be lower and net sediment accumulation will occur.

Sediment resuspension returns previously deposited hydrophobic chemicals back into the water column. The rate at which a chemical is resuspended is calculated as the product of the concentration of the chemical in the sediment and the rate of sediment resuspension.

Resuspension of sediment does occur periodically, e.g., as a result of scouring of a riverbed during storm events. Bioturbation, or movement of sediment caused by sediment-dwelling biota, can also cause resuspension of the surface sediment layer. Anthropogenic activity, such as dredging of sediment and subsequent removal to land or to deep sea, might be considered to be an extreme form of bioturbation. These processes generally are not included in the models considered here, but their effects on long-range transport of persistent organic chemicals can be significant

Importance of sediment–water exchange

In lakes and the lower portions of major rivers and estuaries, there is a net accumulation of bottom sediments. This process of net deposition causes an indirect flux of hydrophobic chemicals introduced into surface waters to the bottom sediments of these systems. Recent surveys of bottom sediments in U.S. freshwater and estuary systems (USEPA sediment survey report 1997) have identified contaminated bottom sediments as a major reservoir and potentially significant source of exposure of hydrophobic contaminants when loading to the overlying water column is decreased.

While the sediment accumulation process represents a net downward flux of solids out of the surface sediments and into deep sediments, the sediment–water interaction represented by deposition and resuspension of the particulate phases can move the chemical either into or out of the sediments depending on the fugacity ratio. This exchange process creates a situation in which bottom sediments can significantly buffer the response time of a water column to changes in chemical inputs.

To demonstrate the potential impact of sediment buffering in determining the response time of a surface water, a simple 2-segment (completely mixed water column with a mean depth of 81.4 m and upper-surface sediment layer with a thickness of 5 cm) mass-balance model (DePinto et al. 1997) was used to calculate the washout of toxaphene from Lake Michigan, with and without sediment–water interactions. The model was run in a time-variable mode. The initial conditions were toxaphene–water-column concentration, 2 ng/L and toxaphene–sediment concentration, 50 ng/g. All external loadings were set to zero, with the exception of an atmospheric gas-phase boundary condition set at 30 pg/m^3. Toxaphene was assumed to have a log K_p of 5.0 and an H of 6×10^{-6} atm-m^3/mol. The lake was assumed to have a steady-state water column suspended sediment concentration of 1 mg/L, gross water column solids settling velocity of 500 m/y, and gross resuspension velocity of 1.7×10^{-3} m/y. These settling and resuspension rates resulted in a net sediment accumulation rate of 3.8×10^{-4} m/y, a solids residence time in the sediments of 130 y, and a solids residence time in the water column of only 0.9 y.

The washout response of toxaphene in Lake Michigan under the above conditions is displayed in Figure 4-5. This shows that the half-time for loss of toxaphene from this system depends greatly on whether or not sediment feedback through resuspension is included in the calculation. Without sediment feedback, the toxaphene washout half-time is approximately 2 y, while including sediment feedback increases the half-time to approximately 15 y. This example illustrates that the parameterization of interfacial transfer and advective transport of a water compartment in a Level III model will be very different depending on whether sediments are included as a separate interacting compartment.

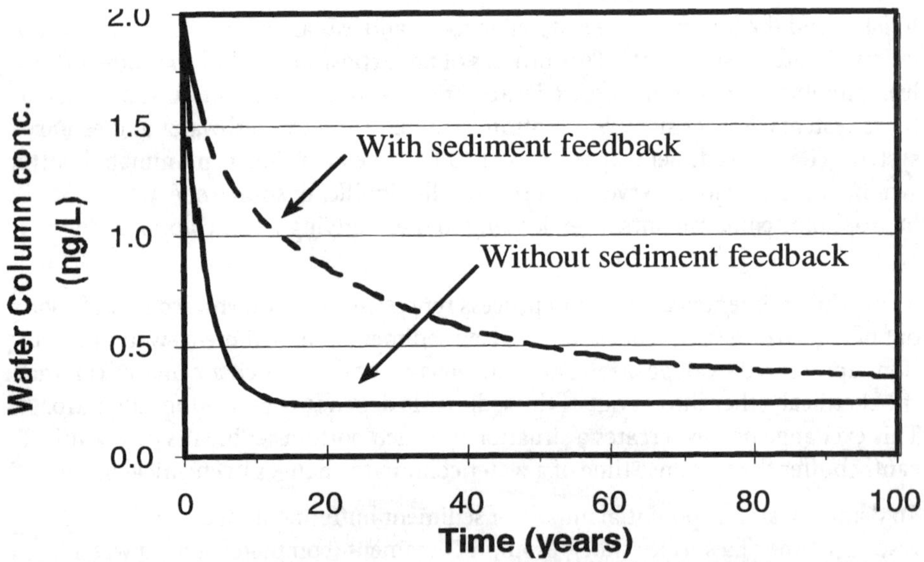

Figure 4-5 Predicted washout response of toxaphene in Lake Michigan with and without sediment feedback (i.e., the sediments are buffering the response to external load reduction. The initial conditions are 2ng/L in the water and 50ng/g in the surface sediment.)

Advective Transport

Persistent organic chemicals can be transported, on regional or global scales, by becoming dissolved in or attached to a moving medium, such as an air current, a flowing river, or an oceanic current such as the Gulf stream. Examples of this advective transport process for chemicals in the air, in rivers, and in oceans are described below.

Atmospheric advection of POPs

Persistent organic pollutants in the atmosphere are carried by the average wind and are well-mixed inside the planetary boundary layer (PBL), which extends to a height of about 1 km above the ground. The net advective transport of POPs is influenced by dry deposition and wet deposition. POPs primarily in particulate phases (e.g., certain PAHs) are expected to have higher deposition rates than do gases and thus to have shorter half-lives for advective transport.

A typical PBL transport velocity is 5 to 10 m/s. However, this value obviously varies with space and time. Today, advanced continental and global meteorological models are available and routinely provide a depiction of local meteorology with model outputs and forecasts available on the Internet. Similarly, oceanic transport models provide credible simulations of ocean currents and transport flows. These advanced meteorological and ocean models have been used often to simulate global climate and expected future changes resulting from increases in greenhouse gases.

Advective transport of POPs in rivers

Persistent organic chemicals have various mechanisms for reaching rivers, following which advective transport can occur. Chemicals can reach a watercourse directly, e.g., via spray drift in the case of agricultural pesticides. Particulate-bound entry of chemicals to a watercourse is also common, the prime examples being chemicals adsorbed on eroded agricultural soils or chemicals adsorbed to suspended solids in wastewater treatment plant effluent.

Once the chemical has entered the watercourse, its transport velocity is limited to the river-flow velocity, which under natural flow conditions depends upon meteorological conditions. Artificial influences (dams, reservoirs, locks, and abstraction points) will regulate river flow and transport velocity in some rivers. Typical flow velocities under natural flow conditions range from less than 0.03 m/s to more than 1 m/s under dry weather conditions (Schnoor 1996), with higher velocities during high flow conditions. Flow velocity, like flow, should follow a Weibull distribution, with a relatively small probability of high velocity events.

As with atmospheric advection, transverse, vertical, and lateral dispersion will occur in rivers and contribute to mixing and dilution. The advective transport length in a river system is necessarily limited to the length of the river. This requirement means that, although advective transport via rivers can be significant on a local or regional scale, its influence on a global scale is limited. Large lakes can also play a part in local or regional advective transport if a chemical is caught in a circulating current pattern. Rivers do provide a route of entry to oceans, whose currents constitute a major, global-scale advective transport mechanism.

Advective POP transport in ocean currents

Two generic cases will be discussed to illustrate this process. In the context of contaminant transport, the surface ocean currents in approximately the top 200 m of the ocean generally will be the most important. Surface currents are typically 10 cm/s and take years to decades to transport water from industrialized, temperate coastal regions to the Arctic Ocean. The flow paths of ocean current and the time scale for them to arrive at various locations are best known for the North Atlantic. The predominant organizing system of surface current here is the Gulf Stream, whose source is in the Gulf of Mexico. This system is capable of transporting large quantities of POPs from near the equator to the European Coast. Substances that are not scavenged by particles entering the sea from the eastern seaboard of North America will likewise be transported by this system, with considerable dilution, to the coast of Europe.

In the Pacific Ocean, the main current system is dominated by a clockwise gyre occupying most of the ocean between 5 to 10 ° N and 45 to 50 ° N. This system moves water from the eastern side of the North Pacific (Philippines, Taiwan, and Japan) across the Pacific Ocean to the west coast of North America at about 45 ° N to 50 ° N latitude. Water is estimated to take 1 to 2 y to transit the Pacific Ocean and perhaps 3 to 5 y to circulate in the gyre. Flow splits at the North American coast with some of the water flowing southward and the remainder northward in surface currents to enter the Gulf of Alaska. From Vancouver Island, it takes approximately 1 to 2 y for water parcels to reach Alaska.

Summary

A formidable and growing body of information exists on the mechanisms of chemical transport in the environment by diffusive, advective, and particle-assisted processes. Robust equations have been developed to describe these processes, but they usually contain parameters that can be highly variable in value, both spatially and temporally. It is probable that inter-media transport rates can be quantified in most cases with an accuracy better than a factor of 3, but situations do exist in which the accuracy is no better than a factor of 10. For some unusual substances or in poorly understood environments, even a factor of 10 may be optimistic. This predictive capability is, however, sufficient that the concepts and equations can be used to quantify the environmental transport of chemicals with enough accuracy to justify the use of mass-balance models as general indicators of chemical persistence and potential for long-range transport.

References

[ACD]. ACD/LogP. 1995. Toronto, ON, Canada: Advanced Chemistry Development.

[ACS] American Chemical Society. 1994. ATOMLOGP [computer software]. Produced by General Sciences Corp. Washington DC: ACS.

Bacci E, Cerejeira MJ, Gaggi C, Chemello G, Calamari D, Vighi M. 1990. Bioconcentration of organic chemical vapors in plant leaves: the azalea model. *Chemosphere* 21:525–535

Baum E J. 1998. Chemical property estimation: theory and application. Boca Raton FL: Lewis/CRC. 386 p.

Bidleman TF. 1988. Atmospheric processes: wet and dry deposition of organic compounds are controlled by their vapor-particle partitioning. *Environ Sci Technol* 22:361–367.

Bierman Jr VJ, DePinto JV, Young TC, Rodgers PW, Martin SC, Raghunathan R. 1992. Development and validation of an integrated exposure model for toxic chemicals in Green Bay, Lake Michigan, Final Report for EPA Cooperative Agreement CR-814885, ERL-Duluth, Large Lakes and Rivers Research Branch, Grosse Ile, MI. 350 p.

Brandes LJ, den Hollander H, van de Meent D. 1996. Simple Box 2.0: a nested multimedia fate model for evaluating the environmental fate of chemicals. The Netherlands: RIVM.

Broto P, Moreau G, Vandycke C. 1984. Molecular structures: perception, autocorrelation descriptor and SAR studies. *Eur J Med Chem Chim Ther* 19:71–78.

Campfens J, Mackay D. 1997. Fugacity-based model of PCB bioaccumulation in complex aquatic food webs. *Environ Sci Technol* 31:557–583.

Chiou CT, Freed VH, Schmedding DW, Kohnert RL. 1977. Partition coefficients and -bioaccumulation of selected organic chemicals. *Environ Sci Technol* 11:475–478.

Clark KE, Gobas FAPC, Mackay D. 1990. Model of organic chemical uptake and clearance by fish from food and water. *Environ Sci Technol* 24:1203–1213.

Cowen CE, Mackay DE, Feijtel TCJ, van de Meent D, Di Guardo A, Davies J, Mackay N. 1995. The multi-media fate model: a vital tool for predicting the fate of chemicals. Pensacola FL: SETAC. 78 p.

Daylight Chemical Information Systems. 1996. CLOGPM Program. New Orleans, LA:

DePinto JV, Raghunathan R, Bierman Jr VJ, Rodgers PW, Young TC, Martin SC. 1993. Analysis of organic carbon sediment-water exchange in Green Bay, Lake Michigan. *Water Sci Technol* 28(8-9):149–159.

DePinto JV, Raghunathan R, Sierzenga P, Xhang X, Bierman Jr VJ, Rodgers PW, Young TC. 1994. Recalibration of GBTOX: an integrated exposure model for toxic chemicals in Green Bay, Lake Michigan . Grosse Ile, MI: USEPA Large Lake and Rivers Research Branch (CR-818560). 132 p.

DePinto JV, Fiest T, Raghunathan R, Smith D. 1997. Analysis of toxaphene behavior in the Great Lakes. *Organochlorine Compounds* 33:280–284.

Devillers J, Domine D, Karcher W. 1995. Estimating n-octanol/water partition coefficients from the autocorrelation method: SAR and QSAR. *Env Res* 3:301–306.

Eadie BJ, Norehead NR, Val Klump J, Landrum PF. 1992. Distribution of hydrophobic organic compounds between dissolved and particulate organic matter in Green Bay waters. *Great Lakes Res* 18(1):91–97.

Finizio A, Mackay D, Bidleman T, Harner T. 1997. Octanol-air partitioning coefficient as a predictor of partitioning of semi-volatile organic chemicals to aerosols. *Atmos Environ* 31(15):2289–2296.

Franz TP, Eisenreich SJ. 1998. Snow scavenging of polychlorinated biphenyls and polycyclic aromatic hydrocarbons in Minnesota. *Environ Sci Technol* 32:1771–1778.

Gailani J, Ziegler CK, Lick W. 1991. Transport of suspended solids in the lower Fox River. *J Gt Lakes Res* 17(4):479–474.

Ghose AK, Pritchett A, Grippen GM. 1988. Atomic physicochemical parameters for three dimensional structure directed quantitative structure-activity relationships. *J Computational Chem* 9:80–90.

Gobas FAPC. 1993. A model for predicting the bioaccumulation of hydrophobic organic chemicals in aquatic food-webs: application to Lake Ontario. *Ecol Model* 1–17.

Hansch C, Leo A. 1979. Substituent constants for correlation analysis in chemistry and biology. New York: Wiley. 339 p.

Hansch C, Leo A. 1995. Exploring QSAR: fundamentals and applications in chemistry and biology. Washington DC: American Chemical Society.

Hansch C, Leo A. 1994. BioByte Corp newsletter: MacLogTM computer software for the MacIntosh. Claremont CA: BioByte Corporation.

Harner T, Bidleman T. 1996. Measurement of octanol-air partition coefficient for polychlorinated biphenyls. *J Chem Eng Data* 41:895–899.

Harner T, Mackay D. 1995. Measurement of octanol-air partition coefficient for chlorobenzenes, PCBs, and DDT. *Environ Sci Technol* 29:1599–1606.

Hippelein M. 1997. Boden/Luft-Verteilung mittelfluchtiger organischer Verbindungem in Abhangigkeit von Temperatur und Bodenfeuchte [Ph.D. dissertation]. Aachen, Germany: Shaker.

Hippelein M, McLachlan MS. 1998. Soil/air partitioning of semivolatile organic compounds: Part 1: method development and influence of physical-chemical properties. *Environ Sci Technol* 32:310–316.

Horstmann M., McLachlan MS. 1998. Atmospheric deposition of semivolatile organic compounds to two forest canopies. *Atmos Environ* 32:1799–1809.

Howard PH, Meylan WM. 1997. Prediction of physical properties, transport, and degradation for environmental fate and exposure assessments. In: Chen F, Schüürmann G, editors. Quantitative structure-activity relationships in environmental sciences VII. Pensacola FL: Society of Environmental Toxicology and Chemistry (SETAC). p 185–205.

Jensen T, Fredenslund A, Rasmussen P. 1981. Pure-component vapor pressure using UNIFAC group contribution. *Ind Eng Chem* 20:239–246.

Jonas R, Heineman K. 1985. Studies on the dry deposition of aerosol particles on vegetation and plane surfaces. *J Aeros Sci* 16:463–471.

Jury WA, Spencer WF, Farmer WJ. 1983. Behaviour assessment model for trace organics in soil I: model description. *J Environ Qual* 12:558–563.

Jury WA, Spencer WF, Farmer WJ. 1984a. Behaviour assessment model for trace organics in soil II: chemical classification and parameter sensitivity. *J Environ Qual* 13:567–572.

Jury WA, Spencer WF, Farmer WJ. 1984b. Behaviour assessment model for trace organics in soil III: application of screening mode. *J Environ Qual* 13:573–579.

Jury WA, Spencer WF, Farmer WJ. 1984c. Behaviour assessment model for trace organics in soil IV: review of experimental evidence. *J Environ Qual* 13:580–586.

Karickhoff SW, Brown DS, Scott TA. 1979. Sorption of hydrophobic pollutants on natural sediments. *Water Res* 13:241–248.

Kerler F, Schonherr J. 1988. Accumulation of lipophilic chemicals in plant cuticles: prediction from octanol/water partition coefficients. *Arch Environ Toxic* 12:1–6.

Kömp P, McLachlan MS. 1997a. Octanol/air partitioning coefficient of polychlorinated biphenyls. *Environ Toxicol Chem* 16:2433–2437.

Kömp P, McLachlan MS. 1997b. Interspecies variability of the plant/air partitioning of polychlorinated biphenyls. *Environ Sci Toxicol Technol* 31(16):2944–2948.

Kömp P, McLachlan MS. 1997c. The influence of temperature on the plant/air partitioning of semi-volatile organic compounds. *Environ Sci Technol* 31:886–890.

Lyman WJ, Reehl WF, Rosenblatt DH. 1982. Handbook of chemical property estimation methods: environmental behavior of organic compounds. New York NY: McGraw-Hill Co. 977 p.

Mackay D. 1991. Multimedia environmental models: the fugacity approach. Boca Raton FL: Lewis/CRC. 257 p.

Mackay D, Patterson S, DiGuardo A, Cowan CE. 1996. Evaluating the environmental fate of a variety of types of chemicals using the EQC model. Environ Toxicol Chem 15:1627–1637.

Mackay D, Shiu WY, Ma KC. 1992–1997. Illustrated handbook of physical chemical properties and environmental fate of organic chemicals: vols. 1-5. Boca Raton FL: Lewis Publishers/CRC.

Meylan WM, Howard PH. 1995. Atom/fragment contribution method for estimating octanol-water partition coefficients. *J Pharm Sci* 84:83–92.

Myrdal PB, Manka AM, Yalkowsky SH. 1995. AQUAFAC 3: aqueous functional group activity coefficients: application to the estimation of aqueous solubility. *Chemosphere* 30:1619–1637.

Neely WB, Blau GE. 1985. Environmental exposure from chemicals: vols. I and II. Boca Raton FL: CRC.

Pankow JF. 1988. The calculated effect of non-exchangeable material on the gas-particle distribution of organic compounds. *Atmos Environ* 22:1405–1409.

Pankow JF, Bidleman TF. 1991. Effects of temperature, TSP and per cent non-exchangeable material in determining the gas-particle partitioning of organic compounds. *Atmos Environ* 25A:2241–2249.

[PMCP] Pomona Medical Chemistry Project. 1987. PC-CLOGP [computer software]. Version 3.32. USEPA Version 1.2. Claremont (CA): Pomona College.

Prausnitz JM, Lichtenthaler RN, de Azevedo EG. 1986. Molecular thermodynamics of fluid-phase equilibria. Englewood Cliffs: Prentice-Hall. 600 p.

Reid RC, Prausnitz JM, Sherwood TK. 1977. The properties of gases and liquids. New York: McGraw-Hill. 688 p.

Rekker RF. 1977. The hydrophobic fragmental constant, its derivation and application : a means of characterizing membrane systems. New York: Elsevier. 389 p.

Rekker RF, de Kort HM. 1979. The hydrophobic fragmental constant: an extension to a 1000 data point set. *Eur J Med Chem* 14:479–488.

Schnoor JL. 1996. Environmental Modeling: fate and transport of pollutants in water, air, and soil. New York: Wiley. 682 p.

Schwarzenbach RP, Gschwend PM, Imboden DM. 1993. Environmental organic chemistry. New York NY: John Wiley & Sons.

Schwartzenbach RP, Westall J. 1981. Transport of nonpolar organic compounds from surface water to groundwater: laboratory sorption studies. *Environ Sci Technol* 15:1360–1367.

[SRC] Syracuse Research Corporation. MPBPVP Programs. 1995. Estimation of melting point, boiling point, and vapor pressure. [computer software]. Version 1.35 for MS-DOS and MS Windows 3.1. Syracuse (NY): SRC.

[SRC] Syracuse Research Corporation. HENRY Programs. 1996. Estimation of Henry's Law constant. [computer software]. MS-DOS and Windows 3.1. Syracuse (NY): SRC.

Thibodeaux LJ. 1996. Environmental chemodynamics: movement of chemicals in air, water, and soil. 2nd edition. New York: Wiley. 593 p.

Thibodeaux LJ, Valsaraj KT, Reible DD. 1993. *Water Sci Technol* 28(8-9):215–221.

Thibodeaux LJ, Nadler KC, Valsaraj KT, Reible DD. 1991. The effect of moisture on volatile organic chemical gas-to-particle partitioning with atmospheric aerosols—competitive adsorption theory predictions. *Atmos Environ* 25:1649–1656.

Thomann RV. 1989. Bioaccumulation model of organic chemical distribution in aquatic food chains. *Environ Sci Technol* 23:699–707.

Tolls J, MacLahlan MS. 1994. Partitioning of semivolatile compounds between air and *Lolium multiflorum* (welsh ray grass). *Environ Sci Technol* 28:159–168.

[USEPA]U.S. Environmental Protection Agency. 1987. PCGEMS user's guide: report prepared under Task 3 –10. Laurel MD: General Sciences Corp. EPA Contract No. 68–02–3970.

[USEPA] U.S. Environmental Protection Agency. 1997. The incidence and severity of sediment contamination in surface waters of the United States: Volume 1, National Sediment Quality Survey. Washington DC: Office of Science and Technology. EPA 823-R-97-006.

Valsaraj KT, Thibodeaux LJ. 1992. Equilibrium adsorption of chemical vapors onto soil surfaces: model predictions and experimental data. In: Schnoor JL, editor.Fate of pesticides and chemicals in the Environment. New York: Wiley.

Warneck P. 1988. Chemistry of the natural atmosphere. San Diego CA: Academic. 757 p.

Welsch-Pausch K. 1998. Atmospharishe deposition polychlorieter dibenzo-*p*-dioxine und dibenzofurane auf futterpflazen [Ph.D dissertation]. Bayreuth, Germany: University of Bayreuth.

Yalkowsky SH, Banerjee S. 1992. Aqueous solubility methods of estimation for organic compounds. New York NY: Marcel Dekker. 264 p.

CHAPTER 5

Persistence and Transport Potential of Chemicals in a Multimedia Environment

Dik van de Meent, Thomas E. McKone, Tom Parkerton, Michael Matthies, Martin Scheringer, Frank Wania, Rich Purdy, Deborah H. Bennett

Persistence in the environment and potential for long-range transport are related because time in the environment is required for transport. A persistent chemical will travel longer distances than will a reactive chemical that shares similar chemical properties. Scheringer (1997) has demonstrated the correlation between persistence and transport distance for different organic chemicals. However, this correlation is not sufficiently robust to predict one property from the other (see Figure 5-1). Specific chemicals that are persistent may or may not exhibit long-range–transport potential. Persistence and long-range transport also present different societal concerns. Persistence concerns relate to the undesired possibility that chemicals produced and used now may somehow negatively affect future generations. Long-range–transport concerns relate to the undesired presence of chemicals in areas where these compounds have not been used.

Environmental policy decisions can be based on either or both considerations, depending on the aim of the regulatory program. In this chapter, definitions and methods for quantifying persistence and transport potential of organic chemicals are proposed that will assist in the development of sound regulatory frameworks.

CHAPTER PREVIEW

Evaluation of Persistence and Long-Range Transport of Organic Chemicals in the Environment. Gary Klečka et al., editors.
 ISBN 1-880611-22-8

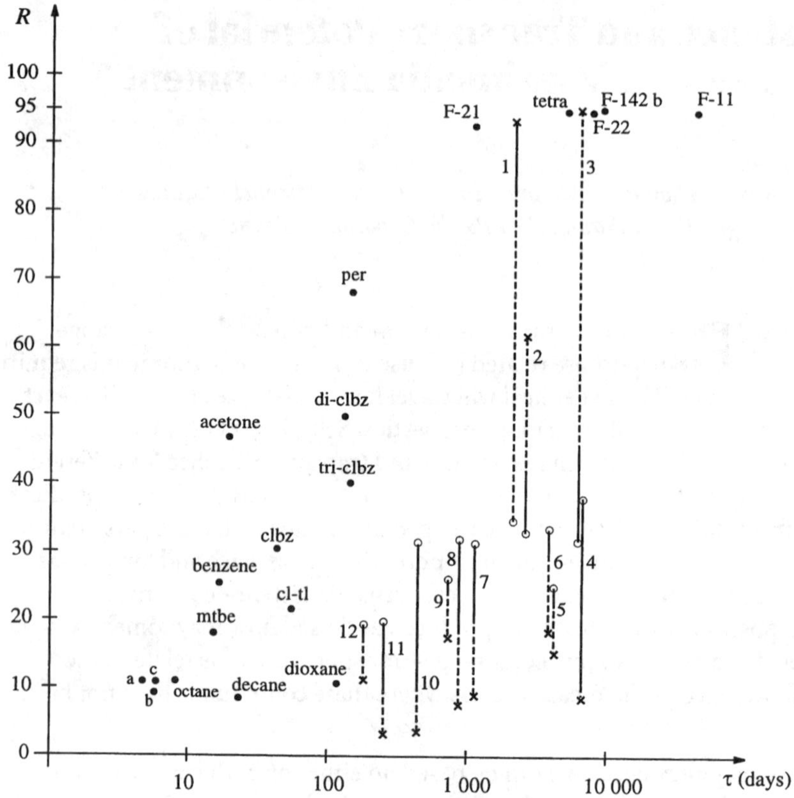

Figure 5-1 Spatial range R (normalized to the earth's circumference) and persistence τ of various chemicals based on the results of the global model proposed by Scheringer (1996, 1997, 1999). 1: hexachlorobenzene, 2: hexachlorobiphenyl, 3: mirex, 4: endrin, 5: DDT, 6: toxaphene, 7: chlordane, 8: dieldrin, 9: TCDD, 10: aldrin, 11: heptachlor, 12: lindane. (With permission from Scheringer 1999, p. 133). See Scheringer, Bennett et al. (2000) for a more detailed interpretation of this plot.

Multimedia Modeling Tools

The models proposed for calculating both persistence and transport potential are based on the standard unit-world fugacity modeling concept as developed by Mackay and co-workers. Mass balance models of this kind have been discussed since the late 1970s (Mackay 1991). Versions and extensions of Mackay's modeling concept have been produced and tested by various authors (McKone 1993; van de Meent 1993; ECETOC 1994; Wania and Mackay 1995; Brandes et al. 1996; Trapp and Matthies 1998). It has been shown that different models in this family yield similar results for primary compartments (i.e., air, water, soil, sediment) if calibrated with the same dimensions and chemical inputs (Cowan et al. 1995; Tell and Parkerton 1997). This model concept is now widely accepted as a useful tool for gaining an understanding of the fate of chemicals in multimedia environments. This concept has also found acceptance in regulatory practice in a number of countries. For example, the European Union uses a Level III model as a regional distribution model for the evaluation of new and existing chemicals (Brandes et al. 1996; Vermeire et al. 1997). The U.S. Environmental Protection Agency has also proposed to use a Level III multimedia model for evaluating persistent, bioaccumulative, and toxic (PBT) chemicals (U.S. Federal Register 1999), while the California Environmental Protection Agency uses a Level III model as a framework for regulating hazardous waste sites (Cal-EPA 1997).

Application of the above models requires parameterization of the following key chemical-specific inputs:

- inter-media partition coefficients describing the equilibrium distribution of the chemical between air–water and soil/sediment–water phases;
- pseudo–first-order transformation rate constants or reaction half-lives of the chemical in air, water, and soil/sediment; and
- mode of entry into the environment (proportions emitted to air, water and soil).

Application of the model implicitly assumes an array of inter-media mass-transfer coefficients.[1]

Persistence in a Multimedia Environment

Definition of persistence

Given a defined rate of chemical release into the environment, concentrations in air, water, and soil are controlled by various chemical and landscape properties, including the persistence of the chemical. Persistence affects the potential for environmental exposure because persistent chemicals exhibit higher concentrations per unit emission and are eliminated more slowly in response to emission reductions.

If emissions are relatively constant over a sufficiently long period of time, concentrations will build up to a steady state, in which releases are balanced by losses. For ordinary chemicals, time scales of practical use are usually sufficiently long to reach a steady state. At steady state, the amounts present in the environment are inversely

[1] Some of the senior scientists stress that the numerical values of the inter-media transfer coefficients used in the present Level III models lack scientific justification. There is considerable uncertainty about the correctness of the assumed values. These assumptions need further testing, especially with respect to gas-phase exchange between the atmosphere and soil and with respect to sediment–water exchange. They plea for use of a Level II model, in which fast inter-media transfer is assumed. While the concern about inter-media transport coefficients was shared by the modeling workgroup, the implicit assumption of zero mass-transfer resistance invoked in Level II suffers from the same drawbacks.

related to the overall transformation rate (proportional to the overall residence time and half-life) in the environment. The overall transformation rate in the environment is a composite of transformation rates for air, water, and soil, weighted by the relative amount of chemical in these media.

If, after a certain period of use, emissions drop to near zero, the amount of chemical in the environment will decrease. The rate of elimination is characterized by the overall transformation rate in the environment. Initially, this is the rate constant at steady state. However, as elimination proceeds, the relative amounts of chemical in air, water and soil generally will change. This will be more pronounced if the differences in transformation rates are great, inter-media mass transfers are slow, and significant amounts of the chemical are present in a "slow" compartment. As a result, the effective transformation-rate constant will decrease with time. Generally, the time necessary for half or near-complete (e.g., 50% or 95%) elimination will be longer than suggested by the steady-state rate constant as illustrated in Figure 5-2. These principles have been used to develop operational definitions of persistence (van de Meent et al. 1992; Webster et al. 1998).

In order to provide a consistent, pragmatic approach, we propose to operationally define "persistence in the environment" as the overall residence time at steady state in a multimedia environment. A simple Level III multimedia mass-balance model is used for this calculation in which only irreversible losses are considered. Irreversible losses consist primarily of chemical transformation of the parent compound but also include unidirectional transport such as transport to deep ocean layers, burial

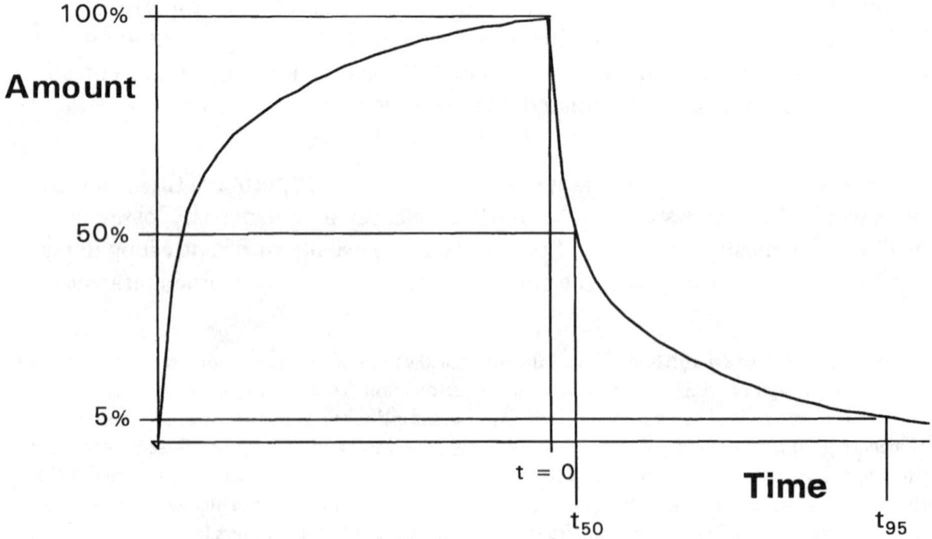

Figure 5-2 Persistence may be defined on the basis of half-life at steady state or time to eliminate x%

in deep sediments, and transport to ground water as well as the upper atmosphere. The dimensions of the model are set to global proportions so as to mimic a closed system, from which no advective losses occur (via air or water flows) as shown in Figure 5-3.

Environmental distribution considerations in persistence assessment

The primary shortcoming of relying on single-medium degradation half-lives in assessing persistence is that each media is given equal importance in the decision process. Thus, there is a possibility that the importance of a half-life is severely over- or underestimated. A half-life in a medium in which the chemical does not partition is of little significance, whereas a half-life in a medium into which a chemical is emitted and from which it is only slowly transferred to other media is obviously of critical importance.

For chemicals that appear principally in only one environmental medium, the decay rates in the remaining media are of no significance and therefore should not be used in persistence classification decisions. Therefore, a tiered screening procedure is suggested. First, chemicals are classified as either single-medium chemicals or as multimedia chemicals. Then, for single-medium chemicals, the half-life in the applicable single medium is used to identify nonpersistent compounds that warrant low priority. For single-medium chemicals with a half-life in the relevant medium that exceeds a prescribed half-life criterion, additional evaluative tiers are considered.

For multimedia chemicals, we propose an initial screening process followed by the use of multimedia models for persistence assessment.

This tiered process for screening chemicals thus begins by first assigning chemicals into one of the following four classes based on partitioning properties:

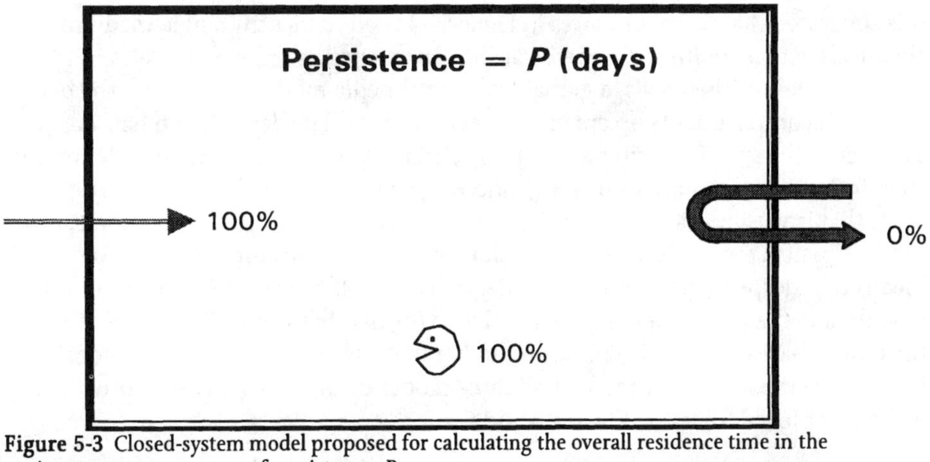

Figure 5-3 Closed-system model proposed for calculating the overall residence time in the environment as a measure of persistence, *P*

- Class A Gas-phase chemicals: Chemicals that partition into the gas phase, regardless of the mode of entry into the environment or the environmental characteristics;
- Class B Aqueous-phase chemicals: Chemicals that partition into the aqueous phase, regardless of the mode of entry into the environment or the environmental characteristics;
- Class C Solid-phase chemicals: Chemicals that partition into the solid phases (soils, sediments), regardless of the mode of entry into the environment or the environmental characteristics; and
- Class D Multimedia chemicals: Chemicals that do not fall into one of the three other categories, i.e., chemicals that partition into more than one environmental medium, at least under some circumstances (i.e., mode of entry).

Only two physicochemical property parameters are necessary for this classification, namely, two out of the following three equilibrium partitioning coefficients:

- octanol–water partition coefficient (K_{OW}),
- air–water partition coefficient (K_{AW}) (Henry's Law constant), or
- octanol–air partition coefficient (K_{OA}).

This categorization can be visualized on a coordinate system mapping chemicals according to partitioning properties (Figure 5-4). Class A chemicals are located in the corner of the diagram with high log K_{AW} and low log K_{OW}, Class B chemicals are in the corner of low log K_{AW} and low log K_{OW}, and Class C chemicals are characterized by high log K_{OA} values. These boundaries are defined such that a classification can be made regardless of mode of entry, i.e., the partitioning is strong enough to overcome the resistance to transfer between phases. The decision on which combination of physicochemical properties leads to a classification into one of these four groups is based on model calculations using the previously described global model.

It is suggested that the boundaries in Figure 5-4 used to identify single-medium chemicals versus multimedia chemicals be constructed as follows. To define the limits for Class A chemicals, a global scale multimedia model is used, and the half-lives in all compartments except air are set to infinity. The degradation half-life in air is set to the specified criteria value, T_{Acrit}. Class A compounds are then defined as chemicals with a combination of K_{AW} and K_{OW} values that yield a calculated persistence that is within a factor of X of T_{Acrit}, i.e., $T_{overall} < X \cdot T_{Acrit}$. X is determined such that $T_{overall}$ never exceeds the desired value for overall persistence. For example, $X = 2$ means that despite being infinite, the degradation half-lives in the other media can only double the overall persistence relative to the half-life in air. This implies that the other phases have such a minor influence that they can be safely neglected. The boundaries must be confirmed for all three modes of entry. An identical procedure is then employed to define Classes B and C.

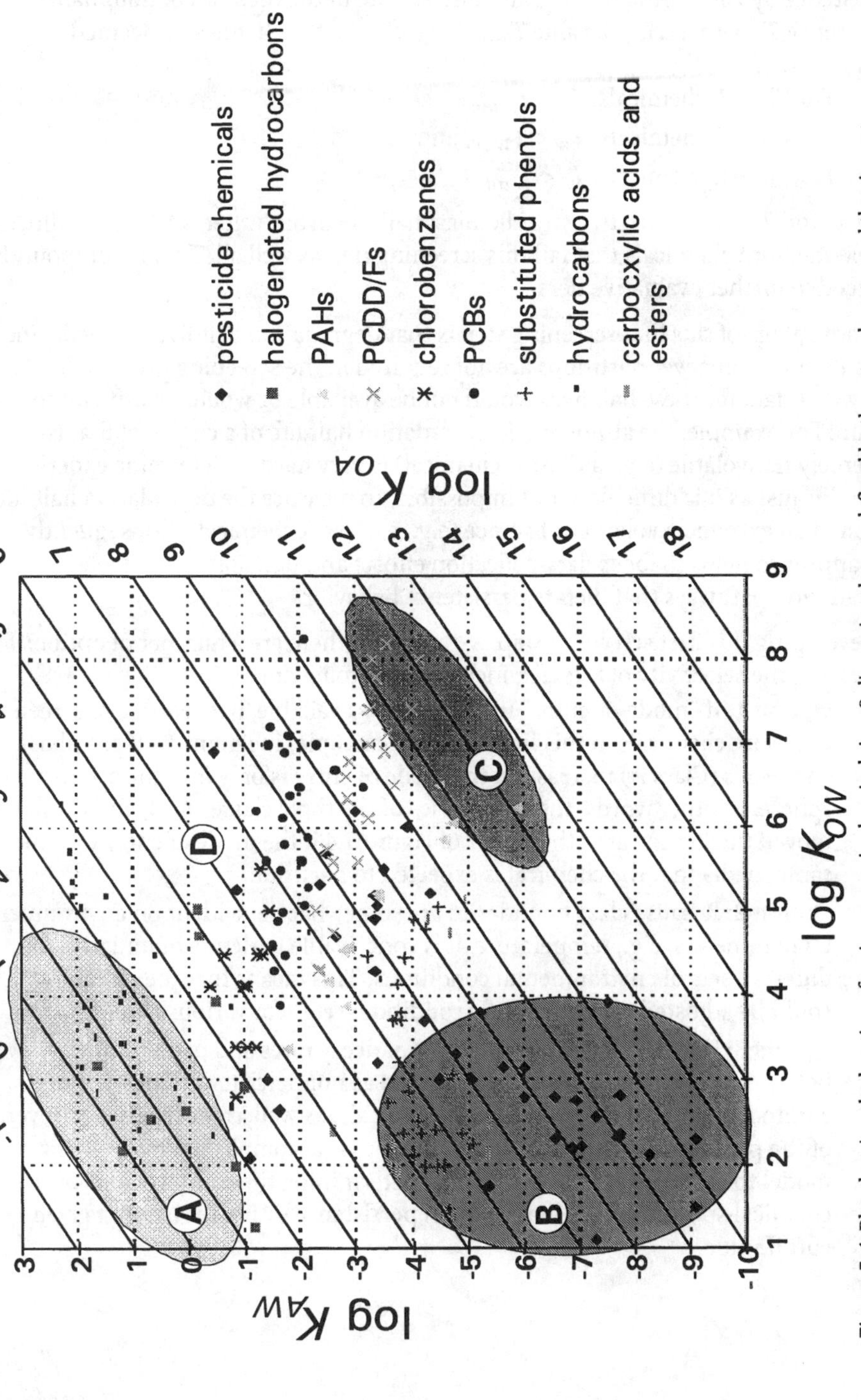

Figure 5-4 Plot showing the location of various organic chemicals in a 2-dimensional space defined by the octanol–water and the air–water partition coefficients log K_{AW} and log K_{OW} (from Wania 1998)

Persistence assessment for single-medium chemicals

Once classified as single-medium chemicals, these compounds can be screened for persistence by comparing the degradation half-life in the medium of dominant occurrence T_X with a criteria value T_{Xcrit}. If $T_X < T_{Xcrit}$, the chemical is deemed nonpersistent:

- For Class A chemicals: $T_A < T_{Acrit}$,
- For Class B chemicals: $T_W < T_{Wcrit}$, and
- For Class C chemicals: $T_S < T_{Scrit}$.

T_A, T_W, and T_S refer respectively to chemical half-life in air, water, and soil/sediment. Single-medium chemicals that fail this screening tier, as well all Class D compounds, proceed to further evaluative tiers.

The advantage of this first screening step is that degradation half-lives in media into which a chemical never partitions are not required in the screening process. It is likely that data for these half-lives would not be available or would be difficult to obtain. For example, the atmospheric degradation half-life of a chemical that is extremely nonvolatile (e.g., a Class C chemical) is very hard to determine experimentally, just as it is difficult, if not impossible, to measure the degradation half-life in soil of an extremely volatile substance (e.g., a Class A chemical). Consequently, this approach helps to focus data-collection efforts and decisions on the relevant degradation pathways that dictate persistence behavior.

In developing this first screening step, we note that there are a number of options for increasing the selectivity of this tier, including the following:

- Option 1: If "mode-of-entry" information is available, it is feasible to screen out more chemicals in this first step. For example, the number of gas phase chemicals (Class A) increases if the mode of entry is only into the atmosphere. In other words, the boundaries of the three classes A, B, and C can be moved further towards the center of Figure 5-4 if the mode of entry is to the same media that the chemical is expected to partition.
- Option 2: Because class boundaries are somewhat dependent on environmental parameters, e.g., temperature, it is conceivable to define boundaries for different sets of environmental conditions. This means that the screening could be adjusted for site-specific conditions, e.g., for various climates.
- Option 3: Under certain combinations of decay rates and partitioning behavior, a chemical can still be classified as nonpersistent. This method eliminates some of the multimedia chemicals, as well as incorporating several of the previously mentioned rules. This can be accomplished by applying a model to calculate persistence and by determining the combinations of chemical-specific inputs that result in persistence estimates below a given cutoff value.

Persistence assessment for multimedia chemicals

For a chemical that is in Class D—a multimedia chemical that can partition in more than one environmental medium at least under some circumstances—the determination of persistence must be based on multimedia partitioning in combination with information on chemical half-lives in the various media. The relative distribution between compartments is influenced by

- equilibrium partitioning of the chemical,
- kinetics of chemical transfer between the media,
- mode of emission into the environment, and
- nature of the environment, as it influences inter-compartmental partitioning and transfer.

Multimedia compartmental fate models (Cowan et al. 1995) address all these issues in combination and thus are uniquely suited for the task of estimating the weighting factors, or chemical-mass distribution. Persistence *P* is calculated as the overall residence time τ_{overall} of the chemical in the multimedia system (Webster et al. 1998):

$$P = \tau_{overall} = \Sigma M_i / \Sigma(M_i \cdot k_i) \quad \text{(Equation 5-1)},$$

where M_i refers to the mass inventory (kg) and k_i refers to transformation rate constant (d^{-1}) in each of the *i* model compartments. The following modeling approaches are suggested as tools for assessing persistence of multimedia chemicals (Figure 5-5).

- Model A: A very simple global-scale model with only three compartments (air, water, soil), five inter-compartmental transfer rates, three degradation loss, and three permanent loss processes, as described by Wania (1998).
- Model B: A global scale model similar to Model A but that additionally includes fresh water and freshwater sediments.
- Model C: A regional scale model similar to Model B, having regional rather than global dimensions (Webster et al. 1998; Beyer et al. 2000).
- Model D: A continental model nested in global model, as described by Brandes et al. (1996). This would address the concern that only a small part of the chemical used on the continental scale interacts with the ocean.
- Model E: A more complex model that includes either more nested compartments or a linked set of multimedia compartments as has been used by Wania and Mackay (1995) and Scheringer (1996, 1997).

These modeling approaches should be evaluated to determine the overall persistence values for a range of chemicals. The output from these different models should be compared based on the following criteria:

- the absolute scale of the calculated persistence values,
- the relative ranking of the chemicals according to persistence,

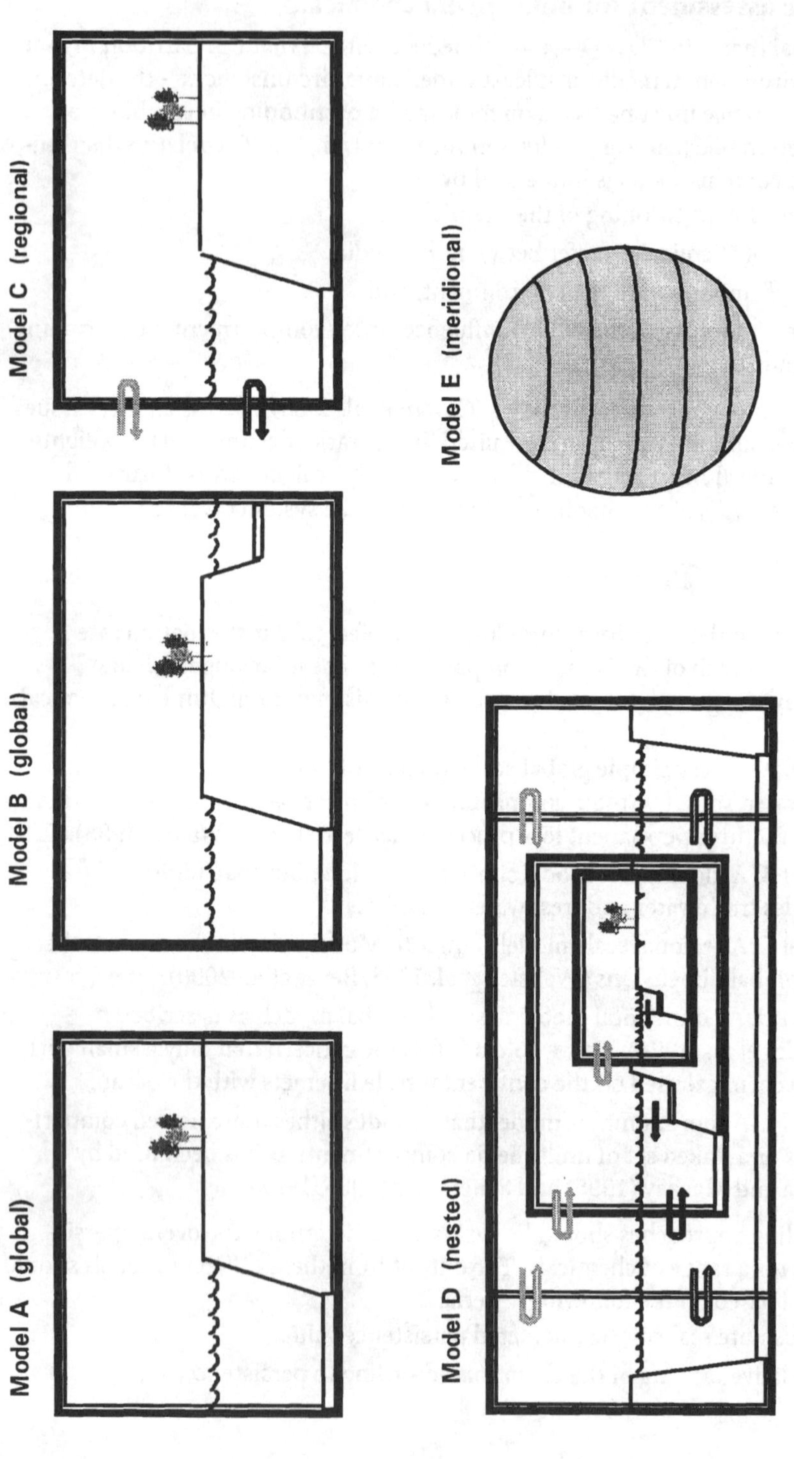

Figure 5-5 Multimedia model approaches for calculating overall residence time in the environment as a measure of persistence

- the dependence of the calculated persistence values on the environmental parameters, and
- the analysis of large deviations of persistence values between the various models, with an aim to identify the potential of a particular model to incorrectly categorize persistence.

On the basis of such a comparative model evaluation, it should be feasible to identify the simplest model approach that will minimize the likelihood that persistent chemicals are misclassified.[2]

Important issues in multimedia model selection

Level II or III: Equilibrium partitioning versus steady state

The main requirement for such a model is to be as simple as possible yet realistically estimate the relative media distribution of a chemical. In principle, several approaches for estimating the relative media distribution exist:

- Assumption of equilibrium partitioning among all compartments—This approach implies that the resistances to inter-compartmental transfer are zero and that there are no effects from advective processes between compartments, often referred to as a "Level II" calculation. There is no need to define the mode of emission or the environmental parameters that influence inter-compartmental transfer, such as deposition velocities and mass-transfer coefficients, but environmental parameters that define size and composition of the environmental media are still required.
- Assumption of a system in steady state, with mass transfer resistances between the compartments—This implies that no change in time occurs and that at any point in time, inputs and outputs are balanced. This is often referred to as a "Level III" calculation. This method is more realistic because the mass distribution reflects the shift from equilibrium resulting from both transport and the mode of entry.

More-complex approaches could be envisaged that require neither the assumption of equilibrium nor steady state. Existing approaches to calculating overall persistence have been either at Level II (Müller-Herold 1996; Müller-Herold et al. 1997) or Level III (Webster et al. 1998; Bennett et al. 1999) or for a pulse input (Scheringer 1996, 1997).

---

[2] In late 1998, F. Wania initiated a comparitive study by modellers to find out how sensitive the computed *P*-value is to model settings. Multimedia modellers were asked to compute *P* with their own favorite model for 180 chemicals, based on a distributed table of properties and half-lives. No results are available at this time.

Model scale

With regard to the scale of the modeled environment, the 2 basic options are 1) a model describing the entire global environment or 2) a defined regional subsection. Each model scale has advantages and disadvantages.

Global scale:

+ System boundaries are well-defined and thus there is only one overall persistence value. This system is naturally closed, i.e., no atmospheric or oceanic advection out of the global system occurs; choice of environmental parameters are constrained to global averages.
– Many chemicals, particularly less persistent ones, do not "experience" the whole global environment but tend to be limited to a smaller scale. Namely, the water volume will tend to be overrepresented in a global model for such chemicals.

Regional scale:

+ Scale, media composition, and environmental descriptors of the modeled system better reflect the environment in which a chemical is likely to occur.
– System boundaries are flexible, so overall persistence value will depend on the selected system. System is open, i.e., atmospheric or oceanic advection occurs. This system can either be artificially closed by setting advective terms to zero or left open with the half-life calculated based on degradation processes only. With an open system, the overall persistence value is dependent on the selected size of the modeled system.

Number of compartments

Although there are numerous possibilities in selecting compartment types, 2 primary alternatives emerge: 1) using a minimum of three compartments, namely, air, water, and soil, whereby "water" lumps the fresh water and marine environment (Scheringer 1996, 1997; Wania 1998) or 2) using a slightly more complex structure that includes fresh water, freshwater sediment, ocean water, and possibly vegetation compartments (in addition to air and soil) (see Chapter 4).

Non-degradation losses

The overall residence time of a chemical in a system is dependent not only on degradation but also on translocation beyond the system boundaries, e.g., by advection in water and atmosphere. Some models simply set any advective transport beyond system boundaries to zero, making degradation the only loss process (Webster et al. 1998). While the artificiality of this assumption is not appealing, the alternative of allowing for advection and calculating an overall persistence based only on the degradation processes is problematic, as this persistence value will be scale-dependent. On a global scale, atmospheric and oceanic advection beyond system boundaries ceases to exist. In order to avoid unrealistically high overall residence times in a global-scale model, it may still be desirable to include processes that lead to loss of chemical out of the sphere accessible to organisms, such as burial in sediments, transfer to the deep sea and the stratosphere, and immobilization in soils (Wania 1998). However, it should be reiterated that overall persistence is always to be based solely on processes that lead to an irreversible chemical conversion.

Environmental parameter selection

What environmental parameters should be employed in the model calculations? Namely, which media volumes (e.g., relative volumes of air, water, and soils or depth of compartments), phase composition (e.g., organic carbon content in soil and water) and kinetic (inter-media mass-transfer coefficients, etc.) and climatic conditions (e.g., temperature) should be used in the calculations? Although there is large choice of environmental parameters, it is believed that the ranking of chemicals according to persistence will not significantly change for a different scale or set of environmental parameters. The lack of dependence on the choice of environmental parameters is demonstrated in the results of the Classification and Regression Tree (CART) analysis presented in an upcoming section.

Transport Potential in a Multimedia Environment

Definition of transport potential

Persistence is a necessary condition for long-range transport: without persistence there is no opportunity for transport. However, persistent chemicals are not necessarily subject to long-range transport unless the residence time of the chemical in the mobile media (air or water) is long enough to allow adequate travel time. With an absence of spatial structure, the closed global model used to determine persistence provides little insight with respect to the spatial range. Chemicals released into the environment from a fixed point or area may be transported away from the source over large distances. Such persistent organic chemicals that exhibit this behavior are now the focus of various national and international regulatory initiatives.

We define the transport potential of a chemical in 3 slightly different ways:

- A scale of an open, mixed-compartment system, such that losses due to advective transport and transformation at steady rate are each 50% of the emission rate. This scale can be found using multi-media model.
- A travel distance of a packet of mobile compartment (air or water), such that a specified percentage of the original mass of chemical in the packet is lost as a result of transformation and inter-media transport. This distance is found using an advective Lagrangian model.

- A spatial range that contains a specific percent of the mass originally emitted. This range can be found using a dispersive Eulerian model.

These three approaches yield similar results and are equvalent options for calculation transport potentials of chemicals. Various percentages can be used in the equations, but we have chosen to use 50 % in all definitions for consistency in this chapter.

Assessment of transport potential

Mixed-compartment approach

In the mixed-compartment approach the transport potential of a chemical is defined by balancing advective transport in air and water with transformation losses in an open system, an approach conceived of by this working group and independently by Hertwich (1999) and Beyer (1999). In such a model, each chemical has a spatial scale at which losses by transformation in the system equal advective transport out of the system. Given a set of system properties (model), this characteristic scale is a function of the properties (inter-media transfer coefficients, rates of degradation) of the chemical. The major forces that facilitate regional and continental transport are air and water flows, e.g., wind and water currents (mobility in soil and ground water are several orders of magnitude slower and thus are not considered). Because advective transport fluxes are controlled by air and water velocities, mobility is not scale-dependent. However, transformation losses are proportional to system volume. Therefore, the proportion of the total chemical budget that is transported out of a system is a function of system scale. This fact provides the mathematical basis for establishing quantitative measures to characterize transport potential.

As a first-tier approach, the characteristic spatial scale can be determined using the mixed-compartment multimedia models described in the previous section. The closed system (as used for calculating persistence in Figure 5-3) has to be opened to account for advective inflow and outflow (Figure 5-6).

The scale represented by balancing the advective and transformation loss results from the non-dimensionalization of the equations describing chemical behavior. Using the global multimedia model, one can "open" the system by allowing for advective transport. Instead of using the global dimensions, the size of the system is reduced to precisely the size that meets the criteria that half of the chemical is transported from the system by advection with air and water and the other half is eliminated by transformation. If only advection in air is considered, this is equivalent to the Lagrangian method, discussed later.

In the mixed-compartment approach, the advective and reactive mass losses can easily be calculated with an open multimedia model (Mackay 1991). Let us assume a standard regional multimedia environment with mass export and degradation. Then, under steady-state conditions, the advective outflow F_{adv} [kg/d] and the reactive loss F_{rea} [kg/d] are constant and their sum is identical to the mass import E [kg/d]. Both mass losses can be calculated from the multimedia model:

$$F_{adv} = u_A \cdot C_A \cdot A_A + u_W \cdot C_W \cdot A_W \quad \text{(Equation 5-2)},$$

$$F_{rea} = k_A \cdot C_A \cdot V_A + k_W \cdot C_W \cdot V_W + k_W \cdot C_W \cdot V_W + k_S \cdot C_S \cdot V_S \quad \text{(Equation 5-3)},$$

where

u_A = wind speed [m/d],
u_W = water speed [m/d],
C_i = concentration in i [kg/m^3],
A_A = cross-section of wind flux [m^2],
A_W = cross-section of water flux [m^2],
k_i = degradation rate in i [1/d],
V_i = volume of i [m^3], and
A, W, S = subscripts for air, water, and soil, respectively.

To calculate the characteristic scale L, F_{adv} is set equal to F_{rea}:

$$k_A \cdot C_A \cdot V_A + k_W \cdot C_W \cdot V_W + k_S \cdot C_S \cdot V_S = u_A \cdot C_A \cdot A_A + u_W \cdot C_W \cdot A_W$$

(Equation 5-4).

If, for simplication, we include advection in air only, convert the equation to a mass-based equation, and substitute

$$k_{overall} = \frac{1}{\tau_{overall}} = \frac{\sum_i k_i M_i}{M_t} = \frac{\sum_i k_i M_i}{M_A} \cdot F_A \quad \text{(Equation 5-5)},$$

where

$k_{overall}$ is the mass-weighted decay rate, mass weighted by the chemical mass distribution in an open Level III system (Bennett et al 1999),
M_i is the mass of chemical in compartment i,
M_t is the total mass of chemical in the system, and
F_A denotes the proportion of the chemical in the atmosphere, equal to M_A/M_t,

we can obtain the characteristic scale, L (=V_A/A_A) by solving Equation 5-4, which yields (Beyer 2000)

$$L = \frac{u_A}{k_{overall}} \cdot F_A = u_A \cdot \tau_{overall} \cdot F_A \quad \text{(Equation 5-6)}.$$

Equations 5-5 and 5-6 enable the effect of inter-media transfer on transport potential to be examined. For Class A chemicals that reside essentially all in air, the effective degradation rate simplifies to the atmospheric degradation rate, since $M_A \cdot k_A \approx \Sigma(M_i \cdot k_i)$. As a result, the characteristic scale is directly related to the atmospheric half-life. However, for chemicals that partition to other compartments,

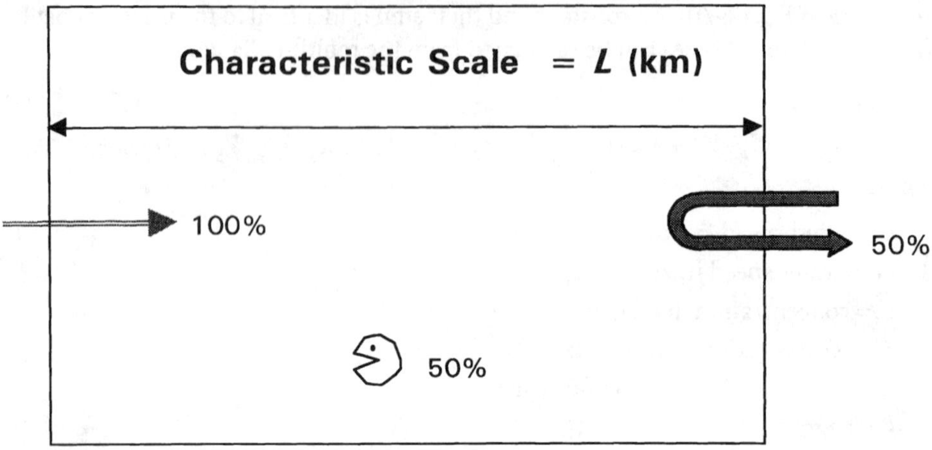

Figure 5-6 A regional, open model proposed for calculating the characteristic scale, L, as a measure of transport potential.

the effective degradation rate increases as transport is retarded, thereby allowing more time for reaction. A similar result is obtained in case of advection with water only (Beyer et al. 2000).

Lagrangian approach

The Lagarangian approach considers the movement of a compound driven by the advective flow, e.g., wind velocity, over a non-moving phase, e.g., soil (Figure 5-7, Bennett et al. 1998). The approach is the same as plug-flow in hydrology (Schnoor 1996; Trapp and Matthies 1998).

During transport, a compound undergoes mass transfer between all media. If the compound is degraded in any of these media, the chemical mass is reduced depending on the travel time and distance. At a steady state, there is one distance at which the concentration in air has been reduced by 63%; at this point, the reaction losses balance the advective transport losses. This distance is defined as the characteristic travel distance (L).

The characteristic travel distance in air (L_A) may be interpreted as the distance to reduce the inital source concentration by 63% (Figure 5-8). From L all other spatial range values can be calculated, e.g., the half-length ($L_{0.5} = \ln 2 * L$) or the distance where 95% (or 5%) of the mass is decayed. While the absolute scale of the different travel distance measure varies depending on the percentage of mass decayed, any of these definitions will provide a consistent basis for persistence classification.

Although conceptually different, the Lagrangian and mixed-compartment approaches yield similar results (Bennett 1999). In fact, the characteristic scale obtained from Equation 5-7 can be interpreted as a Lagrangian travel distance.

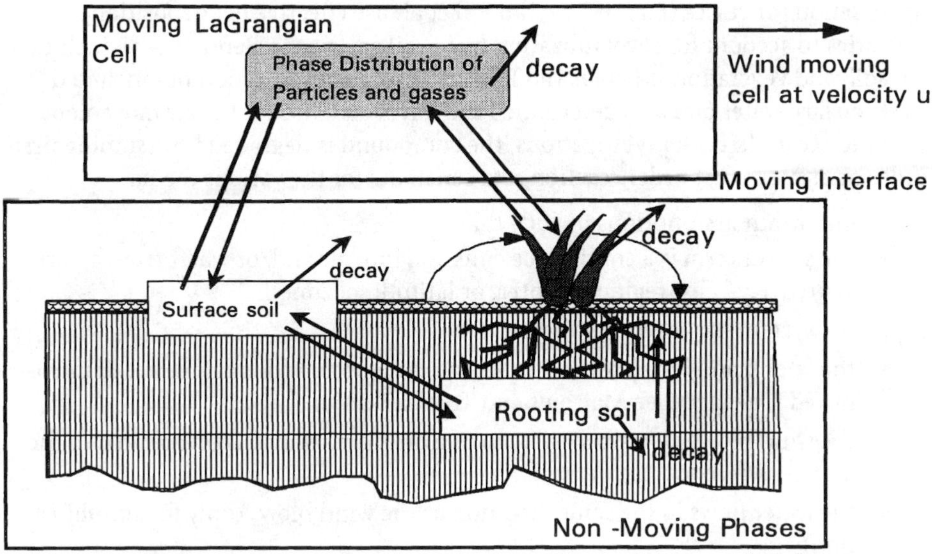

Figure 5-7 Regional multimedia environment used by Bennett et al. (1998) for calculation of the characteristic travel distance

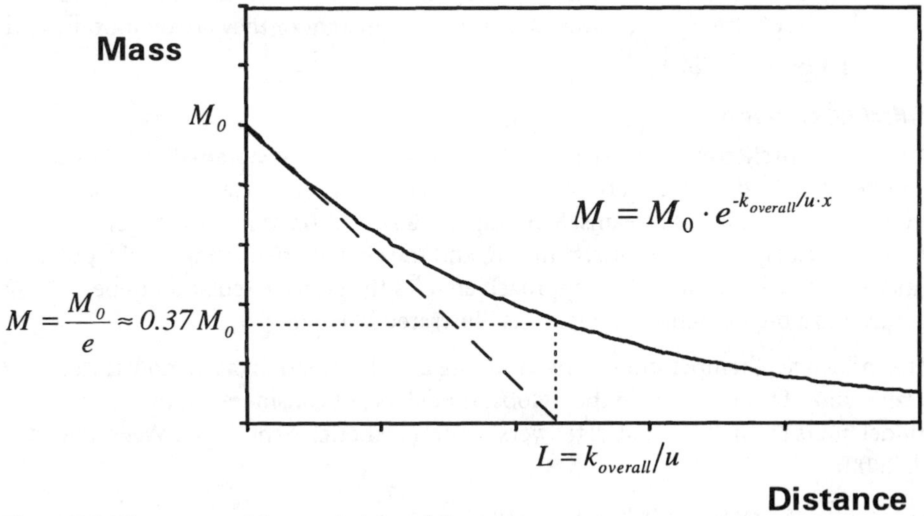

Figure 5-8 The concept of characteristic travel distance

The Lagrangian approach has been used by Bennett et al. (1998) and van Pul et al. (1998) to define a characteristic travel distance (*L*) and to calculate a half-length of travelling, respectively. Although the derivation of these methods differs, the methods are essentially equivalent. While Bennett et al. use a coupled multimedia Level III model to account for the inter-media mass transfer and intra-media degradation processes (Figure 5-7), van Pul et al. use effective net deposition velocities to account for revolatilization from soil and water. Bennett et al. included air, soil, and vegetation, whereas van Pul et al. excluded vegetation but included fresh surface water and also determined the travel distance for the air movement over the ocean. In both investigations, the compound is degraded by assuming first-order or pseudo–first-order reaction rate constants for the various media.

Other important assumptions include that

- the source term is a continuous emission into air and/or water from an area source (e.g., city, region, country, or latitudinal zone),
- the system has reached steady state,
- the lateral and longitudinal dispersion is neglected (this can be easily introduced if needed (see Matthies and Trenkle 1988)),
- the wind blows at a steady state in one direction, which is orthogonal to the area cross-section,
- the water flows in the same direction as the wind blows (only for simplification),
- the landscape properties do not vary spatially or they are spatially averaged, and
- all environmental parameters (wind and water velocity, rainfall, temperature, height of atmospheric layer, etc.) are kept constant or they are temporally and spatially averaged.

Eulerian approach

A more comprehensive approach uses the global circulation system divided into a number of latitudinal (or meridional) zones that exchange matter by interzonal wind and ocean current streams. Scheringer (1997, 1996) used a closed loop of zones and calculated the spatial range, R, and the characteristic time, τ, of a pulse emission into this system. This approach enables the global circulation to be reduced to a one-dimensional model as illustrated in Figure 5-9.

The influence of temperature on partitioning and degradation, as demonstrated by Wania and Mackay (1995) in their global model, is not considered in this circular model but is studied in an updated version of this model (Scheringer, Wegmann et al. 2000).

This model is based on the following concepts:

- each zone consists of homogeneous boxes for air, water, and soil with corresponding first-order degradation rates;
- the zones are connected by air and water flows representing eddy diffusion in the troposphere and ocean;
- the amount M_0 of a chemical is released at time $t = 0$ to one of the zones (pulse emission), and the chemical's partitioning, degradation, and travelling

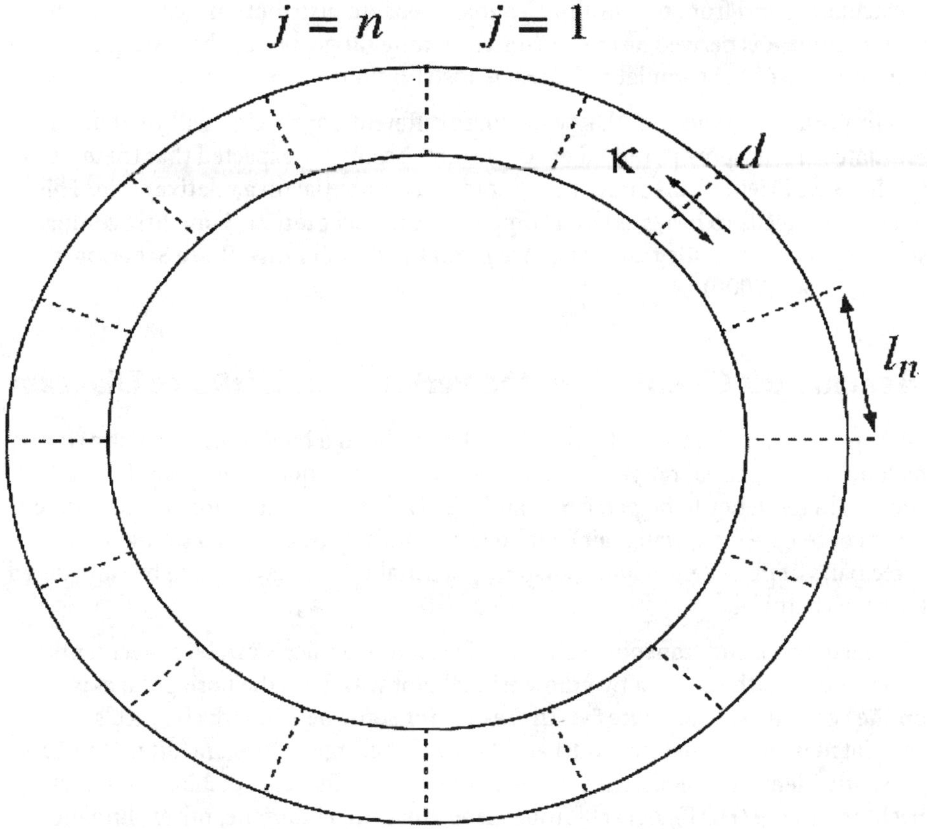

Figure 5-9 Geometry of the global model used by Scheringer (1996, 1997) to determine the spatial range of an organic chemical. Reprinted with permission from Scheringer (1996). ©1996 American Chemical Society.

between the zones are calculated simultaneously in a dynamic, multimedia model; and

- the model parameters (partition coefficients, degradation rates, inter-media transfer parameters such as deposition velocities) do not vary in space and time, i.e., the model in its present state consists of many identical replicates of a dynamic model.

Because this is a dynamic calculation (i.e., pulse input) the chemical's total amount in the system decreases with time. For each compartment (soil, water, air) in each zone (indicated by the index j), the exposure

$$e_j = \int_0^\infty c_j(t)dt \qquad \text{(Equation 5-7)}$$

is calculated, and from the resulting spatial exposure distributions $\{e_j\}_{j=1,...n}$, the spatial range *R* is derived as the 95%-interquantile range, i.e., as the distance that contains 95% of the cumulative exposure distribution.

It is important to point out that while these different approaches will yield different estimates of transport potential on an absolute scale, it is expected that these methods will yield similar rankings of transport potential on a relative scale. This hypothesis needs to be tested by a comparative model exercise. For a first comparison of the characteristic travel distance *L* and the spatial range *R*, see Scheringer, Bennett et al. (2000).

Benchmark Chemicals in the Persistence/Distance Diagram

Persistence (*P*) and transport potential (*L*), as indicated by the characteristic travel distance or the spatial range, are considered to be basic quantities describing a chemical's tendency to be persistent and widely dispersed, i.e., mobile. For different routes of release (soil, water, air), different results may be obtained so that up to three pairs of persistence- and transport-potential values may need to be considered for each chemical.

The persistence and transport behavior of various chemicals can be conveniently visualized with the aid of a two-dimensional plot with P on the horizontal axis and L on the vertical axis. In Figure 5-1, the results for some benchmark chemicals calculated using a dynamic model are shown (Scheringer 1997), including the 12 presently identified persistent organic pollutants (POPs), e.g., dichlorodiphenyltrichloroethane (DDT), hexachlorobenzene, dieldrin, chlordane, mirex, lindane, etc., chlorinated benzenes and halogenated and non-halogenated solvents such as tetrachloromethane (tetra), tetrachloroethylene, CFC-11, HCFC-142b (chlorinated and fluorinated hydrocarbons), cyclohexane, and dioxane (in Figure 5-1 and in this subsection, the symbols *R* and τ are used instead of *L* and *P*).

The diagram shows very high persistence and nearly global spatial ranges for highly volatile and chemically stable compounds such as CFC-11, HCFC-142b, and tetra. The chlorinated benzenes show an increase from intermediate *R* and τ values (monochlorobenzene) to high (di- and trichlorobenzene) and very high values (HCB).

Low values for both *R* and τ are found among the non-halogenated solvents (cyclohexane, butanol, and alkanes). The semi-volatile organic compounds (e.g., DDT) are a special case, with high τ values (more than 100 d) and high or very high spatial ranges (more than 25% of the perimeter of the earth). For the *R* values of semi-volatile organic compounds, intervals are given (full and dotted lines) because the atmospheric degradation rates k_A of these compounds are not well-characterized. This analysis indicates that the atmospheric degradation rate k_A is the crucial parameter that determines long-range–transport potential. Thus, the uncertainty in

k_A leads to considerable uncertainty in R. With respect to the identification of POPs among a variety of semi-volatile organic compounds, the determination of k_A, or at least the range for k_A, is paramount. In particular, the influence of adsorption onto and absorption into aerosol particles on the degradation rates has to be investigated (this is discussed in detail in Chapter 2). In addition to k_A, the Henry's Law constant (H) and the octanol–air partition coefficient (K_{OA}) have a strong influence on the long-range–transport potential because they determine a chemical's volatility, i.e., its tendency to enter the gas phase (Harner and Mackay 1995). These partition coefficients are often known with higher accuracy than are the atmospheric degradation rates, at least in the case of semi-volatile organic compounds.

If cutoff values for persistence and transport potential are assigned, the diagram can be divided into the four regions of mobility/persistence properties. These cutoff values are to some extent arbitrary and may be chosen differently to reflect specific regulatory objectives. To illustrate the categorization of chemicals shown in Figure 5-1, arbitrary criteria of 100 d for P and 1000 km for L are selected. In the next section, CART analysis is used to identify physicochemical properties that lead to exceedence of such criteria.

Application of CART Analysis to Chemical Prioritization

The use of a multimedia framework for classification

As discussed, the task of assessing the persistence and transport potential of both new and existing chemicals would be a major undertaking, requiring either the collection of chemical properties for use in multimedia model simulations for each chemical or the worldwide collection of data in multiple environmental media at each sample site. As an alternative, we present a classification system that considers sets of chemical properties that would result in the classification of a chemical as "persistent" and as having the potential for long-range transport.

A multimedia modeling framework can be integrated with limited data on chemical properties to understand the temporal behavior of a broad spectrum of chemicals. We describe how simple but informative models can be used to develop effective strategies to classify both existing and new chemicals with respect to persistence in the environment.

The framework developed is part of the second tier and follows after the elimination of nonpersistent, single-medium chemicals. The classification scheme used can successfully determine sets of chemical properties that will lead to a nonpersistent chemical. However, those that do not fall into the nonpersistent classification are not necessarily persistent. They need to go into a higher tier of testing, requiring the use of a multimedia model with most of the chemical properties known.

In order to gain broad insight on how chemical properties can be used to classify a chemical as persistent, there is a need to consider how the chemical properties relate to the persistence calculated for a defined environmental system. One method for conducting such an analysis involves the use of a set of binary classification trees. To construct binary classification trees we considered the use of the CART approach. In the next subsection, we provide an introduction to CART and a simple illustration of how CART can be applied to a multimedia model to assess the link between an output classification such as "persistent" and the range of chemical-specific input values that give rise to this classification.

Classification and Regression Trees methodology

The CART methodology is an example of binary, recursive partitioning. The analysis is largely automatic, and the results are concise and easy to understand and are geared toward decision-making. CART is a non-analytic, computer-intensive procedure that leads to classification rules based on inequality constraints applied to individual parameter values or to linear combinations of parameters (Breiman et al. 1984). This technique has been used for more than 20 y in fields such as engineering, public health, medicine, and economics (Breiman 1984). The CART analysis can be used to determine sets of parameters that lead to a nonpersistent classification of a chemical. Often, only a few parameters are used to reach a classification. Identifying these parameters can better facilitate the decision-making process for persistent chemicals.

CART is a set of rules for splitting each node in a binary decision tree, deciding when the tree is complete, and assigning each terminal node to a class outcome. For every model input, CART tests every value as a split point to try to maximize the reduction in variance between the parent node and the resultant two sub-nodes. We have chosen to assign the classification of nonpersistent only when 95% of the chemicals in that parameter grouping are nonpersistent, leading to fewer false-negative decisions. The system is set up to minimize false negatives while allowing for false positives. This implies that a negative result does not necessarily lead to a classification of persistent but identifies a chemical that needs to be evaluated in a higher tier.

Application to a 2-compartment model

In order to illustrate how one would classify pollutants based on their chemical properties, we consider the simple 2-compartment closed system in Figure 5-10 consisting of air (Compartment 1) and water (Compartment 2) at equilibrium with a source to the air, *S* mol/y (a Level II system).

The chemical can be removed from this system only by a transformation reaction in the water compartment. The mass-balance condition for this system is

$$S = f_W V_W Z_W k_W \quad \text{(Equation 5-8)},$$

and the mass in each compartment is given by

$$M_A = f_A Z_A V_A ; M_W = f_W Z_W V_W \quad \text{(Equation 5-9)},$$

where

S is the source term (mol/y),
f_A is the fugacity in air (Pa)
f_W is the fugacity in water (Pa)
V_A is the volume of air (m^3)
V_W is the volume of water (m^3)
M_A is the chemical mass in air (mol),
M_W is the chemical mass in water (mol),
Z_W is the fugacity capacity in water (mol/m^3-Pa), defined as $1/H$, where H is the Henry's Law ratio (Pa-m^3/mol),
Z_A is the fugacity capacity in air (mol/m^3-Pa), defined as $1/RT$, where T is the temperature, and
k_W is the reaction half-life in water (1/y).

Using the above equations with the definition for persistence, Equation 5-1, the persistence P or characteristic time τ for this system is

$$P = \tau = (f_A Z_A V_A + f_W Z_W V_W)/f_W Z_W V_W k_W \quad \text{(Equation 5-10)}.$$

We define the system such that $V_A = V_W$. Because the system is in equilibrium, $f_A = f_W$. K_{AW}, the air–water partition coefficient, which represents the ratio of the fugacity capacities in air and water, is equal to H/RT. Simplifying the above equation, we obtain

$$P = \tau = (1/k_W) \cdot (1 + K_{AW}) \quad \text{(Equation 5-11)}.$$

Under the assumption that $P > 1$ y will result in a "persistent" classification, the above equation was calculated with a broad spectrum of K_{AW} and k_W values to determine which sets of K_{AW} and k_W exceed the cutoff criteria. Solving this analytically, it is apparent that if k_W is less than 1, the chemical will be persistent. If K_{AW} is greater than 1, k_W must be greater than 2 to be nonpersistent, since less than half of the chemical will be in water undergoing transformation processes. The more chemical pollutant there is in the air, the shorter the half-life in water must be for the chemical to remain in the system for less than a year. Regions of chemical properties resulting in a classification "persistent" ($P > 1$ y) are shown in Figure 5-11. The results in this figure can be used to develop a binary classification tree, such as the one shown in Figure 5-12. From this tree, we can determine whether a chemical

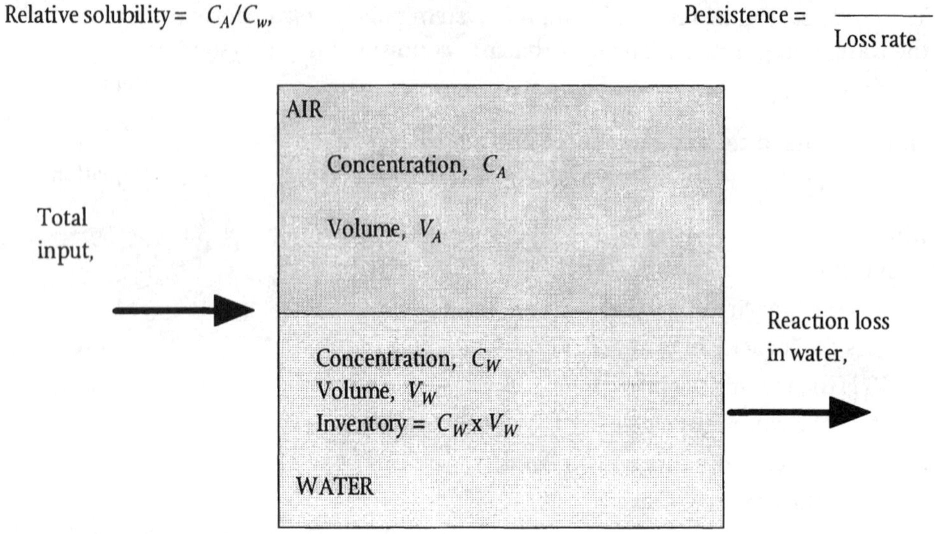

Figure 5-10 Simple 2-compartment system used to illustrate the CART anaylsis

is likely to be persistent based on the series of questions asked in the tree. This example yields a straightforward method for defining and classifying chemicals based on whether or not they persist for more than a year in the defined system. However, as the number of compartments in the model increases and as the number of independent parameters increases, the complexity and tractability of this process increases substantially.

Application of classifying persistence in a multimedia environment

Now that we have demonstrated how CART can be applied to a 2-compartment system, we expand this method to a more complete environmental system. We focus on hypothetical chemicals, with the range of each chemical property falling within the feasible range for organic chemicals. These chemicals are released to the air compartment in a multimedia unit-world as presented in Figure 5-5, Model A, with the proportions of water and soil equaling global averages. A measure of persistence is calculated for each hypothetical chemical. A CART analysis is then completed with the goal of determining which sets of chemical parameters lead to a persistent chemical. The analysis also could be completed for releases into other environmental media.

We calculate the mass distribution in the evaluation unit based on fugacity principles, a common approach for describing partitioning in multimedia systems (Mackay 1991). The steady-state concentrations in each environmental compartment are determined from the interactions between the three environmental compartments and the decay rate in each compartment. The chemical parameter-value ranges, with examples of chemicals near the minimum and maximum of that

Figure 5-11 Dependence of the joint occurrence of k_A and K_{AW} on persistence classification. This figure illustrates that there are large regions in which either the "persistent" or "nonpersistent" classification dominates.

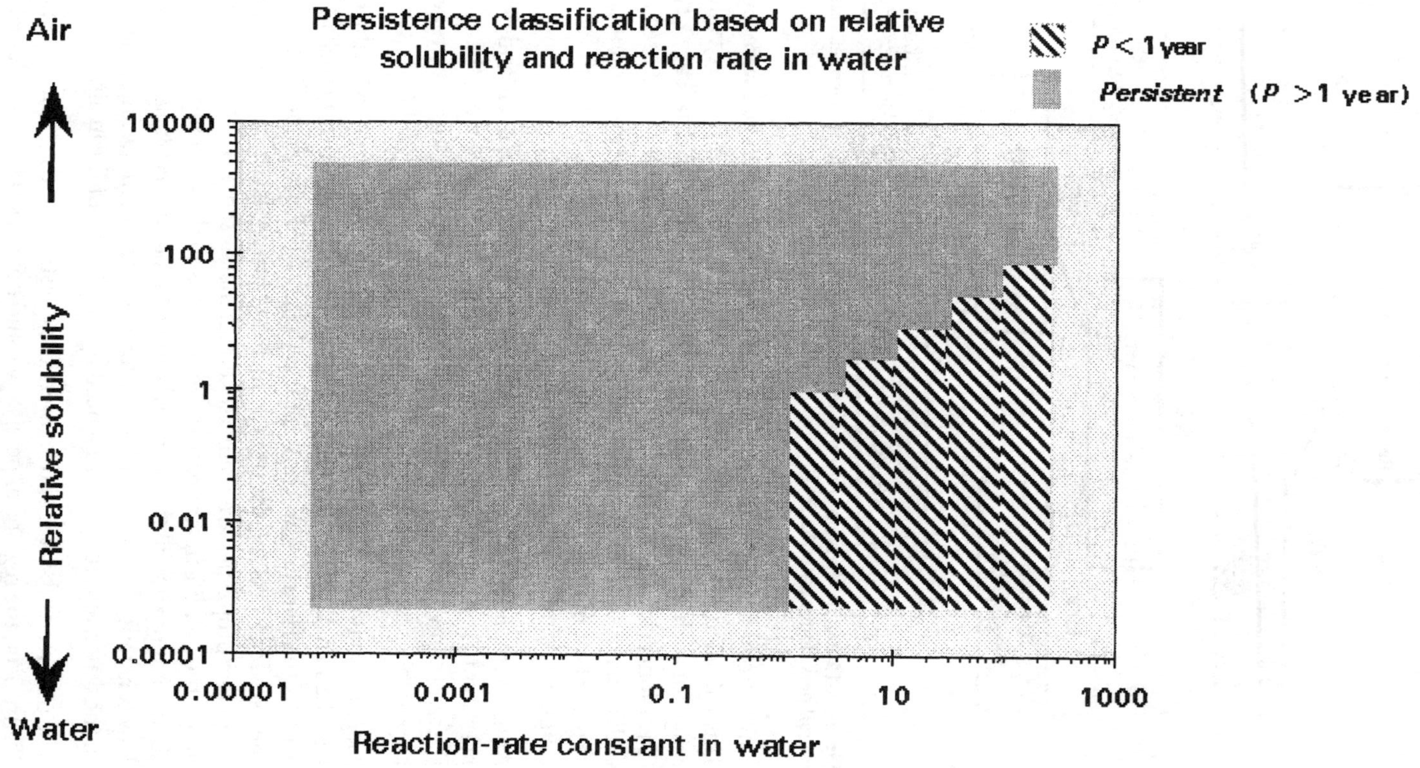

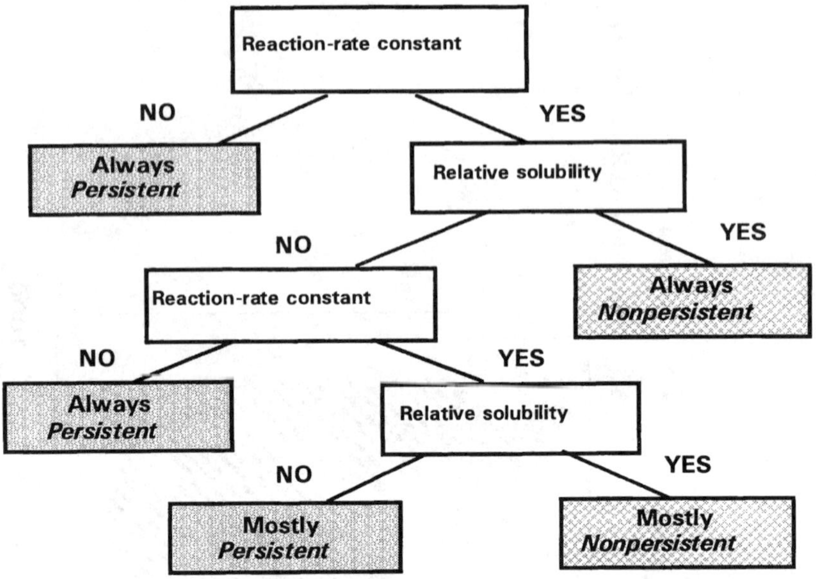

Figure 5-12 Binary decision tree obtained from the results of the 2-compartment case study with the overall persistence criterion set to > 1 y.

range, are presented in Table 5-1. All of the distributions used were log-uniform. This will yield the same number of chemicals having that property value in that range for each order of magnitude. Using the Monte Carlo package Crystal Ball (Decisioneering 1996), 10,000 simulations were completed and used as input to the CART anaylysis. A chemical that remained in the system for 6 months was defined as persistent in this study. Branches not yielding a different result for a significant number of simulations were trimmed.

As an example, we considered a system with source term to air and defined a persistent chemical as one having a characteristic time of more than 6 months. In the tree, shown in Figure 5-13, all of the chemical and landscape properties are used as possible predictor variables.

The first split point on the tree is associated with the question "is the half-life in air less than 26 d?" If the answer is yes, the user follows the left branch of the tree and asks "is the Henry's Law constant less than 0.407 Pa m^3/mol?" If the answer is no, the user follows along the right branch and learns that 98% of the chemicals with these two properties are nonpersistent. It makes sense that the first question is about the half-life in air because the release is to the air compartment, and thus a significant portion of the chemical can be found in this phase. With a high Henry's Law constant, the chemical is more likely to remain in the air, rather than partition to the water, and thus be influenced by the short half-life in air.

The user reads the tree and answers the questions to reach the terminal nodes. In Figure 5-14, the regions of persistence and nonpersistence are plotted based on the

Table 5-1 Ranges of chemical properties

Property	Symbol	Distribution type	Lower end	Upper end	Example of chemical with property at lower end of range	Example with property at upper end of range
Henry's law constant (Pa-m^3/mol)	H	Log uniform	$1x10^{-3}$	$1x10^{5}$	Phenol	Nitrogen gas
Octanol-water partition coefficient	K_{OW}	Log uniform	1	$1x10^{9}$	Butanol, Methylchloride	Di-n-octyl-phthalate
Decay rate in air (1/day)	k_A	Log uniform	$4x10^{-4}$	$1x10^{2}$	Toxaphene, Bromodichloromethane	Benzo(a)pyrene
Decay rate in water (1/day)	k_W	Log uniform	$1x10^{-5}$	$1x10^{2}$	Hexachloroethane	Pyrene
Decay rate in soil (1/day)	k_S	Log uniform	$1x10^{-5}$	$1x10^{2}$	PCB	Anthracene
Vapor pressure in (Pa)	VP	Log uniform	$1x10^{-6}$	$1x10^{5}$	Chrysene, TCDD	Atmospheric pressure
Melting point (K)	Tm	Uniform	100	600	Vinyl chloride	Chrysene, beta-HCH, TCDD
Diffusion coefficient in pure air (m^2/d)	D_A	Uniform	0.2	1.7	Toxaphene	2,4-Dinitrotoluene
Diffusion coefficient; pure water (m^2/d)	D_W	Uniform	$3x10^{-5}$	$1x10^{-4}$	Endrin	Vinyl chloride

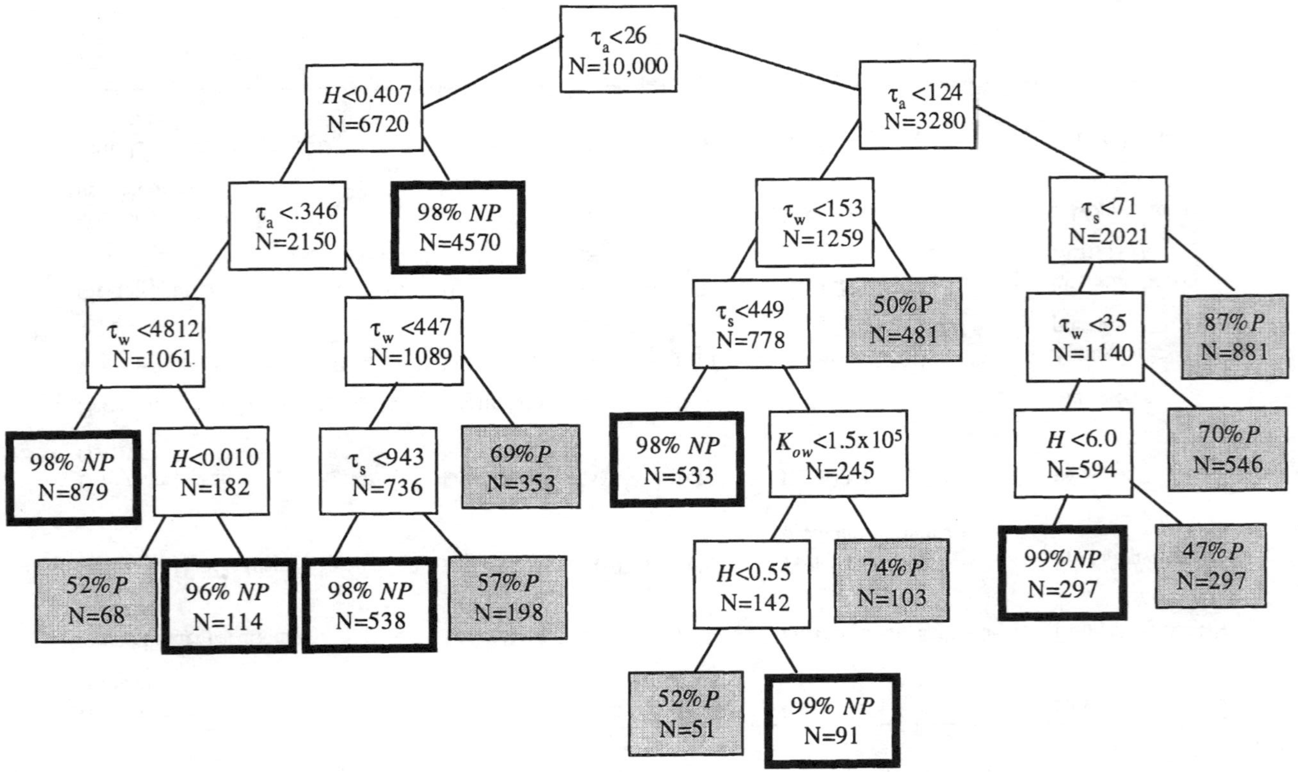

Figure 5-13 CART for emissions to air, persistence defined as 1 y, all chemical and landscape properties used as predictor variables. The tree is read by asking a question regarding the inequality listed. If the answer is yes, the user follows the left branch while if the answer is no, the user follows the right branch. The user follows this procedure until a terminal node is reached, which indicates if the chemical can be classified as nonpersistent or if the chemical needs to be evaluated in a second-tier evaluation. N defines the number of chemical realizations in that node. The terminal nodes indicate the percent of the realizations that are either persistent, P, or nonpersistent, NP.

half-life in air and the Henry's Law constant. Light gray regions are nonpersistent, medium gray regions are nonpersistent if they meet the qualifications listed in that region, and dark gray regions are persistent.

Implications of CART classification example

A screening-level classification system is proposed for determining whether a chemical compound is persistent or nonpersistent. The method also can be used to assess transport potential. This provides a regulatory advantage by allowing the classification of numerous parameters without needing to run a simulation model. Classification of persistence is determined by comparing the overall residence time of a compound in a multimedia evaluation environment to a reference time value. The multimedia model is used with 10,000 simulations to capture a broad range of chemical property sets, allowing us to identify sets of properties that result in a "nonpersistent" residence time in the multimedia model.

The CART analysis reveals that the most important chemical parameters for classifying the persistence of chemicals released to the air is the decay rate in the air. Other important parameters are Henry's Law constant and the decay rate in water. When the source medium is changed from air to soil, both the resulting classification trees and the importance rankings change.

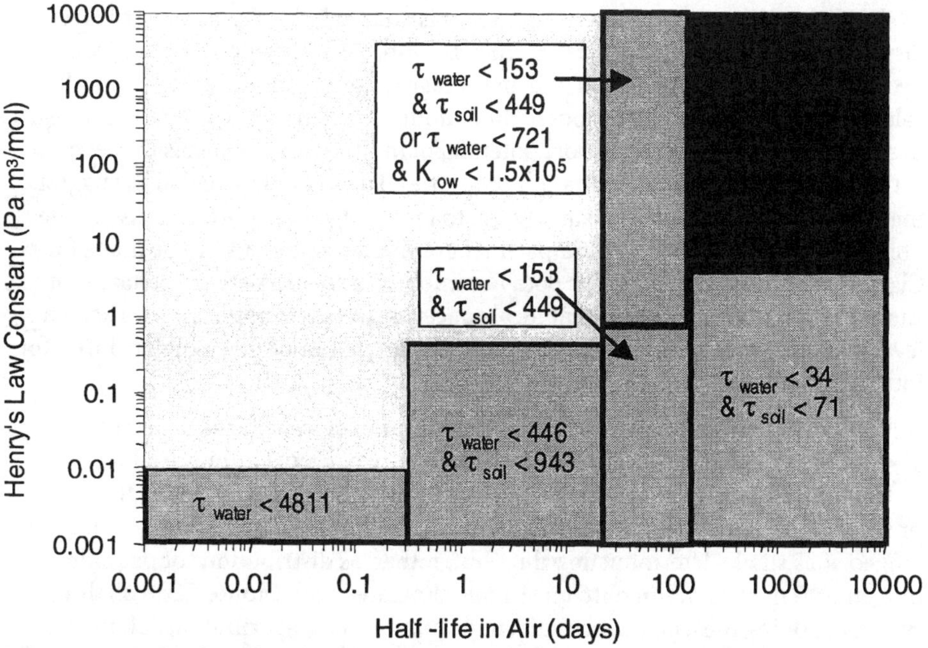

Figure 5-14 Results of CART analysis for emissions to air; persistence defined as 6 months; all chemical and landscape properties used as predictor variables

Although the process of generating the classification trees requires an understanding of multimedia interactions, CART reduces the results to a classification diagram that does not require an understanding of multimedia models for interpretation. In the case study here, the multidimensional response surface of a model can be reduced to a two-dimensional plot with two classifier parameters on the axes. The importance of other parameters shows up in the subregions of this plot.

Use of Persistence and Transport Potential in Decision-making

Application of the modeling tools described in this chapter provide a transparent, scientifically sound approach for characterizing the relative persistence and long-range–transport potential of different chemicals in a multimedia context. These attributes serve as the key attributes that can be used to rank chemicals and thus identify candidate POPs. However, it is important to point out that the models cannot be used to establish recommended criteria for decision-making because criterion selection depends upon the specific regulatory goals. Nevertheless, it is possible to use modeling tools to ensure that proposed criteria are coherent.

For example, consider the proposed air half-life criterion of 2 d. If this criterion is applied to a Class A chemical, application of Equation 5-6 yields a characteristic travel distance of 750 km if an average wind speed of 3 m s^{-1} is assumed. Alternatively, if a half-life criterion in water of 180 d is adopted, the characteristic travel distance for a Class B chemical is less than 0.1 km, assuming a representative hydraulic residence time of 10 d and an average water depth of 3 m. This simple calculation reveals that the proposed air half-life criterion can be used to establish a benchmark for assessing transport potential of multimedia chemicals that exhibit various partitioning and degradation properties. Further, this calculation suggests that Class B chemicals are not susceptible to long-range transport. Because of the non-mobile nature of the soil compartment, a similar argument can be made for Class C chemicals. However, this does not imply that the persistence behavior of such compounds does not present local concerns. Based on application of screening-level modeling tools, however, such chemicals should not be given high priority for international initiatives.

Another important issue related to the classification of a chemical as persistent or susceptible to long-range transport is the influence of model input variability on model output. In comparing calculated persistence/transport measures to a proposed criterion, one should keep in mind that the calculated values are not to be looked at as single deterministic values but, rather, as distributions of possible outcomes. The variance in outcome can be simulated with Monte Carlo analysis. Instead of doing the calculation with single input values, distributions of possible input values that reflect natural variability and parameter uncertainty can be used as input to the model. The outcome would be a distribution, reflecting the operational margins of certainty. In a recent study on persistence assessment, it was suggested that media-specific half-lives be defined in terms of log-normal distributions with defined standard deviations rather than fixed single values. The model result would then be expressed as a distribution of overall persistence values, reflecting more realistically the large variability of degradation potential in the environment. A persistence criterion could then become "overall persistence should be less than 100 d, with a frequency of at least 50%, and 90% less than 200 d" (Webster et al. 1998).

Based upon the above insights, a proposed scheme for identifying candidate POPs subject to long-range transport could rely on the following tiered framework. In Tier I, physicochemical properties are used to classify chemicals. Class A chemicals are compared to the proposed single-medium air half-life criterion. If the chemical exceeds this criterion, the chemical is identified as a candidate POP and moves to the next evaluative tier (e.g., bioaccumulation potential). Because Class B and C chemicals do not exhibit long-range–transport potential, these compounds are dropped from further consideration. Class D chemicals proceed to a second evaluative tier in which transport potential is determined using a multimedia model that allows the variability in media-specific degradation rates to be included. The model calculated distribution of transport potential is then compared to the transport potential criterion that is coherent with the value derived for a Class A chemical and the prescribed air half-life criterion. Alternatively, CART analysis could be used to define the combination of chemical properties for Class D chemicals that exceed the transport-potential criterion. Class D chemicals that fail this decision point move to the next evaluative tier. This scheme is illustrated in Figure 5-15.

It is important to highlight some limitations in assessing persistence and long-range transport using the approaches presented in this chapter. If a chemical is used in only very small quantities and/or in uses that translate into very limited emissions into the environment, this chemical may in fact pose a low concern despite incriminating persistence and long-range potential properties. Similarly, a less persistent and mobile chemical may pose higher concern if emitted in very large quantities. Furthermore, increased persistence does not necessarily imply increased exposure potential. For example, if a hydrophobic chemical is emitted to soil, persistence may be greater than if released to air. However, human exposure potential may be greater because air may serve as a more efficient transport medium to secondary compartments (e.g., plants) that determine indirect exposure pathways (i.e., vegetable ingestion). Consequently, additional information must be examined in higher evaluative tiers (e.g., emission quantity, mode of entry, exposure potential) to ensure effective management decisions. Often, such information is poorly characterized and provides a basis for future data collection, especially for high production volume chemicals.

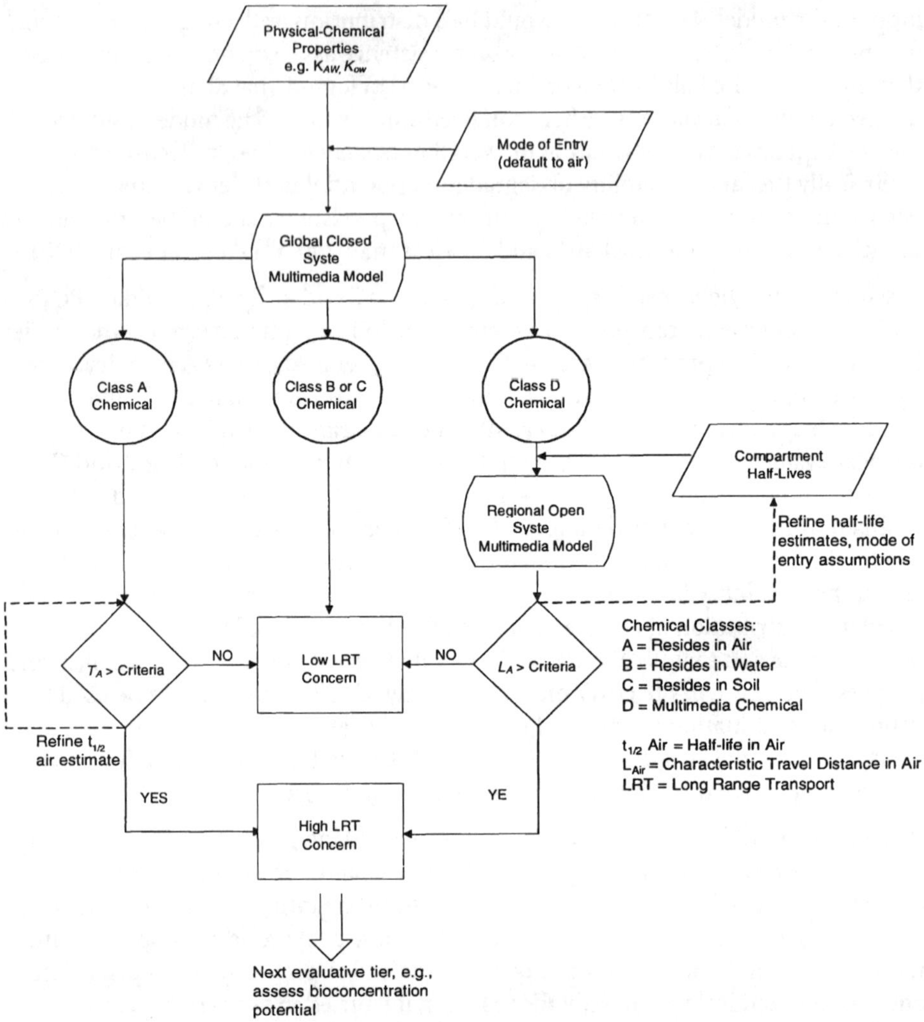

Figure 5-15 Example decision framework for identifying candidate POPs

Higher Tiers

Input variance is only one aspect of the total model uncertainty. To a great extent, the uncertainty in the predicted persistence and transport potential originates from our fundamental doubts about the correctness of the modeling concept applied. Other, equally plausible modeling concepts could have been applied that may have yielded different results. Without further quantitative analysis of the fundamental concepts, this part of the uncertainty cannot be addressed.

The models presented above have been developed for screening and are not applicable for more precise modeling. They are average global models. After a chemical has been identified using criteria including the above global models, it is expected that more data such as toxicity and release modes will be obtained for a risk evaluation. Part of this evaluation is likely to involve the use of models that more accurately reflect reality. The above models are simplified for ease of computation and in order to be useful for a wide variety of conditions. These more specific models can be chosen and parameterized to reflect the conditions of the environment in the locale and/or region of release. The models can be set to reflect levels of UV radiation, wind speed, degradation rates, temperature variations, amount of vegetation, proportion of water, soil types, spatial scale, temporal scale, seasonal changes, marine systems, and other properties. It will likely be useful to use a model such as the global meridional model that explains transport from tropical to polar regions to evaluate the degree of global distillation (Wania and Mackay 1996). The models used can be either the fugacity type or one of the many other types available. The model chosen will depend on the type of chemical, its properties, release conditions, and location of release. The decision of which models to use probably will be made by experts associated with the authority or group that will do the risk evaluation.

It is possible to combine a fate model and effects information into one model that can give an indication of risk. Both the European Union System for the Evaluation of Substances model (Vermeire et al. 1997) and CalTOX (McKone 1993) do this and are already accepted by the European Union and the State of California, respectively.

Caution is advised when deciding to use a more complex model. These models do not necessarily give better information because more complex models include more components which increases the likelihood that part or all of the model may be poorly calibrated. The added components often push the cutting edge of science; thus, there may be insufficient data to adequately calibrate more complex models.

It may be desirable to use one model or set of models for consistency in risk assessment. If that is the case, and the risk assessors consider fugacity modeling to be appropriate, the approaches of Bennett et al. (1998) and Scheringer (1996, 1997) provide a suitable starting point for this task. These papers cover regional transport and global dispersion in complementary ways. These models would be used in two steps. The first step contains the calculation of a chemical's regional travel distance wherein the model contains a moving parcel of contaminated air that is in contact with soil, surface water, and vegetation as stationary phases. If this distance exceeds a threshold of regional dimensions, e.g., 500 to 1000 km, the second step is taken. The second step is to use the global circulation model. This model consists of many replications of the regional Level III fugacity model to form a closed loop that can be looked upon as a one-dimensional representation of the global circulation system (see Figure 5-9) (Scheringer 1996). The model calculates the spread of the chemical along this closed chain of regional models. The overall persistence and spatial range on a global scale are based on macroscopic diffusion in the troposphere and the oceanic surface water. The upper limit of the spatial range is the earth's perimeter, 40,000 km, which is equivalent to a uniform global distribution as observed for

CFCs. The lower limit is given by the size of the regional models for which the loop is made. This "block size" is the spatial resolution of the global model. For a consistent use of both the regional and global model, the maximum dimension of the regional model and the block length of the global model should be of the same order of magnitude. For a proper and consistent application, these models shoud be refined and harmonized.

Conclusions and Recommendations

The following summarize the conclusions and recommendations of the workgroup:

- Separate measures of both persistence and transport potential are needed to classify chemicals with respect to persistence and long-range transport.
- Useful measures of persistence and transport potential can be derived from half-lives in specific media and inter-media partition coefficients by using a Level III multimedia model.
- Given cutoff criteria, the parameter domains that would give rise to undesired persistence and transport potential can be derived using a CART analysis.
- Two-dimensional "Scheringer plots" of persistence versus transport potential for determining agreed-upon chemicals of concern or no concern are a useful pictorial tool.
- Although there is no scientific rationale for criteria selection, science can help in the process of developing criteria that are coherent.

References

Bennett DH. 1999. Semivolatile organic chemicals in a multimedia environment: characteristic travel distance, characteristic time, and human exposure [Ph.D. dissertation]. Berkeley CA: University of California, Berkeley.

Bennett DH, Kastenberg WE, McKone TE. 1999. General formulation of characteristic time for persistent chemicals in a multimedia environment. *Environ Sci Technol* 33:503–509.

Bennett DH, McKone TE, Matthies M, Kastenberg WE. 1998. General formulation of characteristic travel distance for semivolatile chemicals in a multimedia environment. *Environ Sci Technol* 32:4023–4030.

Beyer A. 1999. Assessment of long-range transport of persistent organic pollutants [Diplom thesis]. Osnabrück, Germany: University of Osnabrück.

Beyer A, Mackay D, Matthies M, Wania F, Webster E. 2000. Assessing long-range transport of persistent organic pollutants. *Environ Sci Technol (accepted)*.

Brandes LJ, den Hollander H, van de Meent D. 1996. SimpleBox 2.0: a nested multimedia fate model for evaluating the environmental fate of chemicals. RIVM Report 719101 029.

Breiman L, Freidman J, Olshen S. 1984. Classification and regression trees. Monterey CA: Wadsworth.

[Cal-EPA] California Environmental Protection Agency. 1997. Draft Concept Paper RSU Task D.3, California's Non-RCRA Waste Classification System: Analysis and Proposed Revisions. California Environmental Protection Agency, Department of Toxic Substances and Control, Human and Ecological Risk Division. Available at http://www.cwo.com/~herd1/caltox.htm

Cowan C, Mackay D, Feijtel T, van de Meent D, Di Guardo A, Davies J, and Mackay N. 1995. The multimedia model: a vital tool for predicting the fate of chemicals. Pensacola FL: Society of Environmental Toxicology and Chemistry (SETAC).

Decisioneering. 1996. Crystal Ball. 4.0 ed. Boulder CO: Decisioneering.

[ECETOC] European Center for Ecotoxicology and Toxicology of Chemicals. 1994. HAZCHEM. A mathematical model for the use in risk assessment of substances. Special report no. 8, October 1994. Brussels, Belgium: ECETOC.

Harner T, Mackay D. 1995. Measurement of octanol–air partition coefficients for chlorobenzenes, PCBs, and DDT. *Environ Sci Technol* 29:1599–1606.

Hertwich EG. 1999. Toxic equivalency: accounting for human health in life-cycle impact assessment [Ph.D. dissertation]. Berkeley CA: University of California, Berkeley.

Mackay D. 1991. Multimedia environmental models: the fugacity approach. Chelsea MI: Lewis.

Matthies M, Trenkle R. 1988. Variability of exposure estimations for hazardous chemicals released to the environment. *Chemosphere* 17:1471–1486.

McKone TE. 1993. CalTOX, a multimedia total-exposure model for hazardous wastes sites. Part II: The dynamic multimedia transport and transformation model. Livermore CA: Lawrence Livermore National Laboratory. Report no. UCRL-CR-111456PtII.

Müller-Herold U. 1996. A simple general limiting law for the overall decay of organic compounds with global pollution potential. *Environ Sci Technol* 30:586–591.

Müller-Herold U, Caderas D, Funck P. 1997. Validity of global life-time estimates by a simple general limiting law for the decay of organic compounds with long-range pollution potential. *Environ Sci Technol* 31:3511–3515.

Scheringer M. 1996. Persistence and spatial range as endpoints of an exposure-based assessment of organic chemicals. *Environ Sci Technol* 30:1652–1659.

Scheringer M. 1997. Characterization of the environmental distribution behavior of organic chemicals by means of persistence and spatial range. *Environ Sci Technol* 31:2891–2897.

Scheringer M. 1999. Persistenz und Reichweite von Umweltchemikalien. Weinheim, Germany: Wiley-VCH.

Scheringer M., Bennett DH, McKone TE, Hungerbühler K. 2000. Relationship between persistence and spatial range of environmental chemicals. In: Lipnick RL, Mackay D, Jansson B, Petreas M, editors. Persistent bioaccumulative toxic chemicals: fate and exposure. Washington DC: American Chemical Society (in press).

Scheringer M, Wegmann F, Fenner K, Hungerbühler K. 2000. Investigation of the cold condensation of persistent organic pollutants with a global multimedia fate model. *Environ Sci Technol* (in press).

Schnoor J. 1996. Environmental modelling: fate and transport of pollution in water, air, and soil. New York: Wiley. 682 p.

Tell JG, Parkerton TF. 1997. A comparison of multimedia models used in regulatory decision-making. *SAR and QSAR Environmental Research* 6:29–45.

Trapp S, Matthies M. 1998. Chemodynamics and environmental modeling. An introduction. Heidelberg, Germany: Springer.

U.S. Federal Register. 1999. Persistent bioaccumulative toxic (PBT) chemicals; Proposed Rule, Vol. 64, No 2, 5 January 1999.

van de Meent D. 1993. SimpleBox, a generic multimedia fate evaluation model. RIVM Report No. 672720 001.

van de Meent D, Peijnenburg W, Struijs J, Govers HA. 1992. Towards a QSAR-driven computation of persistence of chemicals in the environment. Poster, 5th International Workshop on QSAR in Ecotoxicology; Duluth, MN.

van Pul WAJ, de Leeuw FAAM, van Jaarsveld JA, van der Gaag MA, Sliggers CJ. 1998. The potential for long-range transboundary atmospheric transport. *Chemosphere* 37:113–141.

Vermeire TG, Jager DT, Bussian B, Devillers J, Den Haan K, Hansen B, Lundberg I, Niessen H, Robertson S, Tyle H, Van der Zandt PTJ. 1997. European Union System for the Evaluation of Substances (EUSES). Principles and structure. *Chemosphere* 34:1823–1836

Wania F, Mackay D. 1995. A global distribution model for persistent organic chemicals. *Sci Total Environ* 160/161:211–232.

Wania F, Mackay D. 1996. Tracking the global distribution of persistent chemicals. *Environ Sci Technol* 30:A390–A396.

Wania F. 1998. An integrated criterion for the persistence of organic chemicals based on model calculations. WECC-Report 1/98. Toronto, Ontario, Canada.

Webster E, Mackay D, Wania F. 1998. Evaluating environmental persistence. *Environ Toxicol Chem* 17:2148–2158.

Chapter 6

Monitoring as an Indicator of Persistence and Long-Range Transport

Davide Calamari, Kevin Jones, Kurunthachalam Kannan, André Lecloux, Mats Olsson, Michael Thurman, Paolo Zannetti

Environmental monitoring is a means of assessing environmental quality. In general, environmental monitoring programs are necessary to develop comprehensive descriptions of environmental quality at spatial and temporal scales and to provide a sound basis for effective measures, strategies, and policies to address environmental problems. The most important objectives of routine environmental monitoring programs are to measure actual concentrations and identification of trends, to evaluate emissions and flows, to test compliance with standards and quality objectives, and to provide early warning and detection of pollution.

Monitoring is the origin of issues pertaining to persistence and long-range transport of chemicals (Goldberg 1979). In the mid- to late 1960s, measurements of dichlorodiphenyltrichloroethane (DDT) and polychlorinated biphenyls (PCBs) in Arctic seals and Antarctic penguins provided the evidence for long-range transportability of these chemicals from the site of their application or usage. While monitoring spatial trends would provide information on the transportability of a chemical, temporal trend studies indicate environmental persistence. Earlier chapters in this book discuss the role of models in predicting persistence and long-range transport. While models are predictive in evaluating these parameters, monitoring studies provide the real evidence of persistence and potential for long-range transport. In this chapter, the role of monitoring data in evaluating persistence and long-range transport is discussed. Various approaches to monitoring and appropriate interpretation of monitoring data are suggested. Illustrations of the use of monitoring data to discern clearance rates and environmental behavior are presented. Persistent organic pollutants (POPs) such as PCBs and organochlorine pesticides such as DDT,

Chapter Preview

Evaluation of Persistence and Long-Range Transport of Organic Chemicals in the Environment. Gary Klečka et al., editors.
 ISBN 1-880611-22-8

hexachlorocyclohexane (HCHs), or chlordanes have been used as examples, where necessary, to describe how monitoring data has been useful in identifying compounds that are persistent and transportable to remote areas from the point of application.

One of the key requirements in monitoring programs is reliable and reproducible data. It is necessary that uniform analytical methodology of the utmost reproducibility and accuracy be used by monitoring laboratories to ensure that analytical results can be correlated and compared on a spatial and temporal scale. Several earlier publications have addressed issues pertaining to data quality, and, more recently, the European Centre for Ecotoxicology and Toxicology of Chemicals (ECETOC) has provided guidance on "data quality" issues such as sampling and storage, analytical quality assurance, and quality control (QA/QC) (ECETOC 1999). A brief discussion of QA/QC follows later in this chapter.

Chemicals, once they are released into the environment, seek out the environmental media (air, water, soil, or biota) in which they are most soluble. For instance, trichloroethylene and benzene are most soluble in air (volatile); thus, they readily partition to air. DDT and PCBs are most soluble in organic matter (e.g., lipid) and therefore tend to bioconcentrate in biota. The dynamics of environmental partitioning of chemicals has been described by multimedia environmental transport models, which help us to understand environmental behavior of chemicals. Despite having relatively low vapor pressures (vapor pressure 10^{-2} to 10^{-5} Pa at 20 °C), PCBs and DDT can reach remote areas through atmospheric transport, which is considered a primary mode of global distribution for organic chemicals. Transport of these compounds to the poles is thought to involve cyclic processes of wet and dry deposition and sublimation and evaporation combined with net atmospheric flux of heat from equatorial regions (Ottar 1981). Monitoring studies have provided evidence that this pole-ward transport can be rapid, based on episodic atmospheric events (Pacyna and Oehme 1988). There is also a close correspondence in time between worldwide production and deposition in areas remote from application (Rapaport and Eisenreich 1988) and parent forms of some compounds. Long-range transport is indirectly involved in inputs into the Arctic Ocean via ocean currents and riverine outflows. Because of their long residence time in the environment, semi-volatile, persistent chemicals tend to partition between various environmental media such as air, water, soil, sediment, and biota depending on the physicochemical properties and climatic features of a geographic region, reaching unexpected destinations in unexpectedly high concentrations. Because of these properties, monitoring a single medium will not provide adequate information about their persistence and transportability. Thus, the need for, and approaches to, multimedia monitoring are discussed in this chapter.

Multimedia monitoring not only helps us to understand partitioning behavior of persistent and semi-volatile pollutants, but also it is essential for the validation of environmental models, since monitoring data is an integral parameter of modeling.

In this chapter, the importance of the complementary relationship between monitoring and environmental modeling, which improves the predictive ability of models of chemical persistence and transportability, is highlighted.

As mentioned earlier, spatial monitoring is one of the means to explain transportability of a chemical. In this chapter, spatial monitoring studies with atrazine, a relatively less-persistent but frequently used herbicide in the U.S., provides an example to describe various factors that influence transportability. Because persistence and inter-media partitioning can vary depending on several environmental parameters (e.g., temperature), trend monitoring studies in the Baltic Sea and tropical Asian countries, which represent temperate and tropical climates, respectively, are discussed to elucidate the importance of emission sources and climatic factors in determining persistence. In addition, the need for proper samples and sampling plans to describe temporal trends using archived samples, ice, and sediment cores is discussed. For certain chemicals, parent compounds may degrade rapidly, while their degradation products metabolites, or derivatives can persist in the environment for a longer period of time. Because models are less likely to predict the persistence and transportability of breakdown products, the importance of monitoring studies to identify persistent degradation of breakdown products is discussed. Despite the progress and widespread application of monitoring studies, there exist several data gaps, which arise from improper and sometimes inadequate use of data. For instance, despite 30 y of monitoring of PCBs and DDT, data generated from different parts of the globe cannot be compared because of the lack of comparability in monitoring programs conducted by different research institutes and because of the lack of data from certain locations. In this chapter, recommendations are made to build global monitoring networks for more effective management of monitoring programs for persistent pollutants in the environment.

Monitoring and Persistence

Definitions

In general terms, monitoring is used to assess the quality of the environment and to provide information for the development of policies and measures to address environmental problems at a local, regional, or global scale. Most often, environmental monitoring is defined as a long-term and standardized measurement, observation, evaluation, and reporting of the environment in order to define status, trends, and mass flows (loads). A few closely related terms often used in conjunction with monitoring are "surveys" and "surveillance." A survey is a sampling program of finite duration, and for a specific purpose, such as an intensive field study or an exploratory campaign to infer semi-empirical relationships and establish tentative theories. Surveillance is a more continuous, specific measurement or observation, with the goal of environmental quality reporting (e.g., compliance with standards and quality objectives) and/or operational activity reporting (e.g., early warning and detection of pollution).

Environmental monitoring is used to define status and trends of environmental pollutants. This means that environmental monitoring studies provide information on the occurrence and movement of pollutants in the environment. However, mere occurrence cannot be considered as a sign of persistence. Persistence is not synonymous with ubiquity (Webster et al. 1998). A chemical may be found routinely over a large area without being persistent. This occurs for a chemical that is continually emitted from a number of sources but is steadily degraded. Several definitions have been suggested for persistence. Environment Canada defines persistence as the length of time a substance resides in the environment. Similarly, persistence has been defined as the tendency of a chemical to persist (survive) in the environment without transformation into another chemical form. In some cases, persistence is defined in medium-specific terms, i.e., the tendency of a chemical to survive without transformation in a specific environmental medium (e.g., soil) and without transfer to another medium (e.g., air). Other definitions have been suggested in previous chapters of this book. A common measure of a substance's persistence is its half-life, i.e., time required for the concentration of a substance to diminish to half of its original value in a particular medium of interest. Many of these definitions narrated above are not specific. For example, Environment Canada's definition does not clearly distinguish between the time a chemical spends in the environment and the time it spends in an individual medium in the environment. Nor does it specify that the reduction in concentration is attributable to irreversible degradation reactions and not merely to a reversible change in speciation or to a loss by transport to another medium. A comparison of high-end thresholds for persistence as suggested by certain regulatory agencies is shown in Table 6-1.

Monitoring data: Representativeness and quality

As mentioned above, monitoring is a key to studying environmental persistence and behavior of a chemical in the environment. However, most monitoring studies lack adequate data quality control and coherency with other monitoring programs. The keys to a cost-effective monitoring program are well-defined objectives, adequate planning, and clear sampling strategy. The objectives must always be defined first so that suitable sampling and analytical methods may be selected. These, in turn, determine the size and duration of the monitoring study and therefore the resources needed. Unfortunately, in many cases the goals of monitoring programs are unclear and the operating conditions are characterized by uncertainty. This generally leads to a lot of data with limited information and content. It is therefore important to take into account some practical considerations when designing a monitoring program. Some of these design considerations relate to the quality and representativeness of the data. Other considerations are more practical and concern the specifics of study design and implementation.

Table 6-1 Comparison of high-end threshold for persistence

Agency	Half-life (d)
United Nations Economic Commission for Europe (1997)	> 2 (air); > 60 (water); >180 (soil and sediment)
Canadian Toxic Substances Management Policy (1995)	> 2 (air); > 182 (water); > 182 (soil); > 365 (sediment)
Canadian Environmental Protection Act	≥ 2 (air); ≥ 182 (water); ≥ 182 (soil); ≥ 365 (sediment)
North American Agreement for Environmental Cooperation – Commission for Environmental Cooperation	2 (air); 180 (water); 180 (soil); 365 (sediment)

The design criteria of a monitoring program should be selected so that the results will clearly represent a well-defined environmental concentration. There are two distinct types of representativeness. One relates to the confidence in the result, i.e., the number of samples and how far apart and how frequently they are taken. The sampling frequency and pattern must be sufficient to adequately represent the "true" concentration at the site selected. The other issue relates to how representative the site is at which the sampling is done of the setting and scenario intended within the framework of the objectives of the program.

The environment is both heterogeneous and dynamic; thus, the concentration of any particular chemical will vary in space and time as a result of variations in its rates of emission, transport, and degradation. Some examples of environmental properties that affect the variability in space and time in different compartments are

- wind speed and direction, turbulence, and precipitation (air);
- current flow rate, turbulence, stratification, depth, and dilution volume (water);
- percolation rate (porosity, rainfall, surface features), organic matter, and moisture (soil); and
- species, feeding habit, age, size, fat content, metabolism, sex, diseases, and season (biota).

Similar environmental factors affect degradation rate in addition to the interaction with other chemicals.

The most useful dataset from a monitoring program is one that captures all of these sources of spatial and temporal variability. The data may then, with proper statistical analysis, be used to estimate probability distributions in space and time. Such probability distributions may be used to choose locations for subsequent sampling programs and, in risk assessment, to estimate the likelihood of reaching a particular concentration at specific locations and times. Knowing the intended use of the data, together with the necessary statistical treatment of the data, is important for planning a representative monitoring program.

Data quality is the key consideration regarding the degree of confidence (validity) in the result of a monitoring study. These quality issues relate to the technical aspects of both sampling and analysis. The quality of the data generated from any monitoring program is defined by 2 key factors: the integrity of the sample and the limitations of the analytical methodology. The 3 major quality features of any analytical methodology are selectivity, sensitivity, and reproducibility. The selectivity of the analysis ought to be such that all of the target analyte, and only the target analyte, is measured and that other related chemicals or contaminants do not interfere with the response. Consequently, no false-positive values will be reported. Confirmation of target analytes with more than one of the several analytical techniques involving mass spectrometry—including nuclear magnetic resonance, infrared, and ultraviolet mass spectral analysis—will improve the reliability of results generated from monitoring studies. The second important feature for a good analytical methodology is that the method ought to be sufficiently sensitive to measure the chemical concentrations expected. The method must have sufficient reproducibility so as to limit the number of replicate samples and to provide information on the actual concentrations among the samples. Numerous published analytical procedures and guidelines are available for environmental analysis of particular chemicals, and these should be utilized whenever feasible.

Data generated by monitoring programs are increasingly being used to evaluate the environmental behavior and impact of chemicals. It is essential, therefore, that the data be used correctly for the purposes originally intended. Many of the problems associated with bias and systematic errors during sampling and analysis may be assessed by a variety of internal quality control measures. The study plan should include the QA/QC procedures to be used. The QA/QC procedures are designed to establish intra-laboratory controls and checks of sample collection and preparation, instrument operation, and data analysis. The QA/QC procedures should be subjected to "Good Analytical Practices" in analogy to "Good Laboratory Practices," and laboratories should undergo to a series of inter-calibration exercises and chemical analysis cross-validations in order to avoid false positives.

There are a number of practical considerations to be dealt with when designing a monitoring program that are generic regardless of the compartment being monitored (Figure 6-1). Some of these design considerations relate to the quality and representativeness of the data. In order to be successful, a monitoring study must be well-planned. The goals and objectives of the study need to be well-defined, as these will determine the site selection, extent of sampling, and types of samples.

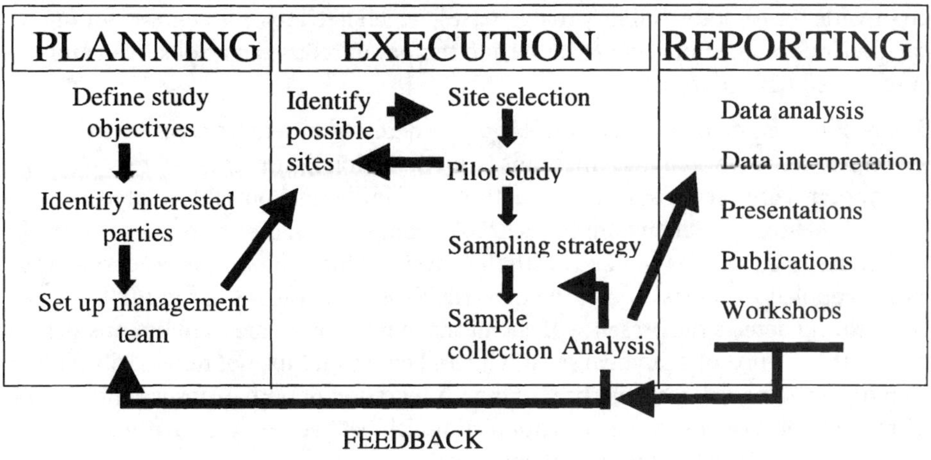

Figure 6-1 Stages in planning and execution of a monitoring program

Use of monitoring data

While the ultimate use of monitoring data is for certain aspects of environmental risk assessment and/or model validation, information gained from monitoring data on persistence and transportability are important in priority-setting of chemicals. It has often been forgotten or overlooked that the ultimate goal of monitoring is to provide information, not data. In the past, many monitoring programs have been characterized by the "data rich, information poor" syndrome.

Based on a number of international inter-calibration studies, it is known that the analytical results of organic contaminants at trace levels (parts per quadrillion to parts per billion) can vary widely among laboratories. The best database tends to originate from a few key research groups worldwide that have developed the sophisticated and validated methodologies needed. These groups are largely based in the industrialized countries in North America, Europe, and Japan, which provides a bias in the spatial coverage for monitoring trace organic pollutants. However, it is important to obtain and compile information on the spatial and temporal distribution of chemicals both regionally and globally because this information provides evidence of their potential for long-range transport and persistence.

It is important to appreciate that semi-volatile contaminants such as PCBs in environmental matrices (notably, air and water) are subject to short-term variability, which can undermine attempts to look at longer-term time trends. Air concentrations of PCBs, for example, respond to ambient temperature, air mass direction and history, and recent depositional events. Consequently, they may vary over time scales of hours, days, and seasons, in addition to underlying year-to-year variations. Concentrations in biota may vary between species, with age and sex, and with "condition" and stage of the life cycle (i.e., starvation/feeding, through lactation, during after illness/disease) of the organism. All of these factors combine together to provide "noise" or variability to databases, which need to be accounted for or factored out before underlying spatial and temporal trends can be observed and interpreted with confidence.

Many of the day-to-day-use chemicals may be detected in remote areas when extremely sensitive analytical methods (with detection limits less than femtogram or attogram) are used. For example, detection of individual polychlorinated dibenzo-*p*-dioxin or dibenzofuran (PCDD/F) congeners in the atmosphere at sub-fg m^{-3} concentrations was achieved with the advent of high resolution gas chromatography coupled with mass spectrometry (HRGC-MS) techniques in the 1980s. Monitoring detects the presence of chemicals, but presence needs to be considered within the context of analytical sensitivity and emission flux, not necessarily within that of hazard or risk, because hazard is an integration of both exposure and adverse effects. In addition, presence does not distinguish between sources (natural or anthropogenic) and local or distant emission.

Several techniques have been used to monitor trends of contaminants in the environment. Archived tissues, snow, ice, and sediment and peat cores have been used routinely to understand temporal trends, while mosses, lichens, or other plant materials (leaves, bark or needles) have been used in spatial trend studies (Jones 1991; Jones et al. 1992; Tremolada et al. 1993; Simcik et al. 1996; Gevao et al. 1997; Ockenden et al. 1998). Monitoring studies using sediment core stratification have the advantage of providing information on the chronologies of accumulation rates of anthropogenic pollutants. This information is important in order to evaluate the rate of emission from probable sources, to relate specific rates of pollutant accumulation and rates of ecosystem response, and to project future patterns and rates of pollutant deposition.

Global inventories and emissions

Information on historical and future trends in use and emissions of chemicals on a global scale is essential for interpreting monitoring data and, eventually, for evaluating chemicals' persistence and transportability. For instance, if the emission rate of a volatile compound is low, it may not be found in remote areas. This is because atmospheric removal rates can exceed emission rates when the latter are small. However, quantitative emission estimates are very uncertain for several pollutants. Even for DDT and HCH, estimates of global production varied by an order of magnitude (Voldner and Li 1995). There are also often data gaps and uncertainties over emissions. For example, several countries do not have toxic substances release inventories. Some contaminants such as PCDD/Fs originate from natural as well as anthropogenic sources (Alcock et al. 1998). Estimation of release inventories is also confounded by the lack of reliable quantification from multiple emission sources (e.g., combustion sources for PCDD/Fs and PAHs).

It is worth noting that despite 30 y of long-term monitoring data, which exist for some POPs like PCBs, reliable interpretation of chemical trends can be complex. Ambient concentrations and declines presumably reflect underlying changes in primary and secondary emissions and dilution processes, such as long-range atmospheric transport from source to sinks. Moreover, removal processes, such as burial in soils/peats/sediments, and degradation processes, such as atmospheric reactions with hydroxyl radicals and biodegradation, sometimes play an important role. Because of the lack of information on usage/emission, the significance of dilution, removal, and degradation processes is unclear or semi-quantitative. Laboratory studies have reported rates of degradation by hydroxyl radical reaction or biodegradation for certain trace organic pollutants (Sinkkonen and Paasivirta 2000a, 2000b).

Global or regional emissions or inventories of chemicals are needed not only to prioritize them in monitoring programs but also for use in modeling (Harrad et al. 1994; Wild and Jones 1995; Duarte-Davidson et al. 1997). While monitoring environmental concentrations in various environmental media of an ecosystem would assist in estimating chemical inventories, this would be a daunting if not impossible task given the number of chemicals in commerce. The following section therefore discusses how monitoring studies are useful in estimating mass balances of PCBs, PAHs, HCH, atrazine, and toxaphene in different ecosystems.

Monitoring Spatial Trends

Issues and complexities

While persistence can be expressed in terms of half-life, for which there exist some guideline values (Table 6-1), an accepted methodology for estimating long-range transport is not available. There is a need to delineate "short-range transport" arising from local contaminant sources including riverine inputs and "long-range transport" arising from distant sources primarily through atmospheric inputs. While global monitoring studies of DDT and HCHs tend to support the view that these compounds are transported to the Arctic from tropical regions (e.g., Iwata et al. 1993), there is no direct evidence to confirm this hypothesis. It can be argued that the origin of these compounds may be primarily temperate North America or Europe, where these compounds were used in large quantities in the past, or local cycles of deposition and re-emission in the Arctic (Oehme 1991). Volatilization from contaminated soils or dump sites in North America or Europe can still contribute to major sources of DDT or HCHs in the Arctic (Bright et al. 1995a). Soil, especially agricultural soil, is likely the largest reservoir of organochlorine pesticides and a major source of emissions. For example, toxaphene was heavily used in the southern U.S. on cotton and soybeans before being de-registered in 1982. Toxaphene concentrations measured in South Carolina air during the mid-1990s were 10 times higher than levels in the Great Lakes region and showed no relationship to air transport direction, suggesting volatilization from regional soils (Bidleman et al. 1998). Moreover, DDT residues in the air over a California farm, where DDT had been applied 23 years ago, were 2 to 3 orders of magnitude higher than those in the Great Lakes region (Bidleman and Falconer 1999). Well-designed spatial monitoring studies can be used to distinguish short-range and long-range transport. For instance, aerial transport from distant sources is expected to exhibit only minimal geographic variation, since atmospheric inputs occur over a large geographic scale. Based on a spatial monitoring study, it was noted that the Canadian Arctic is influenced both by local and distant contaminant sources (Bright et al. 1995b). Short-range transport of relatively polar pesticides, including currently used organophosphorus and carbamates, has been reported recently (Majewski et al. 1998).

Recently, a simple generic procedure has been developed to determine the potential for long-range transboundary atmospheric transport of substances by means of the residence time of the substance in the air (van Pul et al. 1998). The atmospheric residence time is defined as the time period in which the mass of the pollutant in the atmospheric boundary layer is reduced by 50% (atmospheric half-life). An atmospheric residence time of, for instance, 24 h means that at an average wind speed of 5 m/s, half of the substance is transported at least to a distance of more than 400 km. The atmospheric residence time is determined by the total removal rate, resulting from all removal processes from the atmospheric boundary layer such as wet and dry deposition and photochemical degradation. Atmospheric residence time is recommended as an index to consider the spatial extent of long-range transport of substances. While atmospheric residence time provides a base to determine the potential for long-range transport, there are several assumptions and uncertainties involved in its estimations (van Pul et al. 1998). Atmospheric monitoring can be used to measure the half-life of a substance in air under field conditions, which can then be used to estimate the potential for transport. However, further studies are needed on this issue.

Monitoring programs have been established in several countries, which collectively give information on the spatial distribution of persistent organochlorine compounds such as DDT, HCHs, and PCBs in the global atmosphere. Much of this relates to temperate, industrialized regions of the globe and has a bias towards PCBs, PAHs, PCDD/Fs, DDT, and HCHs. These studies demonstrate that urban areas are key sources of PCBs, PAHs, and PCDD/Fs to the atmosphere, reflecting in some cases (e.g., PCBs) emission from past or present usage and in others emission from anthropogenic combustion sources (e.g., PAHs, PCDD/Fs). Agricultural areas of past or present use are the key global source areas for a suite of organochlorine pesticides. Volatilization of toxaphene and DDT from treated soils in the southern U.S. had been unequivocally demonstrated to supply the atmosphere with these compounds and to cause atmospheric transport to regions far removed from the areas of use (Blais et al. 1998); obviously however, it is not the only source of these compounds to remote areas. Maps and datasets have been assembled of the past/ continuing use of γ-HCH (Lindane) and DDT (Wania and Makay 1996), that clearly demonstrate the dominance of tropics as the key global source regions of these pesticides. A major atmospheric monitoring program in the Canadian and Russian Arctic is now providing important information on the supply and concentrations of PCBs, PAHs, DDT, and HCHs to these remote regions.

Global monitoring studies have been conducted by various authors using leaves, bark, pine needles, lichens, mosses, air, or water collected from various locations within a certain period of time (Gaggi and Bacci 1985; Calamari et al. 1991; Iwata et al. 1993; Morosini et al. 1993; Simonich and Hites 1997). Calamari et al. (1991) assembled "global distribution maps" for hexachlorobenzene (HCB), HCH, and DDT by analyzing plant samples collected from a range of locations all over the world. They hypothesized that these organochlorine compounds are supplied to the plants primarily by atmospheric concentrations and deposition and that the plant concentration would directly reflect the atmospheric concentration. Their studies clearly demonstrated that the global atmospheric distribution of HCB, HCH, and DDT differed substantially. HCB, for example, was found at the highest concentrations in plants collected in the polar regions, far from the source regions. In contrast, DDT was highest in tropical plants. This spatial monitoring study illustrates some important features of transport and persistence. A compound's behavior in the environment is clearly controlled by its physicochemical properties. HCB is sufficiently volatile and persistent to move efficiently through the atmosphere. It has a relatively low octanol–air partition coefficient (K_{OA}) and therefore readily partitions from soils or water bodies to air. In contrast, DDT is still found largely in the tropical source regions. It partitions more strongly to soils and atmospheric particles than does HCB. It consequently enters the atmospheric gas phase less readily and has a shorter atmospheric residence time (van Pul et al. 1998). Its movement to remote regions can still occur but occurs less efficiently and less quickly.

Analysis of pine needles has demonstrated another important principle of the global atmospheric distribution of POPs, namely, that source regions and elevated concentrations will be influenced by economic features of different countries. In brief, industrialized countries tend to use and emit a different mixture of POPs than those of nearby less-developed countries; different countries could have different fingerprints of emission that are influenced by the gross national product (Calamari et al. 1995; Simonich and Hites 1997).

As noted above, much of the data on the spatial distribution of POPs has been compiled from the work of different groups, with different analytical methods, conducted in different places at different times. These differences can undermine the detailed interpretation of the datasets or models (Wania and Mackay 1996) and our confidence that the apparent spatial differences are real. Ockenden et al. (1998) therefore set out to make long-term and simultaneous measurements of the atmospheric distribution of PCBs on a latitudinal transect of remote/rural European locations from ~ 47 °N to 79 °N. Mean annual atmospheric temperatures varied between –0.1 and 9.6 °C at their sampling sites. A passive air-sampling technique with semipermeable membrane devices (SPMDs) was used, which was calibrated against the conventional high-volume air-sampling technique. It was found that PCBs demonstrated global fractionation in the atmosphere, but the effect was subtle and would have been difficult to detect unequivocally by a more disparate sampling or data-compilation strategy. The relatively light tetra-chlorinated PCB homolog was evenly distributed on the latitudinal transect. In contrast, the penta-, hexa-, and hepta-homologues were factors of 3, 5, and 7 lower, respectively, in the colder north than the warmer south. Despite the range of chemical properties between the tetra- to hepta-homologues (e.g., K_{OA} 8.0 to 11.6) and the range of environmental conditions, these concentration differences are subtle and are believed to indicate that PCBs are approaching dynamic air–surface equilibrium partitioning conditions (Ockenden et al. 1998).

Iwata et al. (1993) collected air and surface seawater in various seas and oceans during 1989 and 1990 and analyzed for the presence of PCBs, DDT, and HCHs. Among these, HCHs had the highest concentration in air and water, as result of its greater solubility in water than that of DDT or PCBs. The spatial distribution of HCHs revealed the presence of higher concentrations in the northern hemisphere, particularly in cold temperate regions and in tropical countries. While greater concentrations in tropical regions could be explained by recent usage, higher concentrations in cold water bodies was explained by the "cold condensation hypothesis." However, unequivocal evidence of transport of HCH or DDT from tropical countries to the Arctic could not be found. Further, this study estimated mass transfers of HCHs, DDT, and PCBs between air and water. It was found that ocean waters near the tropics and the Arctic could serve as sinks for HCHs and DDT because of the transfer of these compounds from air to water. The magnitude of difference in the flux estimates between tropical and temperate climates is an indication of the environmental behavior of semi-volatile organic compounds.

The results of decreasing south to north gradients of DDT and HCHs obtained from the various spatial monitoring studies mentioned above have been interpreted as an indication of long-range transport from southern sources to the Arctic. However, this does not offer direct evidence of transport from tropical countries to the Arctic. In summary, the studies on spatial trends of selected semi-volatile organic pollutants demonstrate the complexity of their behavior and the difficulties in interpreting spatial monitoring data. The global environment/atmosphere and the other environmental compartments partly reflect past and present usage and local emissions and partly reflect physicochemical properties of the compounds and the ambient climatic conditions. Future distributions will be influenced by these factors, too, but also by their environmental persistence, recycling potential, and susceptibility to the efficiency of global transport. There is a need for harmony in spatial monitoring studies that can be attained by establishing global networks by selecting key stations in the globe representing different climatic regions.

Spatial trend monitoring of atrazine and alachlor: Lessons learned

In order to make an assessment of environmental persistence and transport based on spatial trend monitoring, studies with atrazine and alachlor are used as a model for discussion. Atrazine is one of the most used herbicides in the U.S. at approximately 30 million kg per year. It is applied to corn and other row crops in a preemergent setting in early May across most of a 10-state region, i.e., "The Corn Belt." Similarly, alachlor is an herbicide used in corn cultivation. The fate and transport of the applied atrazine and alachlor have been important environmental issues because of their long history of use, the large quantities that are applied, and suggestions of possible environmental damage.

For these reasons, a detailed monitoring study was carried out to examine the fate and transport of atrazine and alachlor to the atmosphere (Thurman et al. 1991, 1992, 1994; Goolsby et al. 1997; Lerch et al. 1997), which represents a major effort to examine the transport to the atmosphere of a polar herbicide at a regional scale. Furthermore, long-range transport of atrazine from the Corn Belt to Lake Superior was examined in these studies. Atrazine and alachlor are polar and soluble. They do not bioaccumulate because they are relatively degradable, and they have not been found in significant quantities far from their source. They currently are not considered to be of concern as a persistent chemical. However, the detailed study outlined below is instructive for several reasons. First, it shows the detailed nature of sampling that may be carried out to properly examine atmospheric transport. No such detailed studies are available for other classes of compounds. Second, an immunoassay (enzyme-linked immunosorbent assay [ELISA]) was used to analyze 5000 samples for atrazine in order to carry out a large survey over a 25-state area with a reasonable cost. Third, this case study shows that there can be long-range transport even for a compound that is relatively less-persistent when it is applied in large amounts. It also shows that a compound with a low Henry's Law constant may still have considerable transport to the atmosphere (~ 1% of the applied atrazine), resulting in a regional scale distribution. The atmospheric residence time of atrazine (1 d) is shorter than compounds such as ethyl parathion (5 d) or Lindane (242 d) (van Pul et al. 1998). Considering the shorter atmospheric residence time, atrazine may not be considered to have the potential for long-range transport. However, the results of the following spatial trend monitoring studies suggest that atrazine can be transported to as far as 500 km from the point of application.

Spatial and temporal distribution of atrazine and alachlor deposition

Figure 6-2 shows the regional use pattern of atrazine and alachlor in the Midwestern U.S., and the deposition pattern of atrazine and alachlor in 1990 and 1991 is shown in Figure 6-3. The results show that the deposition of atrazine is clearly similar to its use pattern. Regarding the temporal distribution, the large fluctuation in concentration occurred throughout the 18-month sampling period (Figure 6-4). The majority of the deposition occurred during the application period and approximately 2 months after application (May through the end of July). This time period represents approximately 1 to 2 half-lives of atrazine in soil. Atrazine has a 45 to 60 d half life in soils of the Midwestern U.S. This phenomenon is due to the rapid washout from rainfall, which is usually intense during these months. The result is that atrazine is no longer available in the shallow soil for volatilization and transport.

Long-range transport of atrazine

The long-range transport of atrazine occurred during a short window of time, from mid-May until the end of July. During this time, atrazine was found to be transported at least 500 km to the north of the Corn Belt with deposition into the lakes of Isle Royale National Park, which is located in Lake Superior. The highest concentrations were found in early June with washout concentrations of approximately 0.1 to 1 μg/L (Thurman et al. 1991). Washout occurs with low-volume rainfalls in the National Park, i.e., 1 cm or less. This result, however, was considered remarkable given the fact that the distance to the nearest large cornfields was 300 to 500 km to the south. Low-altitude atmospheric transport, perhaps in thunderstorms, which are common during May through July, could have been responsible for this rapid transport. The result of this long-range transport is that concentrations of atrazine may reach 10 to 15 ng/L (ppt) in remote lakes of Isle Royale National Park (Thurman et al. 1991). Although these concentrations may not be toxic to local flora or fauna, this study clearly demonstrated the potential for long-range transport.

Degradation of atrazine

The degradation of atrazine does occur during atmospheric transport, as indicated by the occurrence of metabolites in rainfall samples. It has been shown in laboratory studies that atrazine degrades by both ozone and UV light through the alkyl side chain. An atrazine amide is formed, followed by the dealkylation of the triazine to deethylatrazine (DEA). The compound may be further degraded with ring cleavage. The amount of DEA in rainfall was measured by GC/MS in samples across the 26-state study area, as well as in the long-range transport study in Isle Royale National Park. It was found that DEA occurred in 60% of the samples that contained atrazine, typically at a concentration of about 50% of the parent compound (Lerch et al. 1997, 1999). This ratio of 0.5 for metabolite to parent was remarkably consistent throughout the study areas and is representative of the atmospheric degradation that is occurring in the troposphere. The conclusion reached from this work is that approximately 40% of the atrazine detected had not degraded and probably represented short-range transport or windblown atrazine from nearby fields into the collector. The other 40% of the atrazine was undergoing longer-range transport and was being degraded significantly in the atmosphere. The estimated half-life for atrazine in the atmosphere may be as short as a few days (van Pul et al. 1998), whereas half-lives in water varied from several days (Thurman et al. 1994; Solomon et al. 1996). However, it is still possible to move atrazine 300 to 500 km, despite its short half-life and low Henry's Law constant (2.5×10^{-4} Pa–m^{-3}/mol).

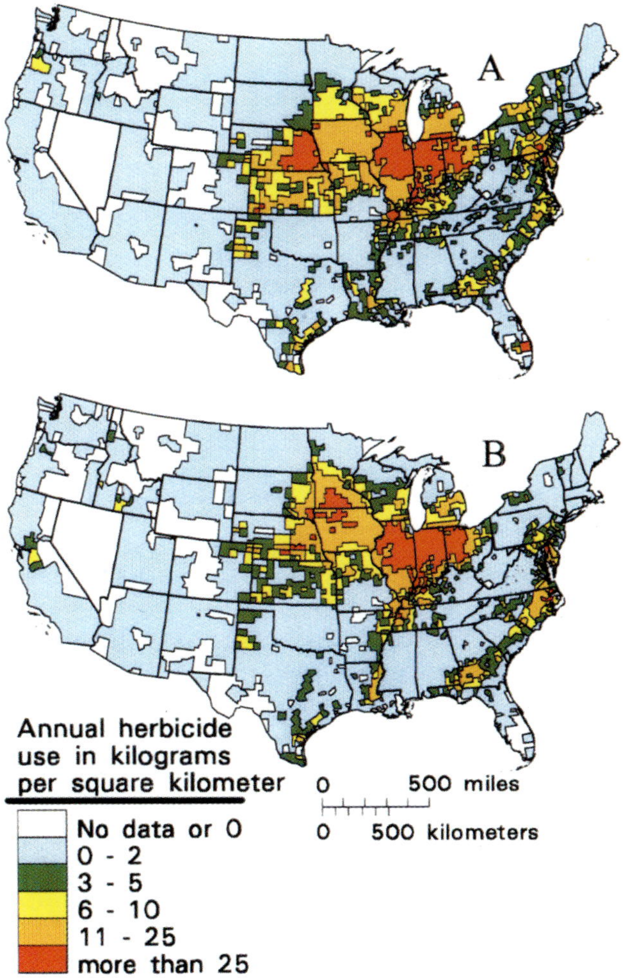

Figure 6-2 Regional use pattern of (a) atrazine and (b) alachlor in the Midwestern U.S. Reprinted with permission from Goolsby et al. (1997) ©1997 American Chemical Society

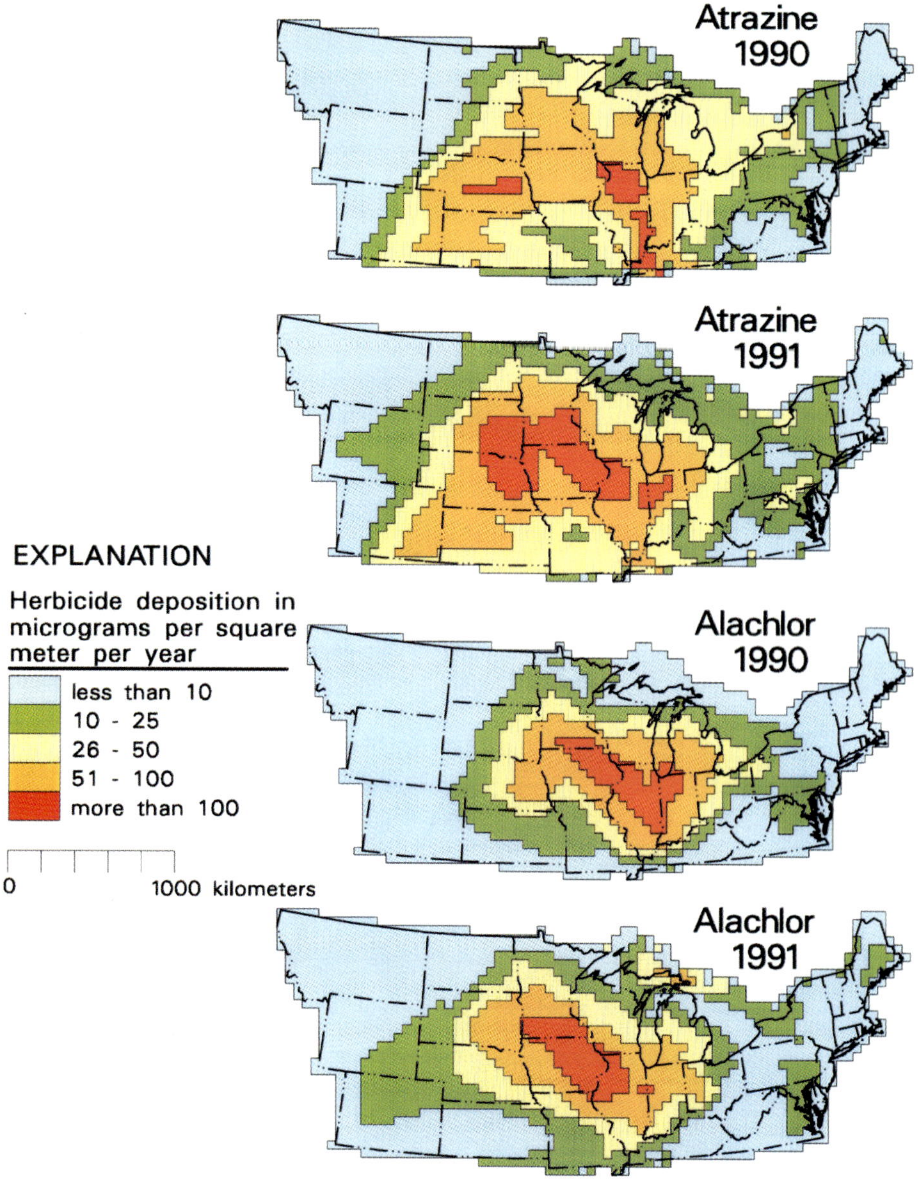

Figure 6-3 Atrazine and alachlor deposition in the Midwestern U.S. in 1990 and 1991
Reprinted with permission from Goolsby et al. (1997) ©1997 American Chemical Society

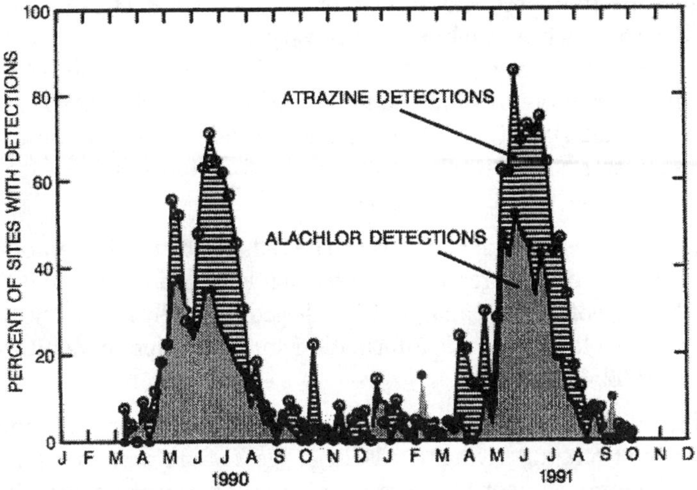

Figure 6-4 Seasonal fluctuations in atrazine and alachlor in precipitation
Reprinted with permission from Goolsby et al. (1997) ©1997 American Chemical Society

Monitoring of Temporal Trends

Temporal trend monitoring studies are valuable because they allow one to evaluate persistence of a substance by the simple concept that an increasing or constant environmental level over a period of time is an indication of persistence. Temporal trend monitoring programs are in existence in several countries, and they monitor heavy metals and organochlorine contaminants such as PCBs, DDT, and HCHs in various environmental media. The international mussel watch program and the National Status and Trends Program of the National Oceanic and Atmospheric Administration are examples of a few (O'Connor 1996). Most temporal trend monitoring studies have primarily used archived biological tissues as matrices to monitor the trends of persistent pollutants. However, in biological samples a number of factors may influence the concentration of contaminants. Before discussing how temporal trend studies are used to evaluate environmental persistence and emissions of a substance, the importance of biological variation in evaluating temporal trend data is discussed.

Short-term variability, random variations, and pooled samples

It is essential that field time-trend monitoring data are available to help elucidate the factors that control the persistence of a substance in the environment and the response of the environment to a change in inputs of chemicals and to test models. These objectives all require that quality datasets be accrued over long time periods and that long-term trends can be distinguished from short-term variations.

Concentrations in various matrices will vary over time. Air concentrations may vary hourly, diurnally, seasonally, or yearly and can be controlled by temperature, local sources, wind direction, precipitation events, ice/snow cover, etc. Concentrations in a biological matrix, such as seal blubber, can vary with age, season, diet, condition, sex, reproductive history, season, habitat, territory, etc. For example, PCB concentrations in the blubber of individual Baltic grey seals varied between 5 and 5000 mg g^{-1} (Bignert et al. 1994). Only when a clear understanding and quantitative assessment of these factors has been achieved can underlying long-term changes in concentration be interpreted.

Random variations between data points can be reduced by careful sampling and analysis. Often for economic reasons samples that have been collected individually are bulked, homogenized, and analyzed. While economically necessary, this approach can lose a lot of valuable information and resolution in monitoring studies (Bignert et al. 1992, 1994).

As highlighted above, information is needed to assess the extent of random and sample variation from underlying trends. Regardless of matrices used (air, water, sediment, soil, biota), comparison of residue levels of persistent contaminant from one year with residue concentrations from another year cannot disclose temporal variations, unless there is reasonable knowledge of the random between-year variation (Bignert et al. 1992, 1994, 1997). Figure 6-5a shows a long-term time trend of PCBs in herring collected from 1972 to 1988. The annual data points are based on individual analyses of 20 specimens/y and are the geometric means of specimens. The fish species selected were sedentary and were of the same age group. When the data was analyzed for the period from 1972 to 1988, a decline in PCB concentrations could be discerned. Using the same dataset, when the study period was separated into 2 groups of 7- to 8-y periods each, no significant change in concentration over time could be seen (Figure 6-5b). In Figure 6-5c, the same time period is covered, but a design based on sampling and analysis every third year is shown. It is apparent that very different conclusions would be drawn depending on the approach used. These issues have recently been discussed for other studies covering other compounds and time periods in relation to the problem of scattered data (Bignert et al. 1995; Olsson 1995).

Use of temporal trend data

Temporal trend studies are useful for determining whether legislative actions taken to reduce the degree of environmental pollution really have had the intended effect. They also help to develop cause–effect relationships between contaminants and affected populations. Monitoring studies can be used to trace sources of certain contaminants. Ratios of parent compounds to metabolites and proportions of components in technical mixtures have been used to infer sources. Atmospheric measurements taken in the mid-1990s at Integrated Atmospheric Deposition Network stations on the Great Lakes show a DDT/DDE ratio of less than 1, suggest-

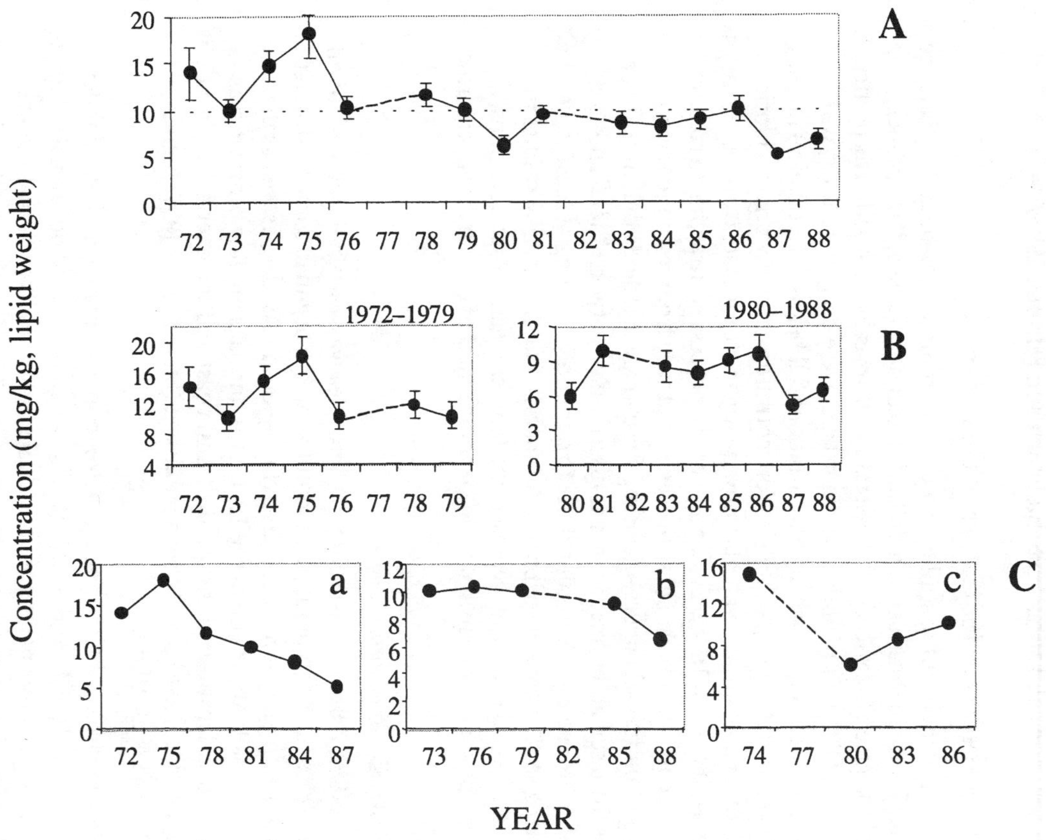

Figure 6-5 (A) Temporal trends of PCB concentrations in herring from the Baltic Sea collected between 1972 and 1988 (B) Trends in PCB concentrations in herring when the data were sepeated for 1972-1979 and 1980-1988 (C) Trends in PCB concentrations in herring when data were taken at 3-y intervals starting 1972, 1973, and 1974.

ing that a substantial portion of recent DDT comes from recycled old residues (Cortes et al. 1998). However, it should be noted that the physicochemical properties of parent compounds and metabolites change considerably, making such interpretations often confounding. Recently, enantiomer (optical isomers) ratios of certain organochlorine pesticides have been suggested to be of use in tracing sources of emissions (Ulrich and Hites 1998; Bidleman and Falconer 1999).

Monitoring studies as indicators of clearance rates

Long-term trend monitoring studies can provide information about the clearance rate of a substance in the ecosystem following ban on its usage. For instance, long-term monitoring on the shores of the Great Lakes revealed that airborne pesticides such as HCHs, DDT compounds, and chlordanes were declining slowly, with "virtual elimination" dates (when concentrations fall below detectability) ranging from 2010 to 2060 (Cortes et al. 1998). Concentrations of many organochlorine pesticides in air and surface water are now nearly in steady state with respect to gas exchange (Hillery et al. 1998). Evidently, certain processes are acting to stabilize atmospheric concentrations. These processes can vary from region to region depending upon several factors, including the climate. Methods that investigate emissions from soil and water, as well as discriminate between "new" and "old" sources of contamination, can contribute to our understanding of these issues. In the following sections, we describe how temporal trend monitoring studies in the Baltic Sea and Asian countries, representing temperate and tropical regions, respectively, provide some insights into clearance rates of certain organochlorine pesticides.

The Baltic Sea in temperate Europe

In the late 1960s, it became clear that the Baltic Sea was a severely polluted marine enclosure. Since then, temporal trend monitoring has begun in some freshwater locations in southern and remote sub-Arctic regions in Sweden, as well as in the marine Baltic itself. At the end of the 1970s, the annual sampling program was enlarged to cover locations at the Swedish west coast and some terrestrial locations. Some of these data have recently been published (Bignert et al. 1997: Bignert, Olsson, Persson et al. 1998).

The Swedish Environmental Monitoring Program uses a range of biota samples, including mammals, fish, and bird eggs. The program is based on annual sampling, specimen banking, and analysis of individual samples selected so that the site-specific matrix will be comparable between years. The sampling locations have been selected to serve as reference sites to study aerially derived contamination, while avoiding local sources. Archiving of the samples, frozen since the time of collection, has allowed cross-checking when analytical methods have changed or when new substances have attracted interest, allowing retrospective analysis. Further information on the matrices used, sampling times and composition, methods of chemical analysis and statistical treatment are given in Bignert et al. (1997) and Bignert, Olsson, Persson et al. (1998). Based on log-linear regression analysis of annual geometric means, the annual decrease in concentrations expressed as a percentage has been calculated. It is thus possible to compare the decreasing rate in various matrices and at various locations for ΣDDT, PCB, HCHs, and HCB in 15 different time series.

Concentration variations differ between different species and sampling sites, with the highest concentrations in the south and the lowest in the north. The concentrations found in the north are similar to those found in corresponding matrices in subarctic and Arctic regions at the North American continent. However, importantly, decreasing rates are similar, irrespective of matrix or sampling sites. For example, the mean rate of decrease for ΣDDT was about 10%, for α-HCH about 20%, for PCB around 7%, and for HCB around 8%. The decreasing rate for PCB in the Baltic region is remarkably similar to that found recently by Hillery et al. (1998) for PCBs in air from the Great Lakes. For pesticides, the variations in rates between species and sites were very small, but for the industrial contaminants PCB and HCB, the variations between matrices and locations varied slightly more. However, neither for the pesticides nor for the industrial contaminants were there any consistent variations between southerly or northerly sites. For α-HCH, for example, rates of decrease were more or less identical in guillemot eggs from the southern Baltic, Arctic char from a remote sub-Arctic lake, and reindeer in the vicinity of the lake. Interestingly, the rates were also similar to those reported in Arctic air for the corresponding time period in the Arctic (Jantunen and Bidleman 1996). In the case of walleye pollack collected from the Bering Sea in 1992, there were 2-, 4-, and 7-fold decreases in the concentrations of PCBs, DDTs, and HCHs, respectively, compared to those of the concentrations reported for walleye pollack collected in 1982 (Kannan et al. 1995).

In another study, the decreasing rates of the pesticides ΣDDT and α-HCH were compared over 10 y after various measures to phase out or decrease use of the compounds (Bignert et al. 1997; Bignert, Olsson, Asplund et al. 1998). Various matrices (7 time series) both from the southern and the sub-Arctic parts of Sweden were used when the percentage annual decrease in concentrations over 1971–1980 for ΣDDT and 1987–1996 for α-HCH were compared. Only minor variations and no significant differences were found when various matrices from terrestrial, freshwater, and marine environments from the north and south were compared. The rate of decrease for ΣDDT was 18% to 19% and 18% to 24% for α-HCH. The differences are remarkably small between the two substances, and there was no indication of a difference between south and north.

In a further study, the time trends were compared in eutrophic and oligotrophic lakes. The rates of decrease were quite similar between the lakes (Bignert, Olsson, Persson et al. 1998) and also between the lakes and the marine environment (Bignert, Greyerz et al. 1998). Despite the differences in sedimentation and degradation rates, similar rates of decrease in the concentrations of organic pollutants is an interesting observation.

In summary, the trend series in terrestrial, freshwater, and marine biota samples from the Swedish monitoring program indicates remarkable similarities in the decreasing pattern for the various matrices and sampling sites. The calculated annual decreases are similar to air measurements on extensive datasets of air samples from the Great Lakes (PCB) and from the Arctic (α-HCH), suggestive of regional and possibly global dispersal/mixing and clearance.

Tropics and the southern hemisphere

The occurrence of high concentrations of POPs such as HCHs, DDTs, chlordanes, and PCBs in the Arctic ecosystem, despite the ban on many of these compounds in most of the industrialized nations of the Northern hemisphere, has prompted the need to identify possible emission sources (Bidleman et al. 1989, 1992; Barrie et al. 1992; Iwata et al. 1993). Investigations conducted by numerous researchers during the last two decades suggest that mid- and low-latitude countries, particularly those in the tropical region, are possible sources of contamination in the Arctic in recent years (Oehme 1991; Tanabe 1991; Barrie et al. 1992). Although this concept is supported by the current high consumption of organochlorine pesticides in tropical developing countries, which enhances food production and eradicates vector-borne diseases (Kannan et al. 1992; Mowbray 1988), there is no direct evidence of transport of POPs from tropical Asian countries to the Arctic. The cumulative production of DDTs and HCHs in India until 1995 was 500,000 t and 1 million t, respectively (Kannan et al. 1995), contributing to a significant proportion of the global usage (Voldner and Li 1995). Concentrations of POPs in environmental matrices in certain Southeast Asian countries, such as Japan, have been slowly declining (Loganathan et al. 1989, 1993). Despite the lack of well-designed time trend monitoring studies in tropical Asian countries, a compilation of information generated by various researchers has indicated that levels have been slowly declining or have been at a steady state depending on the location and matrices. Data from the food-monitoring surveys in India have shown that the concentrations of DDTs have been declining in recent years (Kannan et al. 1997). This is attributed to the ban on the use of DDT in agriculture. However, concentrations in biota did not appear to have declined until recently (Tanabe et al. 1998; Senthilkumar et al. 1999). Concentrations of DDTs, HCHs, PCBs, and chlordanes in dolphins collected from the River Ganges in India showed an increase in concentrations from 1988–1992 to 1994–1996 (Figure 6-6). Iwata et al. (1993) reported that the atmospheric concentrations of HCHs and DDTs in the Arabian Sea were comparable between 1976–1977 and 1989 whereas the concentrations in the eastern Indian Ocean and Southern Ocean have declined since the early 1980s. These results indicate the India has been a major emission source of HCHs and DDTs in southern Asia. Concentrations of PCBs have increased in minke whales from the Antarctic during 1984–1993, while DDT declined slowly (Aono et al. 1997). In contrast, minke whale from the northern hemisphere showed declining concentrations of DDTs, HCHs, and HCB from 1987 to 1994, while demonstrating an increase in PCB concentrations (Aono et al. 1997). These results indicate that PCBs are still being discharged from northern and southern hemisphere countries.

Although the clearance rates of organochlorine pesticides are influenced by climatic and geographic features (Loganathan and Kannan 1991, 1994), field studies conducted in tropical regions suggest that the environmental residence times of POPs are shorter in the tropics than in temperate regions (Sleicher and Hopcraft 1984, Tanabe et al. 1991).

Comparable data between continents and regions of the southern hemisphere are lacking or very limited (Loganathan and Kannan 1994, Connell et al. 1999). A systematic comparison of data for DDTs and HCHs on the northern and southern hemisphere indicated that the northern hemisphere is generally more contaminated than is the southern hemisphere. In the southern hemisphere, concentrations and patterns of POP contamination appear to depend on usage rates and/or methods of disposal. It was also noted that the process of equilibration between the two hemispheres is relatively slow (Connell et al. 1999). The critical factor for further knowledge on the distribution and control of POPs in the southern hemisphere is to obtain comparable measurements between countries and regions using selected environmental substrates from different land uses (e.g., urban, rural, and remote) (Connell et al. 1999).

Monitoring metabolites and degradation products

Chemicals may be broken down in the environment or biota by various degradation pathways, including hydroxylation, dechlorination, carboxylation, hydrolysis, and in some cases, sulphonation. The parent substance may yield similarly (or more) persistent byproducts (e.g., DDT → dichlorodiphenyldichloroethylene [DDE]) or more-polar, less-persistent byproducts (e.g., phenanthrene → hydroxylated byproducts). It can be informative to monitor the ratio of parent:degradation product to gain clues about transport processes and the age of the chemical. For example, DDT (a parent compound) degrades to DDE or dichlorodiphenylethane (DDD) via the loss of chlorine and/or hydrogen. The rate of this degradation is generally understood; thus, the ratio of parent to metabolites gives information on the chronology of input of DDT. Knowledge on metabolism and environmental degradation of POPs is scarce, and a few attempts have been made to evaluate the molecular stability in terms of internal energy (Tremolada et al. 1992, 1993). As mentioned above, certain pesticides such as HCHs are emitted both as technical mixtures containing α-, β-, γ-, and δ-isomers and as the purified γ-isomer. In addition, environmental transformation of γ- to α-isomer is reported to occur (Malaiyandi and Shah 1984), which causes complications in using the monitoring data to interpret the sources of emission.

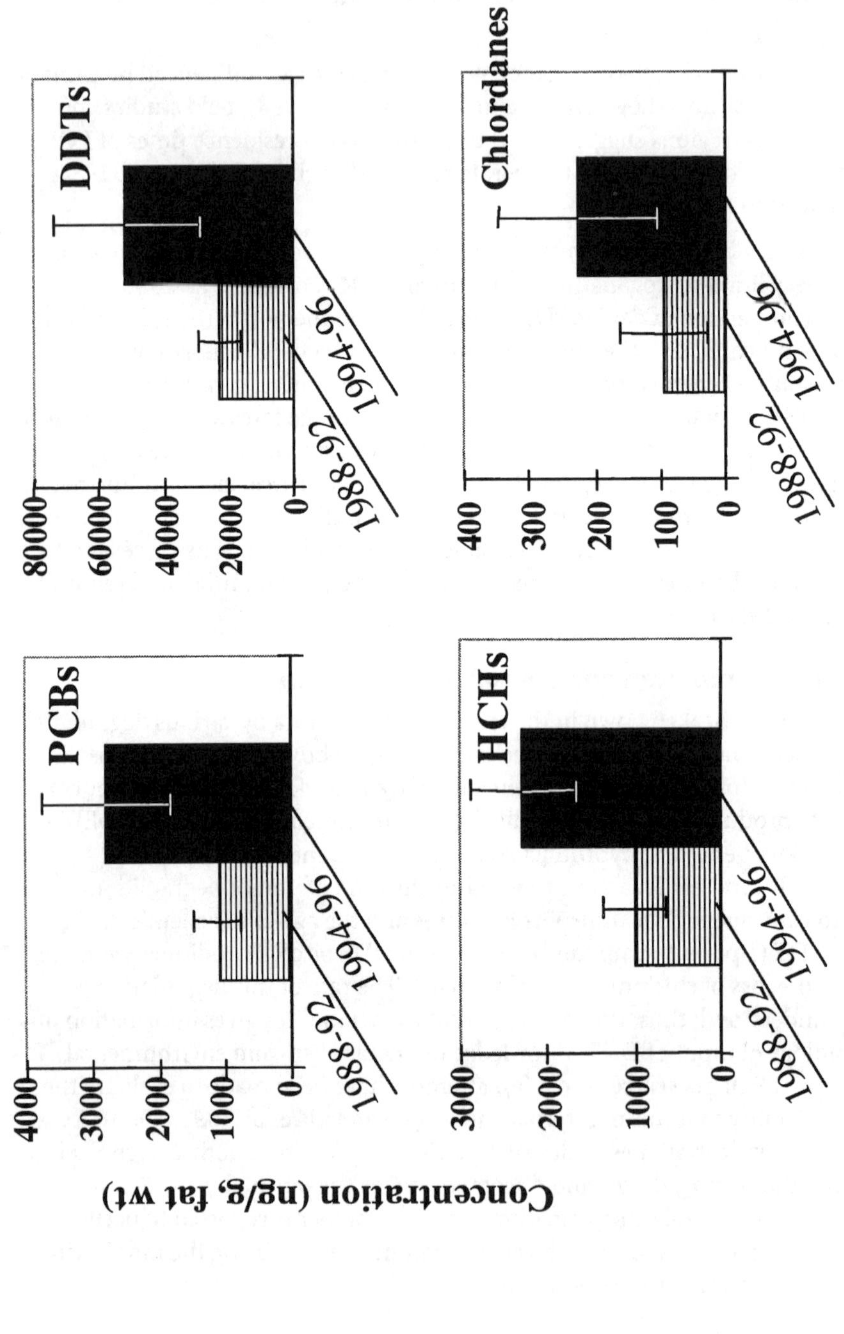

Figure 6-6 Concentrations of PCBs, DDTs, HCHs, and chlordanes (ng/g, fat wt) in the blubber of dolphins collected from the River Ganges, India, between 1988–1992 and 1994–1996 (Senthilkumar et al. 1999)

The major issue is that monitoring studies have routinely focused on tracing parent compounds rather than their metabolites. Some relatively polar, nonpersistent compounds can be degraded to produce relatively persistent metabolites in the environment. Surfactants, particularly nonionic surfactants, are a class of organic compounds that contain both a hydrophobic and a hydrophilic component. For example, nonylphenol ethoxylate (NPE) contains nonylphenol at the hydrophobic end and ethylene oxide on the other end of the molecule. The ethoxylate end degrades rapidly in the environment, leaving a more recalcitrant product, nonylphenol, which is fairly stable in the environment. While physicochemical properties of the parent compound may tend to suggest its lower persistency, the breakdown product can be persistent in the environment. Although it may not be practical to study all metabolites of a compound, some effort should be made to outline important degradation products that contribute to toxicity. Many of the simple box models that are proposed to design criteria for persistence and long-range transport do not consider degradation products. Monitoring can be a valuable means of tracing a persistent metabolite of a nonpersistent parent compound. It is important to understand the chemical degradation pathways of a substance of concern while conducting a monitoring study.

Interpretation of Field Monitoring Data

Often field and monitoring data are presented as simple concentration data. However, monitoring studies have been used recently to interpret environmental behavior. For example, concentrations in air can be related to parallel samples of water or soil, and calculations of the instantaneous net direction of flux or closeness to equilibrium partitioning can be made. Iwata et al. (1993) and Ballschmiter (1992), for example, estimated air:water equilibrium status for a range of POPs compounds in seawater from a wide range of locations worldwide. Similar calculations have been performed for HCH in Arctic air and seawater (Jantunen and Bidleman 1995), for a range of POPs in the Great Lakes (Hornbuckle et al. 1995), and for PCBs in air and soils on a latitudinal gradient in Europe (Ockenden et al. 1998). Efficient use of monitoring data can be made if multimedia monitoring is performed (e.g., Sweetman and Jones 1999). Advantages of multimedia monitoring over single-medium monitoring are discussed below.

Single-medium and multimedia approaches

Most of the environmental monitoring studies of POPs have focused on examining concentrations or trends in a single medium at a given time and location. However, depending on the substance's physicochemical properties and geographic features including edaphic factors and meteorological conditions, it tends to partition to one compartment or another. For example, theoretically a substance with high K_{OW} is not expected to partition into air. At the same time, if its water solubility is low, it will tend to have a high air–water partition coefficient (which is essentially the ratio of vapor pressure to water solubility). *p,p'*-DDT has a very low vapor pressure of 6.14×10^{-5} Pa. Its low solubility in water (6 μg/L) causes its air–water partition coefficient to be substantial (1.46×10^{-3}). Hence, *p,p'*-DDT in water will evaporate fairly rapidly (Lee and Mackay 1995). Lindane (γ-HCH) is more volatile (0.021 Pa), but its air–water partition coefficient is one-tenth that of *p,p'*-DDT because of its high solubility in water. Further, the physicochemical properties that control the partitioning behavior of a chemical can vary depending on the prevailing weather conditions. The vapor pressure of γ-HCH at 15 °C is 0.00472 Pa, whereas at 35 °C it is about 18 times higher (0.0846 Pa) (Lee and Mackay 1995). These properties result in a location- and compound-specific distribution and redistribution among environmental compartments. While environmental monitoring studies provide realistic estimates of half-lives of chemicals under field conditions, monitoring a single medium, and therefore the half-life from a single medium, may under- or overestimate the compound's overall potential for persistence. Therefore, in order to better understand persistence and environmental dynamics, it is essential to monitor the contamination levels in different environmental compartments simultaneously. The following examples illustrate the benefits of multimedia monitoring.

A study conducted in a paddy field in South India examined the half-lives of HCH isomers in air, water, soil, and paddy plants after a single topical application of technical HCH (Tanabe et al. 1991). Air, water, soil, and paddy plants were collected at 5-d intervals for over 40 d after application. About 90% of the applied HCH partitioned to air, leaving less than 10% of the residues in soil, water, and paddy plants in 2 weeks after application. The half-life in air was estimated to be 5 d, whereas the residue concentrations in water and soil decreased by about 50% in 5 and 10 d, respectively. Similarly, it was found in other studies that 50% of the HCH applied to soil in Delhi, India, was lost in 30 to 45 d (Samuel et al. 1988), while 94% was lost 60 d after its application (Kaushik 1989). The half-life of the most stable HCH isomer, β-HCH, was 22 d in a flooded soil in India (Panda et al. 1988). Thus, if the half-life measured in a single medium, (e.g., soil) were to be used to develop criteria for persistence, HCH would not fall under the criteria established by Environment Canada, which requires a half-life of ≥182 d in soil in order to be considered persistent (Environment Canada 1995). On the other hand, if the atmospheric half-lives are considered, i.e., 5 d in the tropical field conditions, HCH would be classified as persistent, as it exceeds the ≥ 2-d criteria of Environment Canada. The half-lives of a chemical in different environmental media vary dramatically depending on the climatic features of a location. The half-life of γ-HCH is only 75 d in warm ocean waters, for example, whereas it is > 100 y in cold ocean waters at 0 °C (Ngabe et al. 1993). These results suggest the need for multimedia measurement of concentrations and half-lives to derive overall environmental half-lives for developing persistence criteria.

Multimedia monitoring and equilibrium partitioning

Multimedia monitoring studies of DDTs, HCHs, and chlordanes in tropical regions have shown that their atmospheric concentrations were several-fold greater than those reported in temperate regions. In contrast, despite the continuing use of these pesticides in tropical regions, their residue levels in sediment and fish tissues are relatively low and are comparable to those found in temperate regions, where these compounds were banned about 25 years ago (Iwata et al. 1994; Kannan et al. 1995). These studies suggest that the persistent chemicals used in tropical areas are rapidly dissipated during their initial transport through inland air and water, leaving lower residues in sediments and aquatic biota. Relatively short residence times of organochlorine pesticides in sediments in the tropics are illustrated by a plot of the concentration ratio of organochlorine pesticides in sediments (normalized to organic carbon) to water (S/W) at different latitudes (Figure 6-7). The plot of Log (S/W) against the latitude of sampling locations indicated smaller S/W ratios at lower latitudes. This suggested that the partitioning of organochlorine pesticides to sediments is less pronounced in the tropics than in temperature regions (Iwata et al. 1994), implying the role of meteorological conditions such as temperature in

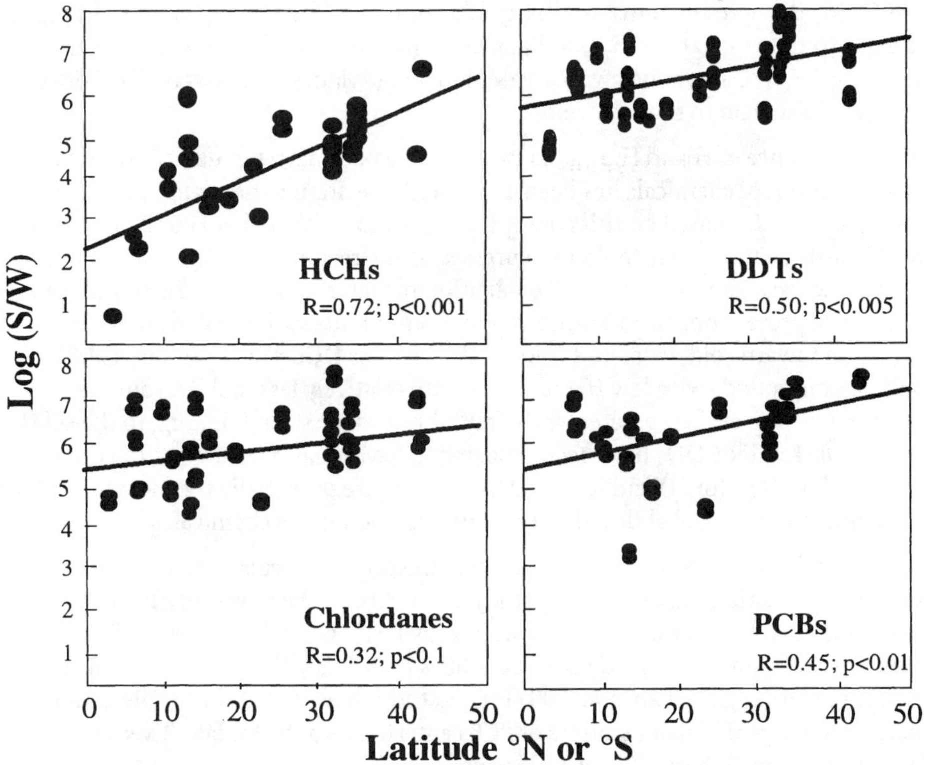

Figure 6-7 Relationship between the concentration ratios of PCBs, DDTs, HCHs, and chlordanes in sediment and water (S/W) and the latitudes of sampling locations (Iwata et al. 1994)

governing the distribution. Therefore, the half-lives of POPs derived from sediment monitoring in the tropics would underestimate their overall environmental persistency.

Multimedia monitoring studies are also useful to evaluate partition coefficients such as K_{OW}, K_{OC}, K_{AW}, and K_{OA}, which are used to examine the factors that influence distribution, dynamics, and fate of organic pollutants in the environment. Monitoring studies designed to elucidate the global distribution of POPs have shown that semi-volatile contaminants like HCB and HCHs have concentrated in Arctic waters, while DDTs and PCBs were higher in low- to mid-latitude regions (Calamari et al. 1991; Iwata et al. 1993). It has been shown that the semi-volatile contaminants released in the tropical low latitude regions would be redistributed to higher latitudes according to contaminant specific transportability (Wania and Mackay 1993; Iwata et al. 1993). Since K_{AW} or the related Henry's Law constant (H) is a measure of equilibrium partitioning of a compound between gas and water phases, this parameter has been used to express the movement of a compound between atmosphere and hydrosphere by comparing the theoretical or equilibrium value of K_{AW} with the calculated ratio of concentrations, using the field measurements of concentrations in air and water (Ballschmitter 1992; Iwata 1994; Mackay et al. 1996). In addition, the transportability of a compound between gas and solid phases is expressed by the ratio of K_{AW} to K_{OW}, i.e., K_{AO} which is usually reported as the reciprocal K_{OA}. Compounds with greater K_{AW}/K_{OW} values tend to transfer more to the gas phase than to the solid phase.

The importance of H and H/K_{OW} as two valuable parameters in describing global redistribution of chemicals has been discussed in relation to data measured in the field (Figure 6-8). Based on this concept, compounds with higher H or K_{AW} values and K_{AW}/K_{OW} ratios that could be expressed as the reciprocal of K_{OA}, for example, chlorofluorocarbons, do not tend to partition to water or to solids. Such compounds tend to be present only in the atmosphere. Compounds such as HCHs tend to partition toward cold water and cold solids, whereas HCB prefers air and cold solids. Compounds with low H and H/K_{OW} are relatively less mobile from the sources because of their greater water solubility and/or strong binding to the solid phase. DDT and PCDDs have been reported to have greater binding affinity to solids. Thus, K_{AW} (or H) and K_{OW} and their ratios are potentially the most important descriptors for the global distribution and partitioning of a chemical.

Multimedia monitoring is useful for tracing the sources of contamination. As an example, a multimedia monitoring study conducted in the sewer district in Buffalo, New York, involved the analysis of soils, atmospheric wet and dry deposition, sanitary flows, and combined sewer overflows to evaluate the sources of PCBs to the Buffalo River (Loganathan et al. 1997). This study showed that the locally contaminated soils were the major sources of PCBs entering into the Buffalo River as opposed to atmospherically derived inputs.

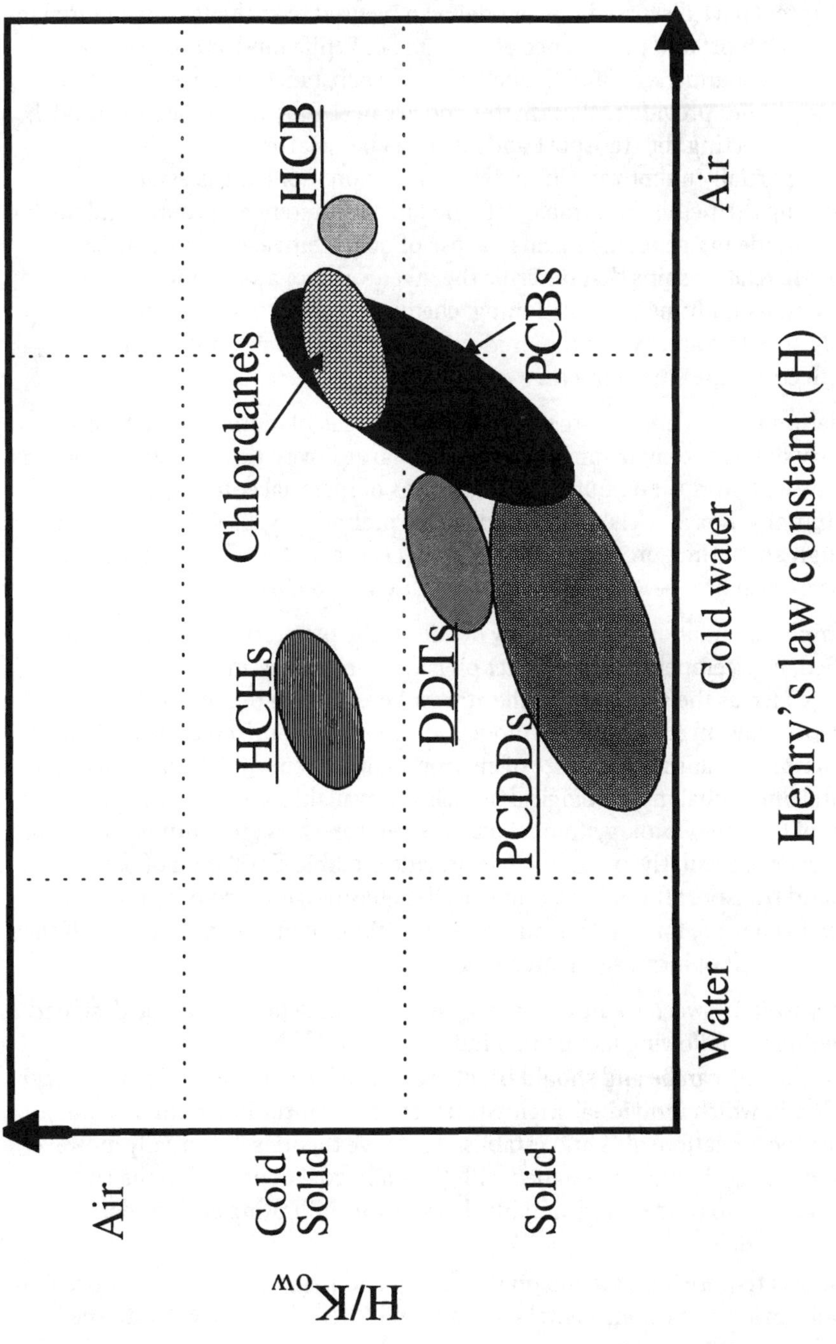

Figure 6-8 Schematic illustration of redistribution of certain organochlorine compounds based on their partitioning coefficients among air, water, and solids and the influence of temperature (Iwata 1994)

Monitoring Data and Environmental Modeling

Earlier chapters have discussed how models can be used to evaluate the potential for long-range transport and persistence of a chemical. While models are useful in various aspects of understanding the behavior of a substance in the environment, monitoring studies provide real estimates and are necessary to validate the models. Mechanisms affecting the transport and fate of a chemical are extremely complex and are still partially unknown. Ultimately, simulation modeling is essential for understanding their emission, transport, and fate in different environmental media. Simulation modeling generally means the use of quantitative, computer-based, deterministic relationships that describe the inventories of a substance in the environment, its multimedia partitioning, chemical reaction and transport to potentially sensitive areas (e.g., the Arctic), and its possible accumulation and adverse effects on the environment.

All simulation models are, in essence, mass-balance calculations. The simplest models may depict the environmental media by large boxes, e.g., the atmosphere as a single box with in-flux and out-flux calculations of material entering and exiting the box. Initially, box models may provide some preliminary, useful information, even though an implicit property of a box model is the uniform concentration of chemicals in each box — a condition that is easily violated for several POPs.

More-complex simulation models allow details of the property of each medium to be specified, e.g., temperature and other physical parameters and advection parameters such as the wind flow in the atmosphere and the ocean currents. Consequently, advanced simulation models are generally more credible in their reconstruction of transport and fate of environmental chemicals. Today, advanced continental and global meteorological models are available and routinely provide a depiction of local meteorology (model outputs and forecasts are routinely available on the Internet). Similarly, ocean models provide credible simulation of ocean currents and transport flows. These advanced meteorological and ocean models have often been used, for example, in simulating global climate and expected future changes as a result of increases in greenhouse gases.

The relationship between monitoring programs and modeling should be designed while keeping the following factors in mind:

- Monitoring can be and should be, at least initially, a tentative, "exploratory" effort in which individual scientists use scientific intuition to infer semi-empirical relationships and establish tentative theories. Eventually, however, monitoring should be designed with the main goal of providing the correct input data to models and suitable datasets for calibrating and validating these models.
- Models require information on the timing and magnitude of emissions. Some monitoring and experimental efforts should therefore be devoted to the collection of these data.

- Models require correct parameterization of physical (e.g., evaporation, ground deposition, sedimentation) and chemical properties (e.g., reaction rates) in the different environmental media. Therefore, some monitoring should be designed in order to collect and measure appropriate environmental data.
- Models provide, as outputs, the concentrations and fluxes in different media. Therefore, one of the main monitoring and experimental efforts should be a well-organized collection of concentrations throughout the world and, especially, in those areas considered the most sensitive and vulnerable. Data should include both regular, continuous measurements and well-focused intensive campaigns. These data should then be used to calibrate, validate, and improve models.

Interactions between monitoring and modeling

Models that predict the physicochemical behavior of chemicals and the fate of chemicals in the environment are a mathematical representation of the real world. In the simplest models, the environmental heterogeneity and temporal dynamics have been substituted by a fixed generic environment and by an assessment of a steady-state situation at local, regional, or global levels.

To provide a more accurate fate assessment, a fundamental interaction between monitoring and laboratory data is needed to parameterize and validate the models for the chemical and environment of interest. Only then can the models be used to extend laboratory and monitoring data and provide a more comprehensive and validated exposure assessment.

Laboratory and field data are frequently used to provide parameters that are necessary to run fate models and develop quantitative structure-activity relationship (QSAR) models that can predict various physicochemical parameters. QSAR models have been successfully used for nonpolar, relatively hydrophobic compounds such as hydrocarbons. However, these models have a high degree of uncertainty when used to predict the physicochemical behavior of surfactants, ionizable compounds, persistent degradation products, and polymers. Experimentally derived data are the most reliable source of the physicochemical parameters for these chemicals.

Models are typically developed for specific chemicals, classes of chemicals, or environments and embody the current understanding of the chemical and the environmental compartment of interest. To ensure accurate predictions, the model results need to be validated by comparing them to monitoring and laboratory data. Each time the model predictions are validated with monitoring and/or laboratory data, confidence in the prediction increases for the environment and class of chemicals considered. However, this does not mean that the model yields accurate or even reasonably accurate predictions for all environments or all chemical classes, unless the validation process has included a wide range of environments and classes of chemicals.

Once models have been validated for the class of chemical and specific environment of interest and parameterized using appropriate field and laboratory data, they can be applied with a certain degree of confidence

- to extend these data to larger geographical areas,
- to provide statistical distributions of concentrations,
- to explore the effect of variations in the environmental and chemical properties (sensitivity analysis),
- to evaluate "what-if" scenarios, and
- to assess responsibility.

The specific chemical under investigation and the processes to which it will be subjected in the environment will determine the minimum parameter set or the minimum number of environmental properties for which information must be given in order for the data to be useful for risk assessment or model validation. More information will be required for some chemicals than for others. For example, if a chemical ionizes or perhaps hydrolyzes at a given pH, data on the pH of the environment will need to be specified. For other chemicals, this information may not be necessary. Alternatively, a chemical may undergo processes in the environment, such as adsorption or biodegradation, whose mechanism may depend on the environmental parameters. Thus, for many organic chemicals, the amount of organic carbon present in soil or sediment will influence adsorption. Additionally, ionic organics may adsorb to charged soil constituents, if these are present at the pH present in the soil. If biodegradation of the compound takes place, the existence of aerobic or anaerobic environmental conditions will be necessary to understand the monitoring result and to correctly position it in either an exposure assessment or a specific modeling context.

The history of chemical release or deposition at the monitoring site must be known in order to correctly position the monitoring data or to determine the appropriateness of the dataset for the modeling purpose. Information is needed concerning the total load emitted and the spatial and temporal emission pattern. This information unfortunately is often missing. This is particularly important in the case of POPs, for which the sources and quantity emitted are generally difficult to define. Background concentrations, as well as the release history of the site, would have some relevance if determination of the total environmental load were the object of the modeling. This is obviously the case for the global modeling of POPs.

The analytical accuracy of an existing dataset may not a priori determine whether it is useful for exposure assessment or model validation purposes. The type of model used, and the accuracy to which the model predictions are either possible or desired must also be considered. Because obtaining monitoring data is much more expensive than is running an existing model, it is probably most cost-effective to accept that all existing environmental monitoring datasets that satisfy the minimum parameter constraints and that meet minimum analytical standards have some degree of use for environmental exposure assessment and model validation purposes. The degree of usefulness of the dataset should be specified in the context of the model itself. However, it must be established that the model is valid for the conditions that pertain to the dataset before the data can be used to validate the model.

In summary, existing data should be used, whenever possible, for environmental exposure assessment and model validation purposes. However, it is essential that the dataset contain sufficient information concerning the chemical and the environment, the release history of the chemical with respect to the monitoring site, and the analytical methods used to determine the concentrations. The definition of sufficiency may vary with the specific chemical and also will vary with the application, i.e., the type of environmental exposure assessment or model validation under consideration.

Recommendations and Data Gaps

Specimen banking

In order to evaluate temporal trends retrospectively, matrices are needed that can provide information about contaminant concentrations in the past. Until now this type of information has mostly been achieved on an ad hoc basis, by analysis of samples collected and saved for other purposes in museum collections or scientific collections. An important question often has been whether the samples have maintained their integrity (i.e., have they been contaminated or lost the analyte of interest during storage). However, some countries have been proactive in establishing Environmental Specimen Banks, with carefully collected, stored, and catalogued specimens preserved for future chemical analysis. Attempts are being made to harmonize the structure of the collections and the format for record-keeping in order to increase international access to the collections. Recommendations have also been made to develop procedures for sampling, preparation, and storage (Olsson and Bignert 1997; deWit et al. 1997). Specimen banks are clearly invaluable tools in enabling scientists to monitor environmental change and investigate new contaminants retrospectively.

Global networking

One of the issues that has emerged from the earlier sections of this chapter is the desirability of comparing trends in different locations. Ideally this would be achieved by cross-checking data from well-studied and well-characterized ecosystems from different parts of the world. Key sampling and environmental monitoring networks should be established in various parts of the globe representing different climatic and geographic conditions that would enable efficient use of funds available for global environmental monitoring.

Oceanic transport

While it has been acknowledged that oceans are the ultimate reservoirs for several manmade chemicals used in the terrestrial environment, movement and transport of these chemicals through ocean currents is less known. Fischer et al. (1991) studied movement of HCH through the water flow in the Atlantic Ocean and suggested that ocean currents also play a role in the transport of persistent pollutants. Similarly, Wania and Mackay (1999) noted the importance of oceanic transport of α-HCH to the Arctic Ocean. Although the atmosphere has been considered to be a major route of transport (Preston 1992), further studies are needed on the role of ocean currents and riverine runoff in the transport of contaminants to remote areas.

Migratory animals

The possibility that migratory birds (Tanabe et al. 1998), fish, or marine mammals could transfer considerable quantities of persistent and bioaccumulative contaminants is uncertain. This pathway is important for exposure of humans and wildlife who consume tissues of migrating birds, fish, or mammals. The relative importance of this issue can be assessed by comparing the mass of chemical in the migratory population (concentrations $\times$ biomass) with inputs via air or ocean currents. Further studies are needed on this topic.

Summary and Conclusions

Monitoring studies and surveys have made essential contributions to the discovery, description, and understanding of the problem with contaminants that have potential for long-range transport and persistence. In the early 1960s, PCBs were first detected in very remote regions of the Arctic and Antarctic. Since then, systematic monitoring studies have shown many substances to be subject to long-range atmospheric transport. Monitoring data obtained in fish and other biota indicated that these compounds are also liable to bioaccumulate into the food chain. This possibility initiated ecological risk assessments for a series of compounds that are now strictly regulated, their use having been severely restricted or banned.

Well-defined, multimedia monitoring programs can demonstrate the effectiveness of the reduction or phaseout measures taken by the regulatory authorities. Monitoring data for DDT or PCBs over recent years collected in different areas of the world generally show reductions in the environmental concentrations. There are underlying reductions in emissions of these organochlorine compounds, which are disappearing at a rate related to their global half-life in the environment.

Monitoring data, if collated in the appropriate way and over a sufficient period of time, can help to define environmental accumulation and removal and clearance times of chemicals. To do so effectively, well-planned multimedia monitoring approaches should be developed, covering significant areas of the world in order to be representative of a regional and/or global pattern. Well-structured monitoring programs should take account of the need for storage/retrospective data and should be designed to test and validate models.

The degree of confidence we can have in using monitoring data for model validation depends not only on the quality and representativeness of monitoring data but also on the quality of the model considered. Simple models can be validated with a limited set of monitoring data, but if the sophistication of the model increases, the amount and the quality of the monitoring data should be adapted accordingly.

Nevertheless, despite the importance of monitoring in assessing persistence and long-range transport, there are severe gaps in the existing monitoring programs. These include the following:

- Only limited retrospective monitoring data are available in all compartments (air, water, soil, sediments, and biota). Furthermore, there are gaps in monitoring of many compounds both in time and space.
- Monitoring has been preoccupied with measuring concentrations rather than with determining data on their global use and release. Thus, there is a lack of geographic information on use and emissions.
- Metabolites have been not emphasized in many monitoring programs. Therefore, we have missed critical information on mass balance, half-life, and fate.
- There is a lack of well-organized monitoring studies in the southern hemisphere. Moreover, gaps exist between modeling and detailed evaluation by monitoring at the global scale. There is a lack of knowledge or a reluctance to accept new monitoring methods and techniques (e.g., ELISA, biosensors, SPMDS).
- Several case studies with multimedia monitoring suggest that Henry's Law constant and octanol–water partition coefficient determine the partitioning behavior of a substance in the environment. These values are important in evaluating environmental movement.
- In addition to the above mentioned parameters, total quantity of global usage and emission are also important in designating a chemical. This information is often the most difficult to obtain, hampering our ability to model.

References

Alcock RE, McLachlan MS, Johnston AE, Jones KC. 1998. Evidence for the presence of PCDD/Fs in the environment prior to 1900 and further studies on their temporal trends. *Environ Sci Technol* 32:1580–1587.

Aono S, Tanabe S, Fujise Y, Kato H, Tatsukawa R. 1997. Persistent organochlorines in minke whale (*Balenoptera acutorostrata*) and their prey species from the Antarctic and the North Pacific. *Environ Pollut* 98:81–89.

Ballschmiter K. 1992. Transport and fate of organic compounds in the global environment. *Angew Chem Int Ed Engl* 31:487–515.

Barrie LA, Gregor D, Hargrave B, Lake R, Muir D, Shearer R, Tracey B, Bidleman T. 1992. Arctic contaminants: sources, occurrence and pathways. *Sci Total Environ* 122:1–74.

Bidleman TF, Falconer RL. 1999. Using enantiomers to trace pesticide emissions. *Environ Sci Technol* 33:206A–209A.

Bidleman TF, Patton GW, Walla MD, Hargrave BT, Vass WP, Ericksson P, Fowler B, Scott V, Gregor DJ. 1989. Supply of chlorinated camphenes to Arctic Ocean fauna by atmospheric transport. *Arctic* 42:307–313.

Bidleman TF, Alegria H, Ngabe B, Green C. 1998. Trends of chlordane and toxaphene in the ambient air of Columbia, South Carolina. *Atmos Environ* 32:1849–1856.

Bignert A, Göthberg A, Jensen S, Litzén K, Odsjö T, Olsson M, Reutergårdh L. 1992. The need for adequate biological sampling in ecotoxicological investigations: a retrospective study of twenty years pollution monitoring. *Sci Total Environ* 128:121–139

Bignert A, Litzén K, Odsj' T, Olsson M, Persson W, Reutergårdh L. 1995. Time related factors influence the concentrations of DDTs, PCBs and shell parameters in eggs of Baltic guillemot (*Uria alge*) 1861–1989. *Environ Pollut* 89:27–36.

Bignert A, Odsjo T, Olsson M, Asplund L, Eriksson U, Haggbert L, Litzen K, deWit C, Rappe C, Asplund K. 1997. The Swedish Environmental Specimen Bank application in trend monitoring of mercury and some orgonohalogenated compounds. *Chemosphere* 34:2059–2066.

Bignert A, Olsson M, de Wit C, Litzén K, Rappe C, Reutergårdh L. 1994. Biological variation: an important factor to consider in ecotoxicological studies based on environmental samples. *Fresenius J Anal Chem* 348:76–85.

Bignert A, Olsson M, Persson W, Jensen S, Zakrisson S, Litzén K, Eriksson U, Häggberg L, and Alsberg T. 1998a. Temporal trends of organochlorines in Northern Europe 1967–1995. Relation to global fractionation, leakage from sediments and international measures. *Environ Pollut* 99:177–198.

Bignert A, Olsson M, Asplund L, Häggberg L. 1998b. Fast initial decrease in environmental concentrations of OCs: A result of atmospheric degradation? Part I. Proceedings from the 18th Symposium on Halogenated Environmental Organic Pollutants: , 17–21 Aug 1998; Stockholm, Sweden, In: DIOXIN–98. Transport and Fate I. N. Johansson, Å. Bergman, D. Broman, H. Håkansson, B. Jansson, E. Klasson Wehler, L. Poellinger and B Wahlström, editors. *Organohalogen Compounds*. 36:373–376.

Bignert A, Greyerz E, Olsson M, Roos A, Asplund L, Kärsrud A-S. 1998c. Similar decreasing rate of OCs in both eutrophic and oligotrophic environments: A result of atmospheric degradation? Part II. Proceedings from the 18th Symposium on Halogenated Environmental Organic Pollutants: 17–21 Aug 1998; Stockholm, Sweden, In: DIOXIN–98. Transport and Fate PI. N. Johansson, Å. Bergman, D. Broman, H. Håkansson, B. Jansson, E. Klasson Wehler, L. Poellinger and B Wahlström, editors. *Organohalogen Compounds*. 36:459–462.

Blais JM, Schindler DW, Muir DCG, Kimple LE, Donald DB, Rosenberg B. (1998). Accumulation of persistent orgnochlorine compounds in mountains of western Canada. *Nature* 359:585–588.

Bright DA, Dushenko WT, Grundy SL, Reimer KJ. 1995a. Evidence for short-range transport of polychlorinated biphenyls in the Canadian Arctic using congener signatures of PCBs in soils. *Sci Total Environ* 160/161:251–263.

Bright DA, Dushenko WT, Grundy SL, Reimer KJ. 1995b. Effects of local and distant contaminant sources: polychlorinated biphenyls and other organochlorines in bottom-dwelling animals from an Arctic estuary. *Sci Total Environ* 160/161:265–283.

Calamari D, Bacci E, Focardi S, Gaggi C, Morosini M, Vighi M. 1991. Role of plant biomass in the global environmental partitioning of chlorinated hydrocarbons. *Environ Sci Technol* 25:1489–1495.

Calamari D, Tremolada P, Notarianni V. 1995. Relationship between chlorinated hydrocarbons in vegetation and socioeconomic indices on a global scale. *Environ Sci Technol* 29:2267–2272.

Connell DW, Miller GJ, Mortimer MR, Shaw GR, Anderson SM. 1999. Persistent lipophilic contaminants and other chemical residues in the southern hemisphere. *Crit Rev Environ Sci Technol* 29:47–82.

Cortes DR, Basu I, Sweet CW, Brice KA, Hoff RM, Hites RA. 1998. Temporal trends in gas-phase concentrations of chlorinated pesticides measured at the shores of the Great Lakes. *Environ Sci Technol* 32:1920–1927.

deWit C, Rappe C, Aslund K. 1997. The Swedish Environmental Specimen Bank application in trend monitoring of mercury and some orgonohalogenated compounds. *Chemosphere* 34:2059–2066.

Duarte-Davidson R, Sewart A, Alcock RE, Cousins IT, Jones, KC. 1997. Exploring the balance between sources, deposition, and the environmental burden of PCDD/Fs in the UK terrestrial environment: an aid to identifying uncertainties and research needs. *Environ Sci Technol* 31:1–11.

[ECETOC] European Centre for Ecotoxicology and Toxicology of Chemicals. 1999. Monitoring and modelling of industrial organic chemicals, with particular reference to aquatic risk assessment. Technical report No 76, European Centre for Ecotoxicology and Toxicology of Chemicals, Brussels.

Environment Canada. 1995. toxic substances management policy: persistence and bioaccumulation criteria. Final report of the ad hoc science group on criteria.

Fischer RC, Kramer W, Ballschmiter K. 1991. Hexachlorocyclohexane isomers as markers in the water flow of the Atlantic Ocean. *Chemosphere* 23:889–900.

Gaggi C, Bacci E. 1985. Accumulation of chlorinated hydrocarbon vapors in pine needles. *Chemosphere* 14:451–456.

Gevao B, Hamilton-Taylor J, Murdoch C, Jones KC, Kelly M, Tabner BJ. 1997. Depositional time trends and remobilization of PCBs in lake sediments. *Environ Sci Technol* 31:3274–3280.

Goldberg ED. 1979. The health of the oceans. Paris, France: UNESCO Press.

Goolsby DA, Thurman EM, Pomes ML, Meyer MT, Battaglin WA. 1997. Herbicides and their metabolites in rainfall: origin, transport, and deposition patterns across the midwestern and northeastern U.S. *Environ Sci Technol* 31:1325–1333.

Harrad SJ, Sewart AP, Alcock R, Boumphrey R, Burnett V, Duarte-Davidson R, Halsall C, Sanders G, Waterhouse K, Wild SR, Jones KC. 1994. Polychlorinated biphenyls (PCBs) in the British environment: sink, sources and temporal trends. *Environ Pollut* 85:131–146.

Hillery BR, Simcik MF, Basu I, Hoff RM, Strachan WMJ, Burniston D, Chan CH, Brice KA, Sweet CW, Hites RA. 1998. Atmospheric deposition of toxic pollutants to the Great Lakes as measured by the integrated atmospheric deposition network. *Environ Sci Technol* 32:2216–2221.

Hornbuckle KC, Sweet CW, Pearson RF, Swackhamer DL, Eisenreich SJ. 1995. Assessing annual water-air fluxes of polychlorinated biphenyls in Lake Michigan. *Environ Sci Technol* 29:869–877.

Iwata H. 1994. Distribution, dynamics and fate of persistent organochlorines in the global environment [Ph.D. thesis]. Matsuyama, Japan: Ehime University.

Iwata H, Tanabe S, Sakai N, Tatsukawa R. 1993. Distribution of persistent organochlorines in the oceanic air and surface seawater and the role of ocean on their global transport and fate. *Environ Sci Technol* 27:1080–1098.

Iwata H, Tanabe S, Sakai N, Nishimura A, Tatsukawa R. 1994. Geographical distribution of persistent organochlorines in air, water and sediment from Asia and Oceania, and their implications for global redistribution from lower latitudes. *Environ Pollut* 85:15–33.

Jantunen LM, Bidleman T. 1995. Reversal of the air-water gas exchange direction of hexachlorocyclohexanes in the Bering and Chuckchi Seas. *Environ Sci Technol* 29:1081–1089.

Jantunen LM, Bidleman T. 1996. Air-water gas exchange of hexachlorocyclohexanes (HCHs) and the enantiomers of a–HCH in Arctic Regions. *J Geophy Res Atm* 02:19279–19282.

Jones KC. 1991. Contaminant trends in soils and crops. *Environ Pollut* 69:311–325.

Jones KC, Sanders G, Wild SR, Burnett V, Johnston AE. 1992. Evidence for a decline of PCBs and PAHs in rural vegetation and air in the United Kingdom. *Nature* 356:137–140.

Kannan K, Tanabe S, Ramesh A, Subramanian A, Tatsukawa R. 1992. Persistent organochlorine residues in foodstuffs from India and their implications on human dietary exposure. *J Agric Food Chem* 40:518–524.

Kannan K, Tanabe S, Tatsukawa R. 1995. Geographical distribution and accumulation features of organochlorine residues in fish in tropical Asia and Oceania. *Environ Sci Technol* 29:2673–2683.

Kannan K, Tanabe S, Giesy JP, Tatsukawa R. 1997. Organochlorine pesticides and polychlorinated biphenyls in foodstuffs from Asian and Oceanic countries. *Rev Environ Contam Toxicol* 152:1–55.

Kaushik CP. 1989. Loss of HCH from surface soil layers under sub-tropical conditions. *Environ Pollut* 59:253–264.

Lee SC, MacKay D. 1995. Displaying multimedia environmental partitioning by triangular diagrams. *Environ Toxicol Chem* 14:1839–1846.

Lerch RN, Thurman EM, Kruger EL. 1997. Mixed mode sorption of hydroxylated atrazine degradation products to soil. A mechanism for bound residue. *Environ Sci Technol* 31:1539–1546.

Lerch RN, Thurman EM, Blanchard PE. 1999. Hydroxyatrazine in soils and sediment. *Environ Toxicol Chem* 18:2161–2168.

Loganathan BG, Kannan, K. 1991. Time perspectives of organochlorine contamination in the global environment. *Mar Pollut Bull* 22:582–584.

Loganathan BG, Kannan K. 1994. Global organochlorine contamination trends: An overview. *Ambio* 23:187–191.

Loganathan BG, Irvine KN, Kannan K, Pragatheeswaran V, Sajwan KS. 1997. Distribution of selected PCB congeners in the Babcock sewer district: A multimedia appraoch to identify PCB sources in combined sewer overflows (CSOs) discharging to the Buffalo River, New York. Arch. *Environ Contam Toxicol* 33:130–140.

Loganthan BG, Tanabe S, Goto M, Tatsukawa R. 1989. Temporal trends of organochlorine residues in lizard goby, *Rhinogobius flumineus* from the River Nagaragawa, Japan. *Environ Pollut* 62:237–251.

Loganathan BG, Tanabe S, Hidaka Y, Kawano M, Hidaka H, Tatsukawa R. 1993. Temporal trends of persistent organochlorine residues in human adipose tissue from Japan 1928–1985. *Environ Pollut* 81:31–39.

Mackay D, Guardo AD, Paterson S, Kicsi G, Cowan CE. 1996. Assessing the fate of new and existing chemicals: A five-stage process. *Environ Toxicol Chem* 15:1618–1626.

Majewski MS, Foreman WT, Goolsby DN, Nakagaki N. 1998. Airborne pesticide residues along the Mississippi River. *Environ Sci Technol* 32:3689–3698.

Malaiyandi M, Shah SM. 1984. Evidence of photoisomerization of hexachlorocyclohexane isomers in the ecosphere. *J Environ Sci Health* A19:887–910.

Morosini M, Schrcitmhller J, Reuter U, Ballschmiter K. 1993. Correlation between C–6/C–14 chlorinated hydrocarbons levels in the vegetation and in the boundary layer of the troposphere. *Environ Sci Technol* 27:1517–1523.

Mowbray DL. 1988. Pesticide use in the South Pacific. UNEP Regional Seas Reports and Studies No. 89 and South Pacific Regional Environment Program, Topic Review No. 26, Nairobi, Kenya: United Nations Environment Program.

Ngabe B, Bidleman TF, Falconer RL. 1993. Base hydrolysis of a– and g– hexachlorocyclohexanes. *Environ Sci Technol* 27:1930–1933.

Ockenden WA, Prest HT, Thomas GO, Sweetman A, Jones KC. 1998. Passive air sampling of PCBS, field calculation of atmospheric sampling rates by triolein-containing semipermeable membrane devices. *Environ Sci Technol* 32:1538–3543.

O'Connor TP. 1996. Trends in chemical concentrations in mussels and oysters collected along the U.S. coast from 1986 to 1993. *Mar Environ Res* 41:183–200.

Oehme M. 1991. Further evidence for long-range air transport of polychlorinated aromatics and pesticides from North America and Eurasia to the Arctic. *Ambio* 20:293–297.

Olsson M. 1995. Ecological effects of airborne contaminants in Arctic aquatic ecosystems: a discussion of methodological approaches. *Sci Total Environ* 160/161:619–630.

Olsson M, Bignert A. 1997. Specimen banking: a planning in advance. *Chemosphere* 34:1961–1974.

Ottar B. 1981. The transfer of airborne pollutants to the Arctic region. *Atmos Environ* 15:1439–1445.

Pacyna JM, Oehme M. 1988. Long-range transport of some organic compounds to the Norwegian Arctic. *Atmos Environ* 22:243–257.

Panda S, Sharmila M, Ramanand K, Panda D, Sethunathan N. 1988. Persistence of hexachlorocyclohexane isomers and carbofuran applied to surface and sub-surface layers of a flooded soil. *Pestic Sci* 23:199–207.

Preston MR. 1992. The interchange of pollutants between the atmosphere and oceans. *Mar Pollut Bull* 24:477–483.

Rapaport RA, Eisenreich SJ. 1988. Historical atmospheric inputs of high molecular weight chlorinated hydrocarbons to eastern North America. *Environ Sci Technol* 22:931–941.

Samuel T, Agarwal HC, Pillai MKK. 1988. Persistence and binding of DDT and gamma–HCH in a sandy loam soil under field conditions in Delhi, India. *Pestic Sci* 22:1–15.

Senthilkumar K, Kannan K, Sinha RK, Tanabe S, Giesy JP. 1999. Bioaccumulation profiles of polychlorinated biphenyl congeners and organochlorine pesticides in Ganges river dolphins. *Environ Toxicol Chem* 18:1511–1520.

Simcik MF, Eisenreich SJ, Golden KA, Liu SP, Lipiatou E, Swackhamer DL, Long DT. 1996. Atmospheric loading of polycyclic aromatic hydrocarbons to lake Michigan as recorded in the sediments. *Environ Sci Technol* 30:3039–3046.

Simonich SL, Hites RA. 1997. Relationships between socioeconomic indicators and concentrations of organochlorine pesticides in tree bark. *Environ Sci Technol* 31:999–1003.

Sinkkonen S, Paasivirta J. 2000a. Rates of biodegradation and metabolism of POPs: fate of POPs in sediment, soil and biota. *Chemosphere* (in press)

Sinkkonen S, Paasivirta J. 2000b. Calculation of photochemical tranformation rates of persistent organic pollutants. *Chemosphere* (in press)

Sleicher CA, Hopcraft J. 1984. Persistence of pesticides in surface soil and relation to sublimation. *Environ Sci Technol* 18:514–518.

Solomon KR, Baker DB, Richards RP, Dixon KR, Klaine SJ, La Point TW, Kendall RJ, Weisskopf CP, Giddings JM, Giesy JP, Hall Jr, LW, Williams WM. 1996. Ecological risk assessment of atrazine in north American surface waters. *Environ Toxicol Chem* 15:31–76.

Sweetman AJ, Jones KC. 1999. Input-output of polychlorinate biphenyls in a long-term study of lactating daily cows. *Environ Sci Technol* 33:104–112.

Tanabe S. 1991. Fate of toxic chemicals in the tropics. *Mar Pollut Bull* 22:259–260.

Tanabe S, Ramesh A, Sakashita D, Iwata H, Tatsukawa R, Mohan D. Subramanian AN. 1991. Fate of HCH (BHC) in tropical paddy field: Application test in South India. *Intern J Environ Anal Chem* 45:45–53.

Tanabe S, Senthilkumar K, Kannan K, Subramanian AN. 1998. Accumulation features of polychlorinated biphenyls and organochlorine pesticides in resident and migratory birds from South India. Arch. *Environ Contam Toxicol* 34:387–397

Thurman EM, Goolsby DA, Meyer MT, Kolpin DW. 1991. Herbicides in surface waters of the midwestern United States: the effect of spring flush. *Environ Sci Technol* 25:1794–1796.

Thurman EM, Goolsby DA, Meyer MT, Mills MS, Pomes ML, Kolpin DW. 1992. A reconnaissance study of herbicides and their metabolites in surface water of the midwestern U.S. using immunoassay and gas chromatography/mass spectrometry. *Environ Sci Technol* 26:2440–2447.

Thurman EM, Meyer MT, Mills MS, Zimmerman LR, Perry CA, Goolsby DA. 1994. Formation and transport of desethylatrazine and deisopropylatrazine in surface water. *Environ Sci Technol* 28:2267–2277.

Tremolada P, Di Guardo A, Calamari D, Davoli E, Fanelli R. 1992. Mass-spectrometry-derived data as possible predictive method for environmental persistence of organic molecules. *Chemosphere* 24:1473–1491.

Tremolada P, Calamari D, Gaggi C, Bacci E. 1993. Fingerprints of some chlorinated hydrocarbons in plant foliage from Africa. *Chemosphere* 27:2235–2252.

Ulrich EM, Hites RA. 1998. Enantiomeric ratios of chlordane related compounds in air near the Great Lakes. *Environ Sci Technol* 32:1870–1874.

van Pul WAJ, Deleeuw FAMM, Vanjaarsveld JA, Vandergaag MA, Sliggers CJ. 1998. The potential for long-range transboundary atmospheric transport. *Chemosphere* 37:113–141.

Villars MT. 1995. Monitoring water quality in the future, Executive summary. Delft Hydraulics, Delft, The Netherlands.

Voldner EC, Li Y-F. 1995. Global usage of persistent organochlorines. *Sci Total Environ* 160/161:201–210.

Wania F, Mackay D. 1993. Global fractionation and cold condensation of low volatility organochlorine compounds in polar regions. *Ambio* 22:10–18.

Wania F, Mackay D. 1996. Tracking the distribution of persistent organic pollutants. *Environ Sci Technol* 30:A 309–A 396.

Wania F, Mackay D. 1999. Global chemical fate of α-hexachlorocyclohexane. 2. Use of a global distribution model for mass balancing, source apportionment, and trend prediction. *Environ Toxicol Chem* 18:1400–1407.

Webster E, Mackay D, Wania F. 1998. Evaluating environmental persistence. *Environ Toxicol Chem* 17:2148–2158.

Wild SR, Jones KC. 1995. Polynuclear aromatic hydrocarbons in the United Kingdom environment: a preliminary source inventory and budget. *Environ Pollut* 88:91–108.

CHAPTER 7

Case Studies on Persistence and Long-Range Transport of Persistent Organic Pollutants

Robie W. Macdonald, Steven J. Eisenreich, Terry F. Bidleman, Jordi Dachs, Jozef M. Pacyna, Kevin C. Jones, Robert E. Bailey, Deborah L. Swackhamer, Derek C.G. Muir

The environmental behavior of a chemical is controlled partly by its properties and partly by the characteristics of the environment. Persistent chemicals have long residence times in the environment and can travel long distances. From the point of view of the receiving environment, persistence can be defined as the opportunity for a chemical to arrive at a given location, enter biogeochemical cycles, and produce a problem such as deleterious effects on biological systems. Historically, persistence has been identified through environmental measurements, particularly of higher-trophic-level organisms that have been affected through food-web magnification. In such cases, target species may exhibit alarming concentrations while only vanishingly small concentrations are observed in the physical environment. Polychlorinated biphenyls (PCBs) in the Arctic provide an exceptionally clear example (Muir and Norstrom 1994).

The notorious dichlorodiphenyltrichloroethane (DDT), with its degradation products dichlorodiphenyldichloroethylene (DDE) and dichlorodiphenyl-dichloroethane (DDD), was one of the first chemicals to be detected in nontarget organisms and in locations far from its application. Rachel Carson's book *Silent Spring* (1962) brought DDT to public attention and spurred the development of monitoring programs for this and many other chemicals in the environment. The dramatic decrease in use of DDT in northern temperate regions since the early 1970s is paralleled in all compartments of the environment (Bishop et al. 1992; Petit et al. 1994; Bignert et al. 1995, 1998; Pekarik et al. 1999). DDT provides the classical

CHAPTER PREVIEW

Evaluation of Persistence and Long-Range Transport of Organic Chemicals in the Environment. Gary Klečka et al., editors.
 ISBN 1-880611-22-8

example of how the environment and a chemical may act together to provide an unanticipated, unpleasant surprise. For DDT, persistence, semi-volatility, and toxicity combined with long food chains in aquatic systems to put top predators at risk. The final property of DDT, its ability to disrupt development and immune systems of organisms, was evident in eggshell thinning, but this particular property has only recently become more widely recognized and understood (compare Repetto and Baliga 1996). Here, we provide a number of case studies that have some similarities to DDT but, more importantly, some differences.

Other chapters in this book discuss models for prediction of the fate and local, regional, or global distribution of chemicals through long-range transport. Models have proven to be extremely useful diagnostic and heuristic tools to help scientists understand why a chemical has become a problem and even explore outcomes of changing emission rates by, for example, banning its use (e.g., Wania and Mackay 1995). However, models have been notoriously poor at predicting a problem in advance of observing it, and the predictions of environmental models are always open to question (Oreskes et al. 1994). Widely used chemicals often have poorly known emission inventories, and it is clear that models cannot foresee every environmental circumstance. Therefore, predictive models of chemical behavior that depend primarily on physicochemical properties will have uncertainty arising from the emission scenario, from components missing from the model as a result of our incomplete understanding of natural systems, and from the fact that natural systems are never closed. For these reasons, we believe that it is crucial to continue to observe the environment for the appearance of new chemicals and to attempt to understand the role of the receptor in the manifestation of persistence. Much can be learned by observing and modeling the response of affected natural systems when the emission of a chemical is curtailed — a point at which we have just arrived for a number of persistent organic chemicals.

Case studies of particular chemicals and locations offer direct insight into the role of the receiving environment or "receptors" in defining persistence. Indeed, we can consider the environment, or subcomponents of it, to provide "living models" at a variety of scales. By definition, these models include complete chemical fluxes, routes of chemical entry, and processes. Often, the output of these natural models is surprising. Our detailed understanding of fluxes, couplings, and processes through case studies then allows us to build more-realistic mathematical models and check them against observation.

The environmental characteristics of receptors include transport pathways within and between reservoirs, the time scales of these couplings, and the quantities contained within the various reservoirs. Therefore, the first task of a case study is to establish these fundamental parameters for the system in question. To predict how a given chemical will behave requires not only a knowledge of the fundamental structure of the system (e.g., mixing rates and depths, exchange rates with external water bodies, exchange rates with atmosphere, particle flux rates) but also a knowledge of how each chemical partitions between media. For example, we now know that some large water bodies were initially important sinks (e.g., the Arctic Ocean for hexachlorocyclohexane [HCH] and Lake Superior for PCB). With a reduction of emissions, these water bodies have now become sources. How long they will continue to be sources is certainly an important component of persistence for the chemicals in question.

For the case studies conducted here, we have selected specific chemicals (Table 7-1) and environments (Figure 7-1, Table 7-2). In most cases, we have selected systems in which major inputs of the selected chemicals have occurred, or continue to occur, through long-range (local, regional, or global) transport and atmospheric deposition. Our choice has been limited to locations in which we have a sufficient understanding of the natural cycles and at least a rudimentary database for the chemical in question. The case studies include the Arctic Ocean (HCH, PCB, chlorobornanes [CHB]), the Mediterranean Sea (PCB, phenanthrene, pyrene, benzofluoranthene, benzo(a)pyrene), the Baltic Sea (PCB), Lake Superior (PCB, CHB), and the Great Lakes (atrazine). In addition, we have examined one terrestrial system (PCBs in Great Britain) and two tropical systems (PCB and HCH in Lake Malawi, Africa and HCH in the Vellar River Estuary, India) in which there are sufficient data to make at least a qualitative study.

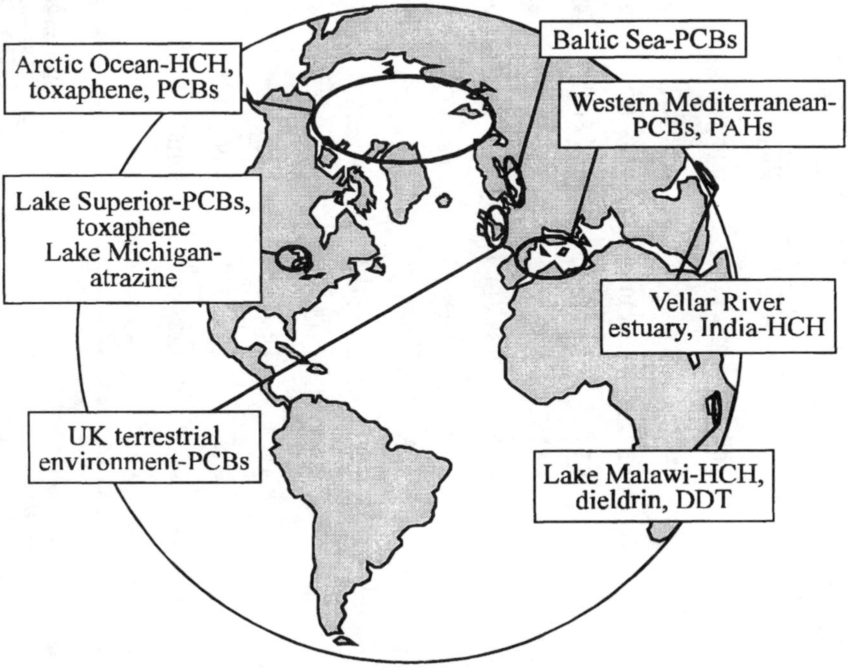

Figure 7-1 Locations for case studies on persistence and transport of POPs

Table 7-1 Properties of the chemicals at 20 °C, except for values from references (l), (m), (n), and (o), which are at 25 °C

Chemical	log K_{OW}	Water solubility mol/m^3 (liq)	Vapor pressure Pa	Henry's Law constant Pa m^3/mol	log K_{OA}	K_b (hydrol) L/mol min	K_{OH} cm^3/s
Toxaphene	5.5 [a]	1.3×10^{-3} [b]	8×10^{-4} [b]	0.30 [c]	9.4 [d]		2.5×10^{-12} [o]
α-HCH	3.8 [f]	1.2×10^{-1} [l]	0.14 [e]	0.42 [c]	7.3 [d]	1.54 [h]	1.4×10^{-13} [g]
γ-HCH	3.8 [f]	1.8×10^{1} [l]	0.039 [e]	0.18 [c]	8.0 [d]	1.09 [h]	1.8×10^{-13} [g]
PCB 28	5.8 [l]	1.3×10^{-3} [l]	1.96×10^{-2} [i]	20.0 [l]	8.22 [j]	0	1.1×10^{-12} [n]
PCB 52	6.1 [l]	4.2×10^{-4} [l]	9.06×10^{-3} [i]	47.6 [l]	8.56 [j]	0	7.3×10^{-13} [o]
PCB 101	6.4 [l]	9.9×10^{-5} [l]	1.83×10^{-3} [i]	35.5 [l]	9.31 [j]	0	3.3×10^{-13} [o]
PCB 118	6.3 [p]	2.33×10^{-4} [p]	6.34×10^{-4} [i]	20 [p]	10.08 [j]	0	3.3×10^{-13} [o]
PCB 138	6.7 [p]	1.6×10^{-5} [p]	2.68×10^{-4} [i]	12 [p]	10.09 [j]	0	1.6×10^{-13} [o]
PCB 153	6.9 [l]	1.6×10^{-5} [l]	3.57×10^{-4} [i]	42.9 [l]	10.04 [j]	0	1.6×10^{-13} [o]
PCB 180	7.0 [p]	1.9×10^{-5} [p]	6.63×10^{-5} [i]	5.4 [p]	10.75 [j]	0	1.0×10^{-13} [o]
Atrazine	2.75 [l]	4.14 [l]	1.19×10^{-3} [l]	2.88×10^{-4} [l]	9.7 [d]	0	2.7×10^{-11} [o]
Phenanthrene	4.57 [l]	3.5×10^{-2} [l]	6.00×10^{-2} [e]	3.24 [l]	7.90 [k]	0	3.1×10^{-11} [m]
Pyrene	5.18 [l]	1.3×10^{-2} [l]	8.19×10^{-3} [e]	0.92 [l]	9.16 [k]	0	5.0×10^{-11} [o]
Benzo(k)fluoranthene	6.00 [l]	4.12×10^{-6} [l]	0.016 [l]	2.5×10^{-4} [l]	11.2 [d]	0	5.4×10^{-11} [o]
Benzo(a)pyrene	6.04 [l]	4.6×10^{-4} [l]	3.65×10^{-6} [e]	0.046 [l]	11.2 [d]	0	5.0×10^{-11} [o]

[a] Isnard and Lambert 1989; [b] Murphy et al. 1987; [c] Jantunen and Bidleman 2000a; [d] Calculated from K_{ow} RT/H; [e] Hinckley et al. 1990; [f] Suntio et al. 1988; [g] Brubaker and Hites 1998; [h] Ngabe et al. 1993; [i] Falconer and Bidleman 1994; [j] Harner and Bidleman 1996; [k] Harner and Bidleman 1998; [l] Mackay et al. 1992, 1997; [m] Atkinson and Arey 1994; [n] Anderson and Hites 1996; [o] Meylan and Howard 1993; [p] Estimated from Mackay et al. 1992–1996 using results for homolog group.

Table 7-2 Physical characteristics of the environmental systems discussed in the case studies

Characteristic	Lake Superior	Lake Michigan	Arctic Ocean	Baltic Sea	Western Mediterranean	Lake Malawi
Surface area (km^2)	8.21×10^4	12×10^4	14×10^6	4.12×10^5	8.5×10^5	28,800
Shelf area (km^2)	~ 45% of SA (< 110m)	~50% of SA (< 100)	30% of Arctic Ocean	—	1.3×10^5	—
Vol deep water (km^3)	1.21×10^4	4.97×10^3	1.5×10^7	—	1.2×10^6	8400
Vol shallow water (km^3)	—	—	2.1×10^6	2.16×10^4*	1.7×10^5	—
Mean depth (m)	145	85	1000	55	1635	292
Maximum depth (m)	405	280	5500	459	2900	706
Precipitation (cm/y)	76	79	19.5	63	20–100	135
River flow (km^3/y)	54	29	3300	480	99.8	29
Flux water in (km^3/y)	—	—	1.5×10^5	2.21×10^4	53,000/38,000**	—
Flux water out (km^3/y)	71	79	1.5×10^5	9.5×10^2	50,500/39,800**	—
Residence time (y)	112	69	10 (surface);30–300 (deep)	25	—	—
Flushing time (y)	177	100	300 (deep)	—	~ 10 (surface)	—

* Total volume of the Baltic Sea is about 2.17×10^4 km^3
** Gibraltar/Sicily

Table 7-2 continued

Characteristic	Lake Superior	Lake Michigan	Arctic Ocean	Baltic Sea	Western Mediterranean	Lake Malawi
Particle flux ($g/m^2 y$)	0.13 ± 0.06	0.2–0.5	~ 1	—	0.05–4	—
Organic carbon flux ($mg/m^2 y$)	40 ± 15	50–100	20	—	7–93	—
Suspended solids (mg/L)	0.1–0.5	0.2–0.7	0.1–15	—	0.1–1.5	—
Sediment accumulation ($g/m^2 y$)	100–800	200–1200	0–5000	0–5000 (shelf); 0.5–1.5 (basin)	400 (shelf); 90 (basin)	—
Sediment accumulation (cm/y)	0.01–0.03	0.03–0.1	0–1 (shelf)	1–3 (shelf); 0.1–0.3 (basin)	0.1–0.65 (shelf); 0.02–0.2 (basin)	—
Watershed area (km^2)	12.4×10^4	12×10^4	10×10^6	1.7×10^6		1.00×10^5
Surface water temperature (°C)	0–15	2–23	−2–0 (ice)	0–16	8–23	
Deep water temperature (°C)	2–4	2–6	−2–5 (open shelf)	1–5	—	—
Ice cover	Partial, ~ 3–4 mo/y	Partial, 2–3 mo/y	$9–16 \times 10^6 km^2$	Partial, 2–5 mo/y	No	No
Primary productivity ($g C/m^2 y$)	100		20 (shelf); 1–10 (basin)	140–160	50–150	
pH	7.8	7.6	7.8	7.1–7.8	7.8	—

For each of these case studies, we present a description of the system. Where possible, we construct a simple budget for the chemical under study using the scheme presented in Figure 7-2 and then summarize the significance of what has been learned from these environmental examples. Finally, we compare and contrast systems in terms of how they receive, cycle, and eliminate chemicals.

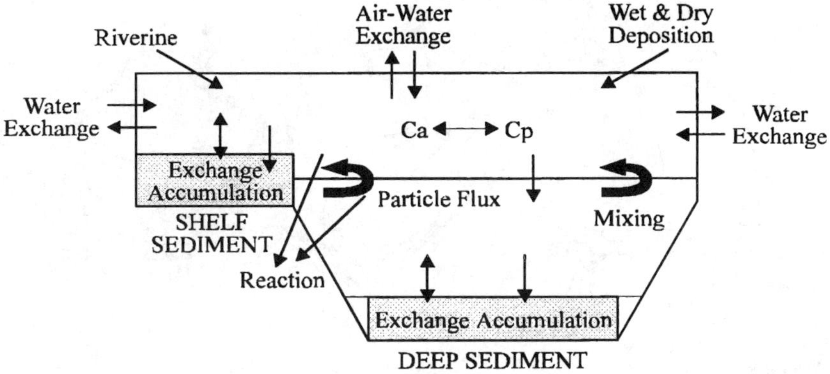

Figure 7-2 Generic model of pathways for POPs in freshwater and marine environments

Arctic Ocean Case Studies

Description of arctic marine system

Recent reviews describe in detail the functioning of the Arctic Ocean as it relates to contaminant transport (Macdonald and Bewers 1996; Macdonald et al. 2000); therefore, only a brief overview is given here. The Arctic Ocean is connected to the Pacific through Bering Strait and to the Atlantic through Fram Strait and the Canadian Archipelago. Flow is predominantly into the Arctic through Bering Strait and out of the Arctic through the Archipelago (Figure 7-3). Large flows go both ways through Fram Strait and the Barents Sea, with a large and uncertain amount of recirculation at the entrance. Freshwater inflow from rivers (Table 7-2) stratifies the upper ocean, thereby limiting vertical mixing. The enormous continental shelves not only process freshwater runoff but also play an active role in the thermohaline circulation of the Arctic Ocean (Aagaard et al. 1985). During winter, ice formation rejects salt to the water and this, together with cooling, may produce dense water that sinks and leaks off the shelves into the interior ocean at a depth appropriate to its density. As a result of this process and the entry of waters of various starting salinities (i.e., densities), the water column becomes vertically stratified with each water mass moving in a restricted depth domain (Figure 7-4). At the surface, the annual cycle of ice formation and melt produces a mixed layer about 50 m deep in winter (polar mixed layer) with a seasonal stratified layer (5 to 10 m deep) contain-

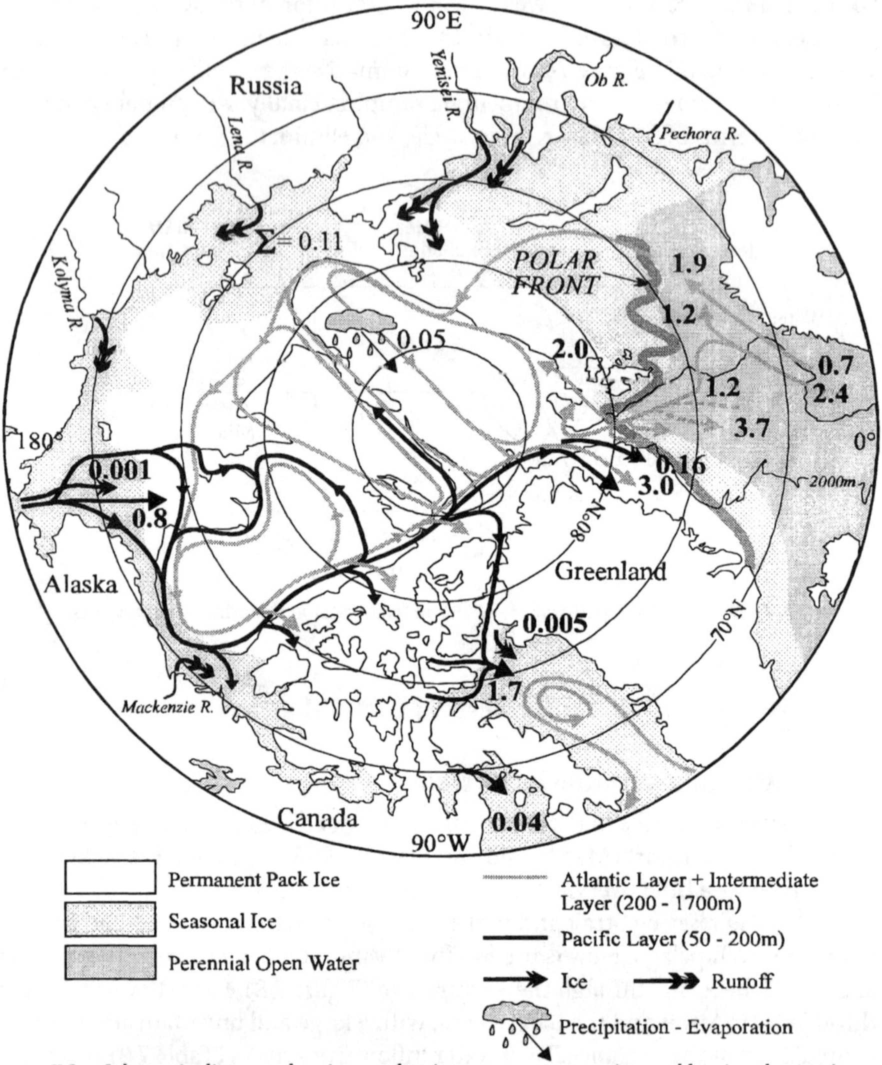

Figure 7-3 Schematic diagram showing predominant currents entering and leaving the Arctic Ocean (after Aagaard 1989; McLaughlin et al. 1996; Rudels et al. 1996). Numbers refer to estimated flow rates in Sverdrups (10^6 m^3/s).

ing ice melt and runoff in summer. Below the polar mixed layer is found a cold halocline, which isolates the surface ocean from the deep ocean with the result that vertical mixing is hindered.

Circulation within the Arctic Ocean is organized horizontally (Figure 7-3) and vertically (Figure 7-4). Because of a strong coriolis effect, currents tend to be constrained to basin boundaries with the result that water motion and replacement

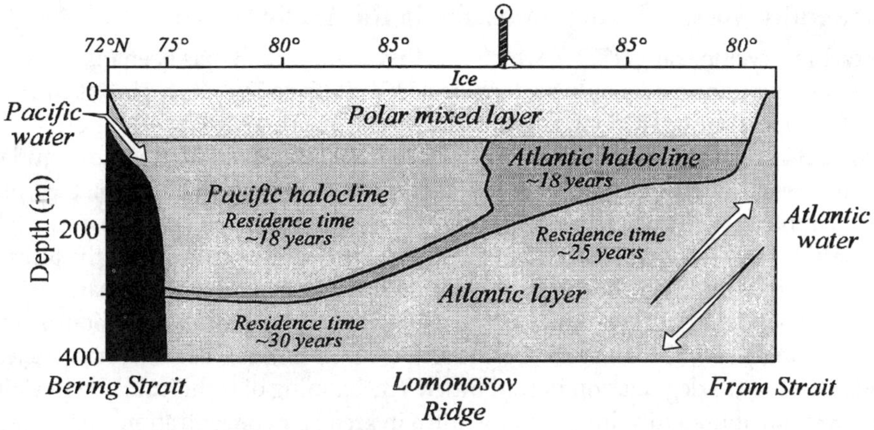

Figure 7-4 Vertical structure of the top 400 m of the Arctic Ocean showing the polar mixed layer (top 50 m), the cold haloclines where stratification occurs, and the Atlantic Layer (estimated residence times are based on freshwater balances).

is faster there than in the basin interiors. The Arctic Ocean contains two predominant domains, one in which the entire water column is dominated by the Atlantic and one in which the Pacific plays an important role (Figure 7-3, Figure 7-4). The separation between these domains occurs approximately at the Lomonosov Ridge, although recent events have caused its position to shift (McLaughlin et al. 1996) with potentially important consequences for property transport (Macdonald 1996). The upper ocean in these two domains behaves differently in terms of circulation and replacement times, and, therefore, each domain must be considered separately in the estimation of budgets. Here, we consider the Canadian Basin, which is dominated by the Beaufort Gyre, to comprise one domain (North American Arctic Ocean [NAAO] at 2.57×10^6 km^2) and the Eurasian Basin, which is dominated by the Transpolar Drift, to comprise the other (Eurasian Arctic Ocean [EAO] with area 7.72×10^6 km^2). Ice cover is permanent over the interior basins but seasonal over the shelves, which tend to clear of ice from about July to September. Ice cover is not 100%, even in winter, and there are cracks through which gas may exchange, although such exchange may be severely restricted. As a consequence of the ocean's structure, contaminants entering the interior ocean at the surface cannot diffuse downward below the polar mixed layer and are also hampered from diffusing back into the atmosphere either by ice cover in winter or stratification in summer. Contaminants that enter at the shelves may be trapped by surface stratification in summer, but they also can be injected into the halocline in winter when dense water is formed. Contaminants entering in water masses from oceans to the south will tend to follow the motion of the water as shown in Figures 7-3 and 7-4.

Case study: Hexachlorocyclohexane in the Arctic Ocean

Hexachlorocyclohexane (HCH), which was first used in 1943, has been manufactured as 2 predominant products: technical HCH (60% to 70% α-HCH, 10% to 15% γ-HCH) and lindane (γ-HCH). It is estimated that by the early 1990s, 10 Mt of technical HCH had been produced globally and used largely in China, India, and the former Soviet Union (Li et al. 1998; Li 1999). Since the 1970s, Canada, the U.S., and most countries in Europe have used pure γ-HCH, the only isomer with insecticidal properties (Walker et al. 1999). The key physical properties that control the ultimate fate of HCH in the environment are high volatility, low Henry's law constant (H), a strong dependency of H on temperature (α-HCH partitions into water about 7 times more strongly at 0 °C than at 25 °C [Kucklick et al. 1991]), low affinity for particles, and resistance to degradation in cold water. The banning of technical HCH in China in 1983 contributed to a dramatic downturn in arctic air concentrations in the early 1980s (Li et al. 1998). Ensuing reductions in HCH use by India and Russia (Iwata et al. 1995; Li et al. 1998) in the early 1990s led to further atmospheric declines (Figure 7-5). The strong partitioning of HCH from air to water means that water phases ultimately become more important reservoirs than is the atmosphere, containing over 90% of the inventory. The result of the dramatic decrease in atmospheric HCH concentration is that the surface water in some oceanic regions became oversaturated and therefore started to outgas (Jantunen and Bidleman 1995).

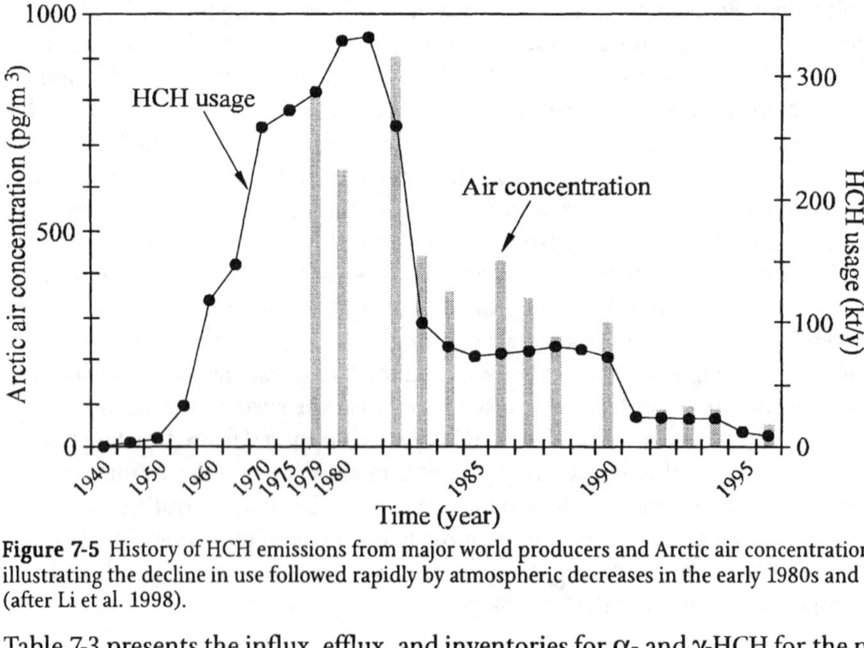

Figure 7-5 History of HCH emissions from major world producers and Arctic air concentrations illustrating the decline in use followed rapidly by atmospheric decreases in the early 1980s and 1990s (after Li et al. 1998).

Table 7-3 presents the influx, efflux, and inventories for α- and γ-HCH for the mid-1990s. The method of calculation and sources of information for Table 7-3 have

Table 7-3 Budget for HCH and PCBs in the Arctic Ocean

Input	Water flow 10^4 km^3/y	Flux (t/year) α-HCH	γ-HCH	ΣPCB	ΣHCH
Bering Strait	2.6	52	12	0.3	64
Norwegian Coastal Current	2.2	10	3.5	0.2	14
Barents Sea	3.8	17	6.1	0.4	23
W. Spitsbergen Current	6.3	29	10	0.6	39
Rivers:					
North America	0.05	0.3	0.17	0.15	0.5
Eurasia	0.3	25	44	15	69
Gas deposition – EAO		34	5.6	38.6	40
Dry fall – EAO		0.06	0.04	6.21	0.1
Wetfall – EAO		10.6	2.8	2.29	13
Gas deposition – NAAO		4.5	0.75	2.05	5.3
Dry fall – NAAO		0.02	0.01	0.89	0.03
Wetfall – NAAO		3.5	0.94	0.32	4.4
Total input		186	86	64	272
Output					
E. Greenland Current[1]	3.2	–28	–6.3	–0.3	–34
E. Greenland Current[2]	7.3	–51	–12	–0.7	–63
Canadian Archipelago	5.4	–193	–28	–0.5	–221
Ice	0.52	–6.8	–1.0	–0.1	–7.8
Volatilization – EAO		–25	–4.0	–0.7	–29
Volatilization – NAAO		–15	–1.2	–18.9	–16
Hydrolysis		–32	–4.0	0.0	–36
Microbial degradation		–204	–23		–227
Sedimentation		–0.06	–0.03	–0.15	0.09
Total output		–555	–79	–21	–634
Net output (input – output)		–369	–7	42	–362
Inventory (water, tonnes)		2910	640	?	3550
Residence time (years)[1]		5.2–15.6	7.4–8.0	?	5.6–13

[1] Polar water [2] Atlantic Intermediate water

been discussed in detail in Macdonald et al. (2000); only salient points will be discussed here.

The river input has been estimated from HCH data together with hydrologic flows for major rivers (Aagaard and Carmack 1989). Data from Canada suggest that HCH concentrations in arctic rivers have responded rapidly to the atmospheric turndown and are presently close to equilibrium (Macdonald et al. 2000). In contrast, a recent analysis of data for the Russian rivers in 1990 through 1996 show a number of sporadic high points and average HCH concentrations well above those found in the North American rivers (Macdonald et al. 2000). Fluxes based on the Russian data amount to 25 t/y α-HCH and 41 t/y γ-HCH, mostly to the Kara Sea (Macdonald et al. 2000).

Atmospheric exchanges take into account rain and snow scavenging of both particulate and vapor phase HCH, dry deposition, air to sea exchange, volatilization from the ocean, and volatilization from ice and snow (Cotham and Bidleman 1991; Macdonald et al. 2000). Values of surface water and atmospheric HCH have been taken from literature. The area of ice cover to correct air–sea exchange estimates for each season was calculated from satellite imagery, which tends to overestimate open water because it includes puddles as sea surface. Presently, the net flux of α-HCH is out of the Arctic Ocean as expected, whereas the net γ-HCH flux is near steady state (Table 7-3).

Ocean exchanges have been calculated by estimating the inflows and outflows from literature (Figure 7-3) and then using HCH seawater data from the regions in question to estimate transport (Falconer et al. 1995; Jantunen and Bidleman 1995, 1996; Macdonald et al. 1996; Hargrave et al. 1997; Harner, Kylin et al. 1999). Seawater exchanges, and not the atmosphere, now dominate the transport of HCH into and out of the Arctic Ocean (Table 7-3, Figure 7-6a). Of these exchanges, the largest outflow was estimated for the Canadian Archipelago (272 t/y α-HCH; 46 t/y γ-HCH).

Particle fluxes contribute at most a minor removal of HCH to the deeper ocean or to sediments (Harner, Kylin et al. 1999). In the Arctic Ocean, two factors play an important role: 1) the particle fluxes are generally small with organic carbon (OC) being efficiently recycled in the upper ocean (Wheeler et al. 1996) and 2) HCH is not strongly scavenged by particles (Hargrave et al. 1988). In support of these low vertical fluxes, HCH is usually not detected in Arctic sediments, although there are exceptions in locations of high productivity (Macdonald et al. 2000).

Degradation occurs within the Arctic Ocean surface layer as evidenced by alterations in the enantiomer ratios of α-HCH (Jantunen and Bidleman 1998; Harner, Kylin et al. 1999). Hydrolysis is also likely to occur for both α- and γ-HCH, but at reduced rates because of the cold temperatures (Ngabe et al. 1993; Harner, Wideman et al. 1999). The half-lives of α- and γ-HCH in seawater at pH 8.1 and 0 °C were estimated at 63 and 110 y (Harner, Kylin et al. 1999). Based on these loss rates

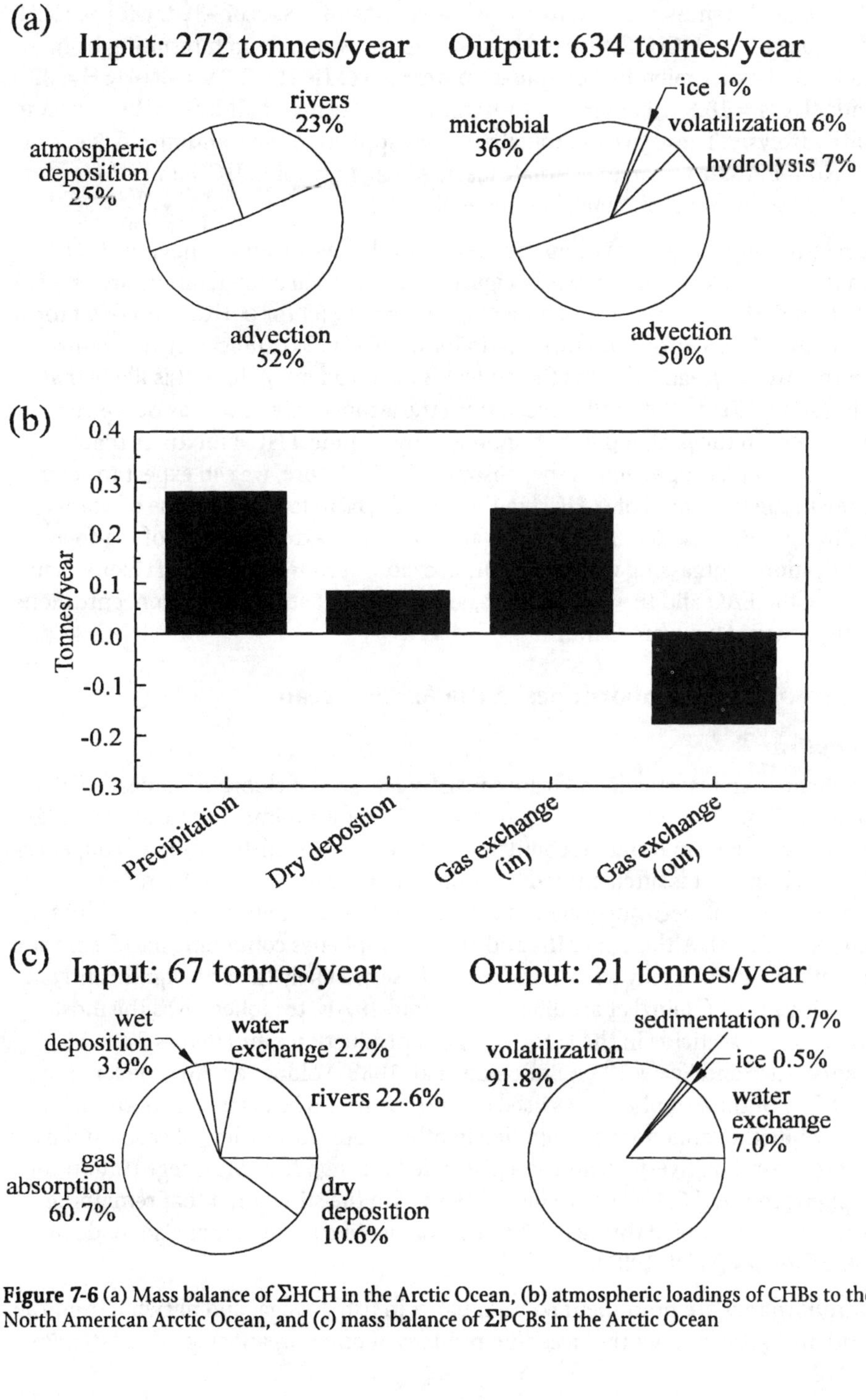

Figure 7-6 (a) Mass balance of ΣHCH in the Arctic Ocean, (b) atmospheric loadings of CHBs to the North American Arctic Ocean, and (c) mass balance of ΣPCBs in the Arctic Ocean

and estimated standing stocks, hydrolysis accounts for losses of –32 t/y for α-HCH and – 4.0 t/y for γ-HCH. Estimated half-lives in the water column resulting from pseudo–first-order microbial degradation were (+) α-HCH = 5.9 y, (–)α-HCH = 23.1 y, and γ-HCH = 18.8 y. These are 3 to 10 times shorter than half-lives with respect to basic hydrolysis. The above microbial rates are applied to the standing stock of HCHs in the Arctic Ocean to estimate losses of –204 t/y for α-HCH and –23 t/y for γ-HCH (Macdonald et al. 2000; Harner et al. 2000).

The relative importance of various processes for delivering and removing HCHs from the Arctic Ocean are shown in Figure 7-6a. The total annual inputs are 186 t for α-HCH and 86 t for γ-HCH, and annual losses are –555 t for α-HCH and –79 t for γ-HCH (Table 7-3). The output/input ratio for α-HCH is 3.0, indicating net removal from the Arctic Ocean, whereas the budget is balanced for γ-HCH. It is likely that loadings of HCHs to the Arctic Ocean from the atmosphere as well as ocean currents were greater in the past and were dominated by technical HCH mixtures that contained a high proportion of the α-isomer. In the future, we can expect to see a decline in the large pool of α-HCH in the Canada Basin largely because of drainage out through the Canadian Archipelago and to a lesser extent because of sea-to-air volatilization. Outgassing will also result in a slow decline of both HCH isomers in waters of the EAO and the regional seas, assuming that atmospheric concentrations continue their downward trend.

Case study of chlorobornanes in the Arctic Ocean

Description

Toxaphene is an insecticidal mixture consisting largely of chlorobornanes (CHBs) with a small proportion of chlorocamphenes. Because a number of toxaphene-like products are made in different countries, and because the distribution of congeners in the environment is often altered from the original products, the term *chlorobornanes* has been suggested as a generic name for environmental residues (Muir et al. 1993). Although CHBs and chlorocamphenes containing 6 to 9 chlorines number over 16,000, only a few hundred of these are likely to occur significantly in the environment (Hainzl et al. 1994). In the mid-1970s, toxaphene was the most heavily used insecticide in the U.S., with peak production estimated as 37 to 55 × 10^6 kg/y (Rice and Evans 1984; Bidleman et al. 1988; Voldner and Schroeder 1989). About 80% of the toxaphene was used on cotton and soybeans in the southern states, while the remainder was applied in other states to a variety of crops and used for pest control on livestock and as a piscicide for rough fish. Most registrations of toxaphene in the U.S. were cancelled in 1982 with the stipulation that remaining stocks could be applied through 1986. Toxaphene-like products are also made in other countries (Saleh 1991).

Chlorobornanes are prominent contaminants in fish, beluga, and narwhal from the Canadian Arctic, frequently exceeding residues of other organochlorine pesticides

(Muir et al. 1997). The widespread occurrence of CHBs in fish (Muir et al. 1997) and sediments (Lockhart and Muir 1996; Donald et al. 1997; Stern et al. 1998) from lakes across northern Canada and in lichens across Ontario (Muir et al. 1993) points to atmospheric deposition as the dominant pathway, although many lakes were treated with toxaphene in the 1960s to remove native fish (Miskimmin et al. 1995; Donald et al. 1997; Stern et al. 1998). Present sources are countries in which toxaphene-like products are still used and volatilization from agricultural soils contaminated by previous use (Harner et al. 2000). It is likely that CHBs reached the Arctic from both North America and Eurasia (Macdonald et al. 2000).

Chlorobornanes in Arctic Ocean surface waters in the mid-1990s ranged from 10 to 18 pg/L in the Bering-Chukchi seas to 60 to 95 pg/L in the northern portion of the Canada Basin (Jantunen and Bidleman 1998). Chlorobornanes at Resolute Bay in the central Canadian Archipelago displayed a seasonal cycle of 30 to 140 pg/L with an annual mean of 85 pg/L in 1993 (Bidleman et al. 1995; Hargrave et al. 1997). Chlorobornanes were determined in air samples from Resolute Bay in June 1992 (6.9 ± 3.3 pg/m^3, n = 6) (Bidleman et al. 1995) and the Bering-Chukchi seas in August to September 1993 (4.1 ± 1.3 pg/m^3, n = 7) (Jantunen 1997).

Atmospheric deposition into the North American Arctic Ocean

There are no measurements of CHBs in the eastern Arctic Ocean, and lack of this information precludes formulating a budget for the entire Arctic Ocean, as was done for HCHs. Instead, we present an estimate of the atmospheric loadings to the NAAO, based on input (wet and dry deposition, gas exchange) and output (gas exchange) pathways.

Atmospheric deposition and volatilization of CHBs were estimated in the same manner as for HCHs (Macdonald et al. 2000). Vapor pressure and the Henry's Law constant were calculated as functions of temperature (Hoff, Muir et al. 1993; Jantunen and Bidleman 2000a). These properties were used to model wet and dry deposition of particulate and gaseous species and air–water gas exchange. Precipitation scavenging and dry deposition were estimated by calculating the partitioning of CHBs between the aerosol and gaseous phases (using aerosol properties typical of the arctic environment) and applying scavenging ratios or dry deposition velocities according to relationships presented by Cotham and Bidleman (1991). Gas exchange was based on the two-film model with fugacity definitions, taking into account fractional ice cover in different months. Surface water concentrations in the NAAO were assumed to be 60 pg/L.

Atmospheric loadings to the NAAO (t/y) are estimated as follows: precipitation 0.28, gas exchange 0.25, and particle dry deposition 0.083, for a total of 0.61. Volatilization removes 0.18 t/y. The net loading is depositional, 0.43 t/y. Unlike the HCHs, all atmospheric processes are significant for CHB loadings (Figure 7-6b). Dry deposition is highest during winter haze season when about 90% of the CHBs are predicted to be associated with aerosols (Figure 7-7). Contribution from precipita-

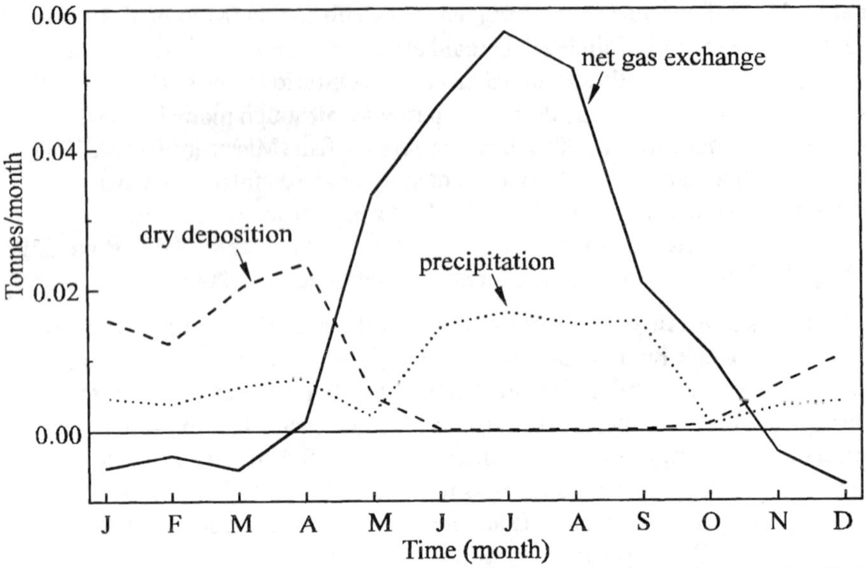

Figure 7-7 Effect of season on dry deposition, precipitation, and net gas exchange to the CHB budget of the North American Arctic Ocean

tion is fairly constant over the year because of particle scavenging during winter and gas scavenging during summer. The water–air fugacity ratio of CHBs varies from 0.4 to 0.7 in spring and summer to 5 to 13 in winter. Lower concentrations of CHBs in air during winter lead to fugacity ratios > 1.0 and net volatilization. Little outgassing actually takes place, however, because of the ice cover. During summer the atmospheric concentrations are higher, fugacity ratios reverse to < 1.0 (net deposition), and gas exchange maximizes because of the expanded area of open water.

Case study of polychlorinated biphenyls in the Arctic Ocean

Description

Polychlorinated biphenyls have been used in major quantities in electrical transformers and capacitors, and also have been used in plasticizers, paints, and in copy paper. Since their introduction in 1929, ~ 5.7×10^8 kg of PCBs have been produced, of which ~ 35% remains in mobile environmental reservoirs including aquatic systems, plants, and soils (compare Tanabe 1988). Polychlorinated biphenyls were banned from new uses in 1979 in the U.S. and Canada. Of the 209 possible congeners, only about 110 congeners are routinely found in the environment. Aqueous solubilities of PCBs (S_{aq}) vary from 10^{-4} to 10^{-7} mol/m^3, subcooled liquid vapor pressures ($P°_L$) range from 10^{-1} to 10^{-4} Pa, H values range from 10^2 to 10^0 Pa m^3/mol, log K_{OW} range from 4.5 to 9, and log K_{OA} range from 4 to 10 (Mackay et al. 1997). The temperature dependence of $P°_L$, H, and K_{OA} contribute significantly to their distributions and transport in the environment.

Mass-balance estimates for PCBs in the Arctic Ocean are confounded by a lack of measurements of seawater concentrations at key locations and by order-of-magnitude differences in concentrations in river water suspended particulate matter and in seawater between North American and Norwegian and Russian waters. There is very limited information on PCB congener profiles in seawater and bottom sediments, although substantial information is available on air concentrations of individual congeners.

Polychlorinated biphenyl budget for the Arctic Ocean

Total PCB (ΣPCB) was used as a basis for a preliminary mass balance given in Table 7-3. The average vapor pressure and H for tetra- and pentachlorobiphenyls, which predominate in air and seawater, were used. Extrapolations to Arctic temperatures (0 °C for water) for subcooled liquid vapor pressure versus temperature were made using slopes for CHBs and for Henry's Law constant ($\text{Log } H = 12.77 - 3416/T$) based on the work for PCBs by Tateya et al. (1988).

ΣPCB atmospheric concentrations based on monthly averages for 1993 for Alert (NAAO) and Ny Alesund (EAO) and water concentrations of 10 pg/L imply air-to-water fugacity ratios (FR) ranging from 0.1 to 4.0. Gas phase PCB concentrations are well-documented, although they are not collected at the same time or location as water sampling; thus, uncertainty in the direction of flux results mostly from uncertainty in water concentrations. Much lower FR values would result if PCB concentrations of 200 to >1000 pg/L determined for the Russian seas (Melnikov and Vlasov 1995) and the North Pole (Strachan et al. 1998) were used. However, the higher values are difficult to reconcile with those of Chernyak et al. (1995) for the Barents Sea (38 pg/L), Hargrave et al. (1988) at the Canadian Ice Island (7 pg/L), or Iwata et al. (1993) (6.6 to 9.3 pg/L) in the Chukchi Sea. Therefore, we have used 10 pg/L as an ocean-wide average and 25 pg/L for inflowing waters of the EAO (Table 7-3). Atmospheric deposition of PCBs via dryfall, gas exchange, and precipitation is estimated to equal volatilization. Outgassing of PCBs would be expected to be especially important in the Russian seas, assuming relatively high water concentrations there. Sedimentation is relatively more important for PCBs than for HCHs and CHBs.

Using average concentrations of 5 ng/L PCBs in Eurasian rivers (derived from concentrations on river suspended particulate matter [Melnikov and Vlasov 1995]), fluxes of 15 t/y are estimated, or about 23% of the total input estimated for PCBs (Table 7-3, Figure 7-6c). Although concentrations of 5 ng/L are high, it should be noted that PCB continues to be used in Russia because of the lack of funds to replace electrical transformers and other PCB-containing systems. Thus, elevated PCBs in Russian rivers flowing through urban areas (e.g., Ob River) may not be unreasonable. It should be noted that even if the concentrations were one-tenth as large (i.e., equivalent to North American rivers), they would still give rise to significant inputs. Thus, accurate determination of Eurasian riverine concentrations is critical for accurate estimates of loadings to the Arctic Ocean. By comparison, ocean current inputs are less significant because of low surface-water concentrations.

Removal of PCBs by sinking particles is potentially more significant than it is for HCHs and CHBs (Table 7-3) because of greater hydrophobicity. However, because of low particle fluxes in the interior Arctic Ocean, sedimentation and burial of PCBs are not predicted to be major removal processes compared to outflows via ocean currents. Sedimentation may, however, be particularly significant over the enormous shelves of the EAO whose sediments could, at least temporarily, sequester particle-bound PCBs. Sediment concentrations are generally higher in the EAO than in the NAAO, although measurements in the latter region are limited. For the budget, average concentrations of 0.5 ng/g dw were used for EAO sediments and 0.25 ng/g for the NAAO.

The PCB budget is highly uncertain. Further measurements, especially of seawater concentrations, and confirmation of high PCBs in Russian seas and rivers are critical to improve the PCB budget for the Arctic Ocean. The implication of this budget is that the Arctic Ocean is not in steady state and that inputs presently exceed outputs by a factor of 1.5 (i.e., this ocean is still in the loading phase).

Lessons learned: Persistent organic pollutants in the Arctic Ocean

One of the most extensive datasets of all the POPs is available for HCH both in terms of its environmental distributions and emission inventories. We can document with some confidence the events that led to the present situation where the world's highest HCH aquatic concentrations are to be found in the surface waters of the Canada Basin, far from where HCH was used. Indeed, it is these surface waters that will set the upper bound for persistence in global surface-water masses (Macdonald et al. 1997). From the 1940s until about 1980, the atmospheric concentrations of HCH increased (Figure 7-5). Hexachlorocyclohexane partitions overwhelmingly into cold water, and the atmosphere is effective in transporting HCH to the Arctic within days from regions where it was applied (Macdonald et al. 2000).

A global model of the distribution and fate of α-HCH was recently constructed, which considers source apportionment, mass balance, and trend prediction (Wania and Mackay 1999; Wania et al. 1999). The model shows that most of the α-HCH found in the Arctic originated in the northern temperate and boreal zones; transport from tropical regions was minor (Wania and Mackay 1999). Concentrations in the upper Arctic Ocean increased until about 1985 and then declined slowly. Atmospheric concentrations declined about 10-fold during the period 1980–1995. Water–air fugacity ratios indicated net deposition until about 1990, when the system was close to air–water equilibrium, and indicated net volatilization afterwards. These predicted trends and their magnitudes agree with the measurements made in arctic air and water (Wania et al. 1999).

In most of the Arctic Ocean, the HCH-containing water is sandwiched between an ice cover or seasonal stratification at the surface and a halocline at depth. As a result, volatilization losses are relatively minor (Figure 7-6a) and HCH is removed mainly by leaking out in the surface water exiting through the Archipelago or Fram Strait (Carmack et al. 1997; Macdonald et al. 1997, 2000) and by degradation (Harner, Kylin et al. 1999; Macdonald et al. 2000). What initially began as long-range atmospheric transport has now been taken over by oceanic transport. Influxes to the Arctic Ocean presently occur predominantly through ocean currents, and the Archipelago has become the downstream recipient of the large reservoir that has grown into the Canada Basin surface waters over the decades from 1950 to 1980. The clearance time for HCH from the Canada Basin is set predominantly by the oceanic replacement times, which are on the order of decades, and by in situ degradation rates.

Comparison of the well-constrained HCH budget with the more speculative one for PCBs illustrates the value of obtaining high-quality and high-resolution concentration data.

Gas exchange dominates the atmospheric loadings of CHBs and PCBs, but precipitation and dry particle fluxes also make important contributions. This underscores the importance of haze aerosols in scavenging CHBs and PCBs with relatively low vapor pressures and delivering them to the Arctic.

The low Henry's Law constant (H) of CHBs (comparable to that of HCH; Table 7-1) produces high equilibrium concentrations in cold water. By contrast, PCBs with relatively high H and a greater tendency to adsorb to sinking particles should not attain as high concentrations in the upper water column. Air–sea exchange depends directly on the area of open water in the Arctic Ocean, and improvements in our knowledge of ice cover is critically needed. Recent changes in ice cover or stratification (Macdonald et al. 1999) could exert a large influence on gas exchange of CHBs and other POPs.

Inputs of PCBs from Russian rivers are highly uncertain but appear to be much higher than from North American rivers, possibly because of the continued use of PCBs in Russia. Future decisions on use of PCBs in Russia are likely to have a large impact on the persistence of PCBs in the Arctic Ocean.

One of the main outflows of surface water from the Arctic Ocean is through the Canadian Archipelago. Even though international controls will stop or severely restrict new usage, the reservoir of elevated CHBs and HCHs in the Canada Basin is likely to buffer the biological systems in the Archipelago against declines for years or decades.

Lake Superior Case Studies

Description of Lake Superior limnology

The physical setting of Lake Superior (Eisenreich et al. 1981; Baker et al. 1991), its food web, and its contamination by PCBs (Kucklick and Baker 1998) have been well described.

Lake Superior is a large, dimictic lake located in a subarctic temperate zone in the mid-latitudes of North America (Table 7-2). It is the largest freshwater lake on earth by surface area and second in water volume after Lake Baikal. The lake stratifies in May or early June, overturns in November to January, experiences partial ice cover for 4 to 5 months a year, and full ice cover occurs approximately 1 in 10 years. Surface currents are generally counterclockwise and are especially pronounced along the Keweenaw Peninsula. Deep water temperatures are typically 2 to 4 °C, and the upper few meters of surface water of the open lake in summer can reach 15 to 20 °C. The average stratification depth, about 15 to 20 m, is sufficient to dampen deep water currents. Episodic benthic currents during poorly stratified periods redistribute sediments throughout the lake. Resuspension of deep sediments throughout the water column occurs at water depths less than 100 m. The lake is oligotrophic but receives erosional inputs in the extreme western end. The major source of particles to the lake is erosion in the western basin with additional inputs from primary productivity, resuspension, and aeolian input. Suspended solids concentrations average 0.1 to 0.5 mg/L during the summer and the total OC concentration averages 1.6 mg/L in the open lake. The fractional OC content of surface water particles range from 0.2 to 0.4, with lower levels in the western arm as a result of dilution by eroded red clays. Lake Superior has a small watershed to lake area ratio and receives more than half of its hydrologic inputs from direct precipitation (0.76 m/y). Atmospheric deposition supplies a substantial fraction of nutrients and pollutants to the system. The mass sediment accumulation rate in Lake Superior is 100 to 800 g/m^2 y or 0.01 to 0.03 cm/y (Eisenreich 1987) with a mean of ~ 350 ± 250 g/m^2 y in the depositional basins. The fractional OC content of surface sediments is typically ~ 0.01 to 0.02. Sediment accumulates over ~ 45% of the lake area and at water depths greater than 110 m in the open lake. Settling particle fluxes in Lake Superior are 0.13 ± 0.06 g/m^2 d and OC settling fluxes are 40 ± 15 mg C/m^2 d. Primary productivity is about 50 mg C/m^2 y.

Case study of polychlorinated biphenyls in Lake Superior

Description

Lake Superior has received most of its PCB input from precipitation, gas exchange, and to a lesser extent, dry particle deposition. Recent decreases of PCB in water and fish concentrations attest to a response to regulatory action. It has been hypothesized that PCB concentrations in Lake Superior are predominantly controlled by atmospheric exchange. Thus, the long-term trend in water-column and biota concentrations may very well be determined by atmospheric concentrations that continue to be supported by emissions from terrestrial soils and vegetation. This theme is common to the future of many of the historical POPs.

Polychlorinated biphenyl mass budget for Lake Superior

Essential features of the PCB mass budget in Lake Superior and the internal biogeochemical cycle may be found in Capel and Eisenreich (1985), Jeremiason et al. (1994, 1998), and Baker et al. (1991). Unique to Lake Superior is the availability of PCB concentration data for the water column measured by a single laboratory using similar methodologies since 1978. In addition, there are considerable data available on PCB concentrations and fluxes in atmosphere, water column, and sediment, permitting the construction of PCB mass budgets (Table 7-4) and the unlocking of the internal biochemical cycle.

Polychlorinated biphenyl concentrations in the water column of Lake Superior were measured from 1978 to 1992 to ascertain long-term trends (Jeremiason et al. 1994). Based on 35 congeners, PCB concentrations decreased from 0.99 ng/L in 1980 to 0.09 ng/L in 1992 (Figure 7-8a). Extrapolating to 82 congeners that were measured in later studies, concentrations decreased from 2.4 ng/L in 1980 to 0.18 ng/L in 1992. Polychlorinated biphenyl concentrations measured in the spring of 1993 (Anderson et al. 1998) suggest a further decrease to 0.083 ng/L. The water-column PCB concentration therefore decreased with a pseudo–first-order rate constant of ~ 0.20/y, corresponding to a half-life of about 3.5 years (95% CI, 2.5 to 6.3 y). The loss of PCBs from the water column of Lake Superior from 1980 to 1992 amounted to an impressive 26,500 kg. Sediment cores collected from 1980 to 1992 suggested that the total sedimentary inventory amounted to 4900 kg since the time of PCB production began. Thus, partitioning of PCBs to aquatic particles followed by delivery to the sediment and burial cannot come close to accounting for the missing PCBs.

Sediment cores collected in 1986 and 1992 were used to estimate the accumulation rate of PCBs. Surficial sediment PCB concentrations varied from 8 to 18 ng/g in cores for which sediment accumulation rates varied from 70 to 340 g/m^2y. The PCB accumulation rate and inventory for each core was divided by the ^{210}Pb-determined sediment focusing factor (0.85 to 1.8) to yield a lake-wide sediment inventory of 50 g/m^2 or 4100 kg, and a lake-wide PCB accumulation rate of 110 kg/y. Riverine inputs were estimated as the product of an average tributary concentration of 2 ng/L and the annual volumetric flow. Other discharges directly to the lake contributed ~ 41 kg/y. Atmospheric inputs were estimated from measured volume-weighted rain concentrations and annual rainfall of 0.76 m/y as 125 kg/y. Dry particle deposition of PCB was estimated by Strachan and Eisenreich (1988, 1990) as 32 kg/y. Outflow fluxes, determined by multiplying the lake-wide water-column concentration by the outflow via the St. Mary's River, amounted to a loss of 60 kg/y.

Table 7-4 Annual budgets in Lake Superior for PCBs (1986)[1] and CHBs [2]

	ΣPCBs		CHBs	
	In (kg)	Out (kg)	In (kg)	Out (kg)
Rivers	110	60	38	77
Atmosphere – gas absorption	??	—	348	—
Volatilization	—	1900 (net)[3]	—	390 (net)[3]
Precipitation	125	—	12.5	—
Dry particle deposition	32	—	6.3	—
Direct discharges	41	—	0	—
Settling particle flux	3000	2890	—	—
Net sediment accumulation	—	110	—	90
Degradation (Transform)	Nil	Nil	—	?
Water-column loss	—	1800	—	—
Total	308	2070	57	557
	Inventory (kg)	Time scale		Time scale
Atmosphere	200	—		—
Water column	10,000	4.8 y[4]		21 y[4]
Sediments	4900	—		—
Total	15,100	—		—
Water concentration decrease($\tau_{1/2}$)	—	3.5 y[5]		—
Fish concentration decline($\tau_{1/2}$)	—	5–10 y		Not observed
Air concentration decrease($\tau_{1/2}$)	—	3–10 y		—

[1]Jeremiason et al. (1994); [2]Swackhamer et al. (1999); [3]Net volatilization = inputs – outputs + first-order loss rate; [4]Residence time for water column (inventory/outflow); [5] First-order rate constant for water-column loss = –0.20/y

In the mass balance (Table 7-4), volatilization was assumed to account for the difference between inputs and outputs of PCBs. Thus net volatilization (volatilization – absorption) was estimated as ~ 1900 kg/y or ~ 63 ng/m^2 d (net vol = inputs – outputs + M_{lost}). Jeremiason et al. (1994) constructed a model that predicted that volatilization exhibited a rate constant of –0.24/y, and sedimentation of –0.004/y, supporting the finding that volatilization predominated over sedimentation as the major loss pathway of PCBs from the lake. This mass balance implied that oligotrophic lakes in the mid-latitudes do not always act as sinks for hydrophobic

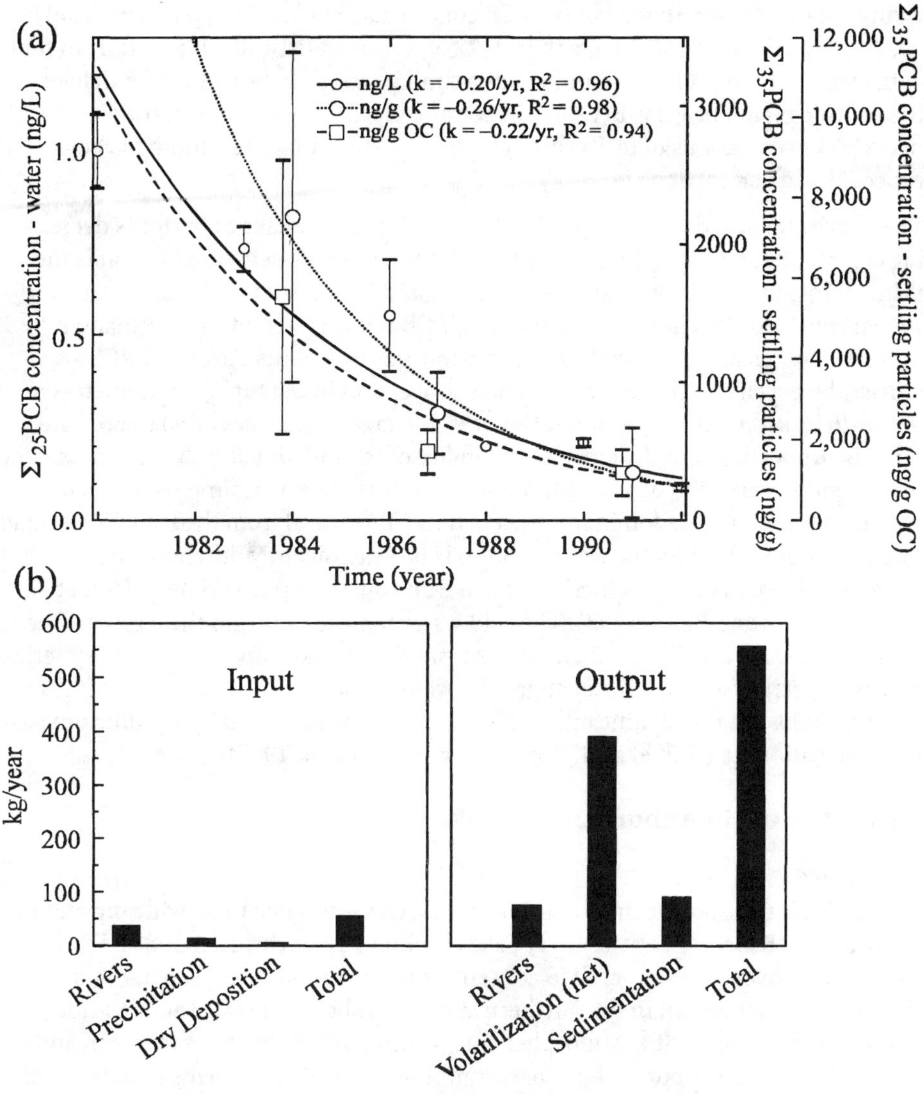

Figure 7-8 a) Decrease in ΣPCB concentrations and particle fluxes in Lake Superior (after Jeremiason et al. 1988) and b) CHB inputs and outputs for Lake Superior during the mid-1990s (after Swackhamer et al. 1999)

chemicals such as PCBs but presently represent sources and therefore slow the rate of decrease of PCBs in the atmosphere. Hornbuckle et al. (1995) calculated the net volatilization flux of PCBs from Lake Superior in 1992 as ~ 250 kg/y. This is similar to the ~ 400 kg/y loss predicted by application of the Jeremiason model to 1992, assuming an overall water-column first-order rate constant of –0.20/y.

Figure 7-8a shows the change in the PCB concentration of settling particles from 1984 to 1991. The rate of change (k = –0.22/y) is not statistically different from that of the water-column PCB loss and yields a similar overall half-life for PCB delivery to the benthic region of ~ 3 y. Because the water concentrations support the aquatic particle concentrations of PCBs through phase partitioning, this finding is both expected and comforting.

The above scenario demonstrates that the PCB burden in Lake Superior is decreasing exponentially with a rate constant of –0.20/y and that net volatilization is the primary loss process. Both Baker et al. (1991) and Jeremiason et al. (1998) have shown conclusively that sediment burial of PCB is unimportant as a permanent loss process: biologically produced natural organic matter and its burden of PCBs is efficiently recycled in the benthic boundary layer and in the top few millimeters of the surficial sediment. Recycling ratios of PCB congeners, values of 50 to 500, are directly proportional to the aqueous solubility of the individual PCB congeners. The inefficient preservation of autochthonous organic matter in sediments leaves net volatilization as the predominant process for PCB removal from the lake. The overall rate of loss of PCBs from the water column is not significantly different from regional decreases in atmospheric and fish PCB concentrations (Hillery, Hoff et al. 1997; Kucklick and Baker 1998). The cold water temperature and the large surface area available for exchange enhance the sensitivity of Lake Superior to atmospheric inputs, especially air–water exchange. The water-column concentrations of PCBs in the mid-1990s are not significantly different from that predicted by equilibrium gas-phase partitioning (Hoff et al. 1996; Mackay and Bentzen 1997).

Case study of chlorobornanes in Lake Superior

Description

Atmospheric transport from the southern U.S. was estimated to provide most of the historical CHB loadings to the Great Lakes (Voldner and Schroeder 1989). During the years of toxaphene usage, CHB concentrations in ambient air were far higher in the southern states than in the northern states and showed a clear south-to-north gradient (Rice et al. 1986). Atmospheric deposition in the Great Lakes region and eastern Canada is supported by concentration–time profiles of CHBs in dated peat cores collected during the mid-1980s (Rapaport and Eisenreich 1986). Peak deposition took place in the mid-1970s, and the shape of the profiles tracked the production records. Sediment cores from the Great Lakes show similar profiles in some locations (Howdeshall and Hites 1996; Pearson et al. 1997), but suggest that non-atmospheric inputs may have taken place in Lake Michigan (Pearson et al. 1997).

The concentration of CHBs in Lake Superior (900 to 1100 pg/L [Jantunen and Bidleman 2000b; Swackhamer et al. 1999]) is higher than in any of the other Great Lakes (170 to 470 pg/L in Lakes Huron, Michigan, Erie, and Ontario [Swackhamer et al. 1999]). Furthermore, in contrast to PCBs and DDT, concentrations of CHBs in fish from Superior did not decline between 1982 and 1992 (Glassmeyer et al. 1997). Possible reasons for this include 1) Lake Superior has (or had) point sources of CHBs resulting in higher inputs than in the other lakes and 2) Lake Superior has slower output rates and/or a greater capacity for CHBs because of its physical differences compared to the other lakes. Lake Superior has the largest volume, surface area, and depth of all the lakes as well as lower suspended particle concentration and lower average temperatures for most of the year (Table 7-2). Toxaphene use in the Great Lakes drainage basin was estimated to have been approximately 40×10^6 kg, accounting for less than 4% of the U.S. total use (VonRumker et al. 1974; Voldner and Schroeder 1989; Swackhamer et al. 1999), and the overall agricultural drainage basin for Lake Superior is small. It is therefore unlikely that agriculture within the drainage basin explains the high CHB concentrations in the lake. The mass of CHBs needed to increase the water concentration above that in the other lakes is sufficiently large that a spill or other point source is also unlikely.

Modeling loadings and retention of chlorobornanes in Lake Superior

Swackhamer et al. (1999) investigated why CHB concentrations differ between Lakes Superior and Michigan using a Stella II dynamic mass balance. Inputs and outputs of CHBs were modeled from 1950 to the present with monthly or seasonal time steps. Inputs included precipitation, dry particle deposition, and gas absorption. Losses included sedimentation, volatilization, and outflow through St. Mary's River. Because historical atmospheric concentrations of CHBs are not available, they modeled the atmospheric profile over time from measurements made in 1989 (Hoff et al. 1992a, 1992b; Hoff, Muir et al. 1993) extrapolated back in time to match the shape of the known production curve for toxaphene (Rice and Evans 1984). Thirty-year average surface temperatures from National Oceanic and Atmospheric Administration satellite data, the Henry's Law constant for technical toxaphene as a function of temperature (Jantunen and Bidleman 2000a; Table 7-1), and the air-side mass-transfer coefficient for gas exchange (Hoff, Muir et al. 1993) were used to predict historical water concentrations. The predicted water concentrations for the mid-1990s were very similar to those measured in Lakes Superior and Michigan, and the sediment profiles predicted for Lake Superior were similar to observed sediment core measurements. Observed sediment profiles for Lake Michigan did not match predictions, suggesting that some non-atmospheric input of CHBs had occurred.

The model indicated that CHBs in the two lakes increased up to the late 1970s, then declined. However, the decline in Lake Superior was slower predominantly because of a lower sedimentation rate of particle-associated CHBs and less volatilization from the lake as a result of its colder temperature. While the average temperature of Lake Superior is only a few degrees colder than that of Lake Michigan, this slight difference has a large effect on the fate of CHBs over several years. Despite considerable uncertainty in the model, it accurately portrayed the relative differences between the two lakes and highlighted the importance of site-specific characteristics (temperature, particle flux) in controlling fate and transport of POPs.

Present-day transport and air–water exchange of chlorobornanes

Despite the banning of toxaphene, atmospheric transport continues. Hoff et al. (1992a, 1992b; Hoff, Muir et al. 1993) noted that high levels of CHBs in air of southern Ontario occurred during transport events from the southern and central U.S. Chlorobornanes were widespread and uniformly distributed in lichens across Ontario (Muir et al. 1993). Average concentrations of CHBs in ambient air of South Carolina and Alabama measured between 1994 and 1996 were about 10 times higher than in air over Lake Superior (Bidleman et al. 1998; Jantunen et al. 2000). Soil residues of CHBs in the 50 to 2000 ng/g dry weight range were found in a 1995 survey of agricultural fields in Alabama, and a fugacity model predicted emission of CHBs and atmospheric concentrations within the range of monitored values (Harner et al. 2000). It is therefore likely that volatilization from contaminated soils will supply CHBs to the atmosphere for years to come.

Concentrations of CHBs in Lake Superior surface water collected in August 1996 and May 1997 (Jantunen et al. 2000) averaged 900 pg/L, in agreement with concentrations of 1100 pg/L determined by Swackhamer et al. (1999). Atmospheric concentrations in 1996 to 1997 were 6 to 22 pg/m^3 over Lake Superior (Glassmeyer et al. 1999; Jantunen and Bidleman 2000b) and were slightly lower over Lake Ontario (Shoeib et al. 1999), showing a decrease since CHBs were measured at Green Bay, Wisconsin in June 1989 (59 pg/m^3) and over the lower four Great Lakes in August 1990 (33 pg/m^3) (McConnell et al. 1998). Water–air fugacity ratios, calculated from air and water concentrations and the temperature-dependent Henry's Law constant for technical toxaphene (Jantunen and Bidleman 2000a), averaged 1.8 in August and 1.3 in May. Thus, CHBs are close to being in air–water equilibrium, with a slight potential to volatilize.

Mass budget for chlorobornanes in Lake Superior

The standing stock of CHBs in the water column is 12,100 kg, based on a water concentration of 1000 ng/m^3 (Jantunen and Bidleman 2000b; Swackhamer et al. 1999) and the lake volume (1.21×10^{13} m^3). The sediment inventory has been estimated at 6500 kg (Pearson et al. 1997). Riverine input of 38 kg/y was calculated from the annual inflow (5.4×10^{10} m^3) and the assumption that river water has the same CHB content as precipitation (700 ng/m^3; see Figure 7-8b). Loss of 77 kg/y by advection through the St. Mary's river was calculated from the concentration of CHBs in lake water at the outflow volume (Swackhamer et al. 1999). Sedimentation of ~ 90 kg/y was estimated from the present accumulation rate of 970 to 1400 ng/m^2 y (Pearson et al. 1997) and the lake area is 8.21×10^{10} m^3.

Input and removal of toxaphene by atmospheric processes were estimated by Swackhamer et al. (1999) according to the scheme of Hoff et al. (1993b). Processes considered were wet and dry deposition of gaseous and particulate CHBs and air–water exchange.

Gas exchange (Table 7-4, Figure 7-8b) is by far the dominant loading process for CHBs in Lake Superior. The net gas exchange varies seasonally, from 398 ng/m^2 (deposition) in summer to –2964 ng/m^2 (volatilization) in fall. On an annual basis, the net exchange is volatilization and accounts for removal of –390 kg CHBs (70%) from the lake (Swackhamer et al. 1999). Removal by sedimentation and outflow from the St. Mary's River account for 16% and 14%, respectively, of total losses (Swackhamer et al. 1999). The residence time of CHBs in the water column, estimated from the inventory and removal, is about 21 y.

Lessons learned: Polychlorinated biphenyls and chlorobornanes in Lake Superior

The story of CHBs in Lake Superior illustrates how the lake's limnology can have a large influence on the accumulation and clearance of POPs. The cold water temperature and the large surface area available for exchange enhance the sensitivity of Lake Superior to atmospheric inputs, especially air–water exchange. The model of Swackhamer et al. (1999) attributes the 2- to 5-fold higher concentration of CHBs in Lake Superior to a lower sedimentation rate and a colder average temperature than those of the other Great Lakes. Outgassing of Lake Superior, demonstrated for PCBs (Jeremiason et al. 1994), is also an important loss mechanism for CHBs (Figure 7-9), but on a longer time scale. The half-life for PCB decline in Lake Superior water is 3.5 y (Jeremiason et al. 1994), whereas the residence time is 12 y for CHBs. Clearance of CHBs by volatilization would be faster were it not for atmospheric burdens buffered by continued volatilization of CHBs from agricultural soils in the southern states upwind from Lake Superior (Bidleman et al. 1998; Jantunen et al. 2000; Harner et al. 2000).

The PCB burden in Lake Superior is decreasing exponentially with a rate constant of –0.20/y, primarily by net volatilization. Although PCBs are more strongly sorbed to sinking particles than are CHBs, sedimentation does not provide a permanent loss because of the efficient recycling of organic matter and its burden of PCBs in the benthic boundary layer and in the top few millimeters of the surficial sediment. Because PCBs are not efficiently buried in sediments, air–water exchange becomes the predominant loss mechanism. The overall water-column decrease in PCBs parallels regional decreases in atmospheric and fish PCB concentrations (Hillery et al. 1997a–1997c; Kucklick and Baker 1998). The water-column concentrations of PCBs in the mid-1990s are not significantly different from those predicted by equilibrium gas-phase partitioning (Hoff et al. 1996; Mackay and Bentzen 1997).

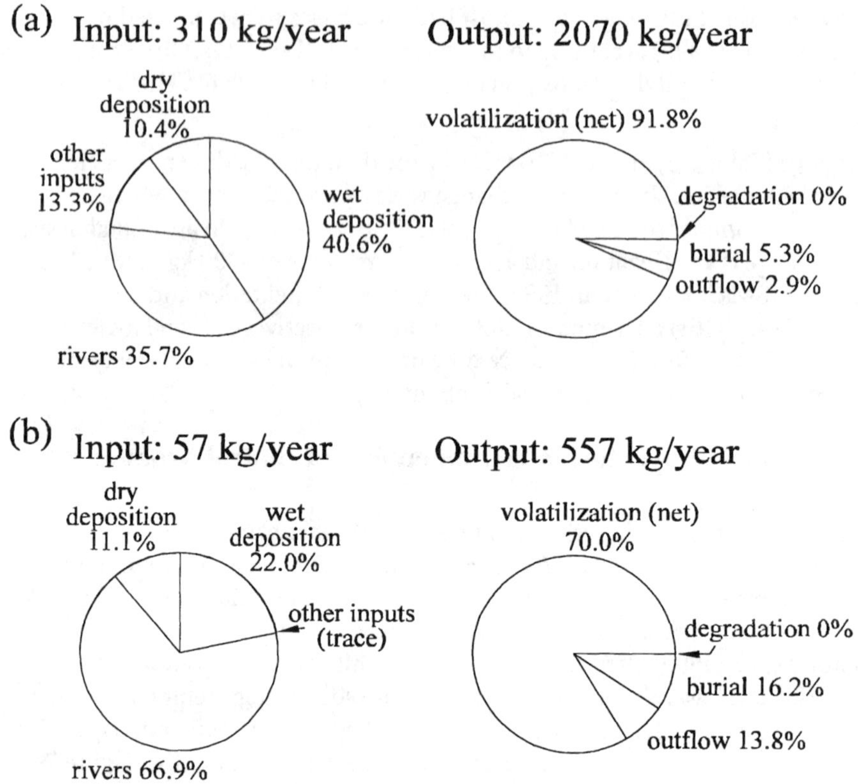

Figure 7-9 (a) ΣPCB inputs and outputs in Lake Superior (Jeremiason et al. 1994) and (b) CHB inputs and outputs in Lake Superior (Swackhamer et al. 1999)

Case Study of Atrazine in Lake Michigan

Description of Lake Michigan limnology

Lake Michigan, located in the mid-latitudes of North America, is a large oligo- to mesotrophic, dimictic lake with a water-residence time of ~ 69 y and a flushing rate of ~ 100 y (Table 7-2). The southern basin is heavily populated on its southwestern shores and receives the majority of river drainage from agricultural lands. The northern basin is deeper and colder and is removed from most urban activities. Water circulates generally south along the western shore and north along the eastern shore. This phenomenon in the southern basin results in resuspension of sediments on the western shelf and deposition of sediment in the eastern part of the southern basin. The major outflow through the Mackinac Straits in the north results in significant south-to-north transport of water, sediment, and chemicals, although sediment is efficiently trapped in the deep northern basin. Organic matter in the depositional zones of the north and south basin is preserved such that OC concentrations in sediments are 0.02 to 0.05 g/g dw. Sediment cores reflect sedimentation rates ranging from 0.03 to 0.1 cm/y on a whole-lake basis. The distribution of organic contaminants in bottom sediment cores and the water column point to efficient lateral transport of water and sediment. Recent data from the Lake Michigan Mass Study show that atrazine, PCB, and Hg concentrations in the water column vary little from north to south, east to west, and top to bottom. The lake experiences partial ice cover in especially cold winters, but harbors and embayments usually freeze over each year for 1 to 2 months. The lake has not completely frozen over in the last decade.

Description of atrazine and its metabolites

The physicochemical properties of atrazine and its metabolites determine persistence in large lakes. Several degradation pathways have been identified for herbicides, including aerobic and anaerobic microbial degradation (Giardina et al. 1980, 1982; Giardi et al. 1985; Winkelmann and Klaine 1991), aqueous and soil photolysis (Rejto et al. 1983; Pelizzetti et al. 1990; Ciba-Geigy Corporation 1994), hydrolysis (Muir and Baker 1978; Solomon et al. 1996; Lerch et al. 1997), and oxidation (Beltrán et al. 1996). Major transformation products of atrazine that are present in water are deethylatrazine (DEA), deisopropylatrazine (DIA), deethylhydroxyatrazine, and hydroxyatrazine (Pereira and Rostad 1990; Kruger et al. 1993; Ciba-Geigy Corporation 1994; Thurman et al. 1994; Lerch et al. 1997), although the hydroxyatrazine and deethylhydroxyatrazine are prominent in soil (Lerch et al. 1997). The physicochemical properties of atrazine and other related herbicides that create a potential for accumulation in lake water include high water solubility, moderate hydrogen bond acceptor and donor capability (Welhouse and Bleam 1993), low vapor pressure, and low partitioning to organic matter (Mills and Thurman 1994; Seybold et al. 1994). After application in the spring, herbicides are washed from soils in a large pulse, depending on the amount and timing of rainfall (Thurman et al. 1991, 1992; Pereira and Hostettler 1993; Richards and Baker 1993, Schottler et al. 1994; Muir and Grift 1995). Volatilization, particulate transport, and wet deposition also may deposit herbicides to lakes, especially in the warmer months and following application. Long-range transport and atmospheric deposition may be important in lakes removed from agricultural activity such as Lake Superior (Glotfelty et al. 1990; Schottler and Eisenreich 1994; Cromwell and Thurman 1995; Muir and Grift 1995; Goolsby et al. 1997).

Mass balance of atrazine in the Great Lakes

Atrazine and the major atrazine metabolites, DEA and DIA, were measured in Great Lakes waters from 1990 through 1995 (Schottler and Eisenreich 1994, 1997; Hines 1996). Concentrations were ~ 3 to 4 ng/L in Lake Superior, 35 to 40 ng/L in Lake Michigan, ~ 20 ng/L in Lake Huron, ~ 30 to 110 ng/L in various basins of Lake Erie, and 70 to 95 ng/L in Lake Ontario. Inventories of atrazine plus the major metabolites exceeded 600 t, with anticipated water-column residence times on the order of years. Atrazine and its metabolites were approximately the same concentrations spatially and temporally within each system with the exception of Lake Erie, attesting to efficiency of mixing and to an apparently long water-column residence time.

Two mass balances have been undertaken. The first was intended to estimate the in situ transformation rates and the relative importance of sources and sinks and to assess future conditions given various scenarios for atrazine application in the Midwest Corn Belt. The second was a mass balance constructed on Lake Michigan based on an intensive series of water-column measurements, tributary inputs, and atmospheric deposition. Atrazine contrasts starkly with most other POPs under consideration because it has a low vapor pressure and relatively high water solubility, which lead to particle transport in the atmosphere, little tendency to volatilize, and little loss via sedimentation. Furthermore, atrazine is known to transform to a variety of metabolites at rates unknown in lakes.

The concentration of atrazine in a lake (or basin) can be described by the expression (Schottler and Eisenreich 1997)

$$\text{Concentration} = \int_{t=0}^{t} (\text{Mass In} - \text{Mass Out})/\text{Volume} \qquad \text{(Equation 7-1)}.$$

The premise of Equation 7-1 is incorporated into a dynamic mass-balance model constructed on the basis of

$$\text{Mass}_t = \text{Mass}_{(t-dt)} + (\text{Inputs}_t - \text{Outputs}_t)\ dt \qquad \text{(Equation 7-2)},$$

where Mass_t is the mass of herbicide in the lake at a particular time, Inputs_t and Outputs_t (kg/y) are the inputs and outputs at any particular time, and dt is the time step. Solutions to this general equation for a selected time period are solved iteratively using a fourth-order Runge Kutta algorithm.

Water concentrations were calculated by combining the premise of Equation 7-1, into the format of Equation 7-2 by using a time step dt = 0.001 y. Tributary and precipitation loadings were varied monthly based on known data, and tributary loading was varied annually based on trends in atrazine use (see Table 7-5). Air–water exchange and dry deposition were varied monthly using seasonal estimates of atrazine air concentration. Losses through internal transformation and sedimentation varied with water concentration. Atrazine concentrations in Lakes Huron, Michigan, Ontario, and Superior showed no significant spatial trends from 1991 to 1995 and were, therefore, treated as one-box, well-mixed systems. Lake Erie, showing a strong west-to-east concentration gradient, was divided into 3 basins according to hydrology, morphology, and existing basin models of Lake Erie. Tributary loading to the Great Lakes was estimated from application records of atrazine use in watersheds of each of the lakes (compare Bodo 1991). An essential feature of tributary loading captured the average fraction of total agricultural

Table 7-5 Inputs and outputs of atrazine in the Great Lakes (tonnes/y)

	Superior	Michigan	Huron	Erie-Central	Erie-East	Ontario
River inflows	0.03	9.0	4.3	0.8-3	0.45–1.6	1.5–3
Precipitation	0.82	2.6	1.2	0.65	0.11	0.38
Dry deposition	0.12	0.16	0.17	0.15	0.005	0.16
Air–water exchange	0.035	0.03	0.03	0.25	0.001	0.001
Connecting channel			3.1	10–13	13–20	14–21
Outflow	0.26	2.9	3.5	13–20	14–21	16–19
Transformation	0.45	8.9	5.5	3–5	1.1–1.5	8–9
Sedimentation	0.003	0.025	0.015	0.06	0.025	0.03

Source: Schottler and Eisenreich (1997)

application that reaches rivers (~ 0.5% to 2%) and the spring flush phenomenon (60% to 90% of annual atrazine flux in rivers occurs in first 90 d after application).

Precipitation inputs were based largely on seasonal data by Goolsby et al. (1997). Fluxes varied from 10 $\mu g/m^2$ y for Lake Superior to ~ 40 to 60 $\mu g/m^2$ y for the Lake Erie basins. Dry particle deposition was estimated from atmospheric measurements of gas-and particle-phase concentrations, although most measurements over the lake indicated that greater than 90% of atrazine is in the particle phase. Air–water exchange was estimated monthly for each lake based on calculated distributions between the gas and particle phases and the spatially dependent concentrations in the water column of each lake.

Sedimentation was assumed to be negligible for atrazine, since > 95% is in the dissolved phase in the water column. Outputs and inputs via rivers and connecting channels were estimated from lake water concentrations and annual volumetric flows. Internal transformations were estimated by solving the dynamic mass balance and calibrating to the measured water column concentrations.

The dominant inputs of atrazine to most of the Great Lakes are tributary or connecting-channel flows followed by atmospheric deposition (Table 7-5). Only in Lake Superior is atrazine input dominated by atmospheric processes. The dominant outputs are outflows and transformations with losses to sediments and the atmosphere being unimportant. For example, tributaries contribute 76% of atrazine to Lake Michigan, and transformation accounts for 76% of the annual loss (Figure 7-10). Because inputs of atrazine are similar in magnitude to outputs, the concentrations in Lake Michigan have remained constant since 1991 and the residence time of atrazine relative to all processes is on the order of years. Inputs range from 1 t/y for Lake Superior to 16 to 25 t/y for Lake Ontario (Table 7-6). Transformation (mostly to DEA and DIA) ranges from 1% of the inventory per year in Lake Superior to 10% per year for Lake Erie. The comparatively slow transformation in Lake Superior water probably can be assigned to exceedingly low concentrations, cold water (> 95% of lake volume is hypolimnetic water and always less than 4 °C), and very low biological productivity (leading to low bacterial populations).

An atrazine transformation half-life can be calculated from the transformation rate and the first-order rate law such that

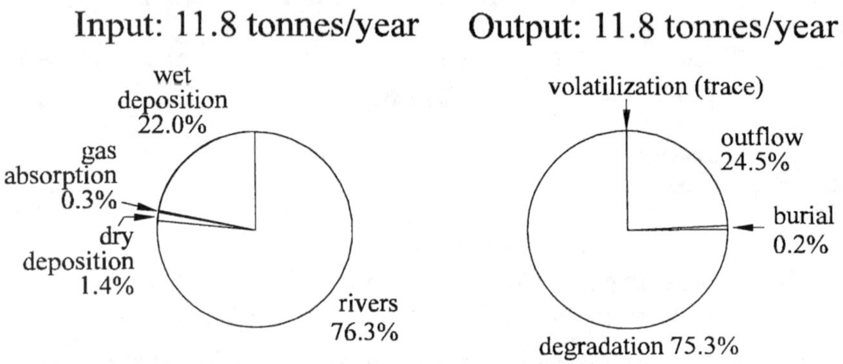

Figure 7-10 Input, output, and transformation of atrazine in Lake Michigan

Table 7-6 Comparison of atrazine inventory in the Great Lakes to the annual input of atrazine, and the annual loss though internal transformation

	Inventory[1] (t)	Annual input (t/y)	Annual input divided by inventory (%)	Percent of inventory lost annually by transformation
Superior	36	1.03	3	1
Michigan	176	11.8	7	5
Huron	75	8.78	12	7
Erie[2,3]	30–60	10–25	30–40	10
Ontario[2]	125–140	16–24.5	12–18	6

[1] Inventory was calculated from average lake-wide concentrations (Schottler and Eisenreich 1994,1997) and volume of the lake. [2] Concentrations and loadings to Lakes Erie and Ontario vary yearly; thus, a range of values is presented. [3] The 3 basins of Lake Erie were combined to give total annual inputs and loss by transformation and a lake-wide inventory.

An atrazine transformation half-life can be calculated from the transformation rate and the first-order rate law such that

$$t_{1/2\ \text{transformation}} = \ln 2/k_t \qquad \text{(Equation 7-3)},$$

where $t_{1/2\ \text{transformation}}$ is the pseudo–first-order atrazine transformation half-life. An overall atrazine loss rate constant is calculated as the sum of the individual loss rate constants:

$$k_{total} = k_t + k_f + k_s \qquad \text{(Equation 7-4)},$$

where k_{total} is overall atrazine loss rate constant (y^{-1}), k_t is the in situ transformation rate constant,

k_f is the loss rate constant due to outflow (reciprocal of lake residence time [Table 7-6]), and

k_s is the sedimentation loss rate constant.

In the Great Lakes, sedimentation accounts for <<1% of annual atrazine loss; thus, k_{total} is essentially $k_t + k_f$. Finally, the bulk water-column residence time for atrazine is simply

$$R_{atz} = 1/(k_{total}) \qquad \text{(Equation 7-5)},$$

where R_{atz} is the atrazine water-column residence time. The bulk water-column residence times vary from 1 y in Lake Erie to as much as 25 y in Lake Michigan (Table 7-7). Internal transformation rates for atrazine estimated from the dynamic mass-balance model are on the order of 0.07 to 0.13/y. Lake Michigan exhibits an overall atrazine half-life of 10 to 27 y, with the in situ transformation rate accounting for perhaps 75% of the chemical residence time, forming mostly DEA and DIA based on metabolite distributions.

Table 7-7 Summary of loss rates and water column residence time of atrazine in the Great Lakes

Lake	Internal transformation[1] $t_{1/2}$ (y)	Internal transformation[1] k_t (y^{-1})	Flushing rate[2] k_f (y^{-1})	Water-column residence time[3] $(k_t + k_f)^{-1}$ (y)
Superior	> 7	< 0.1	0.005	> 10
Michigan	13.9	0.05	0.016	10–27
Huron	9.2	0.075	0.048	6 –11
Erie West	2.3	0.3	7.7	< 0.13
Erie Central	5.5	0.13	0.61	1.1–1.5
Erie East	7.7	0.09	1.25	0.72 –0.77
Ontario	9.9	0.07	0.13	4.3–5.9

Source: Schottler and Eisenreich (1997)

[1] Internal transformation values represent the best estimate; a range of values is discussed in the text.
[2] Flushing rates are the reciprocal of lake residence times. [3] Range for atrazine water column residence time was based on the range of k_t's determined at high and low estimates of inputs discussed in text.

Mass balance of atrazine in Lake Michigan

Tributary loads of atrazine to Lake Michigan in 1994 and 1995 were calculated from measured water concentrations at 11 river mouths over the hydrologic year (Hines 1996). Because the 11 tributaries sampled represent 78% of the hydrological tributary inputs, the loading of atrazine to Lake Michigan using a mass balance was proportionally increased from 1640 kg/y to 2100 kg/y. The average concentrations of atrazine, DEA, and DIA in Lake Michigan tributaries were 73, 30, and 24 ng/L, respectively, while corresponding concentrations in lake water were 39, 26, and 16 ng/L, respectively. The net direction of air–water exchange for atrazine was from air to water. A mass balance (Table 7-7) was constructed using a steady-state approximation given the constant water-column concentrations:

$$\frac{d[C]_{atz}}{dt} \approx 0 = \text{tributary load} + \text{atmospheric load} - \text{outputs} - \text{transformation} \quad \text{(Equation 7-6).}$$

The tributary loadings for atrazine assessed through extensive sampling of river waters in this study were less than one-half those estimated by Schottler and Eisenreich (1997) based on application and runoff rates. The discrepancy is not large given potential year-to-year variations in rainfall, tributary flow, and runoff rates. What is significant is that the in situ transformation rate of atrazine in Lake Michigan estimated by Schottler and Eisenreich (1997) permits the steady-state mass budget of Hines (1996) to balance. Atrazine and its metabolites continue to accumulate in Lake Michigan with expected decreases only years after atrazine usage is minimized or stopped in the basin. An additional observation of importance is that atrazine and its metabolites drain from the northern basin subwatersheds of Lake Michigan even though there is no documented use of the herbicide there. Atmospheric transport and deposition is implicated as the source.

Lessons learned

Although atrazine has a 30- to 90-d half-life in soils and soil-water in the upper 1 to 2 m of the soil horizon, transport out of this zone into receiving waters (rivers and lakes) leads to significantly longer half-lives. Lake Michigan and other large aquatic systems are most sensitive to tributary inputs of atrazine, while the long-term response is controlled by long hydraulic residence times and slow transformation rates. Despite the fact that perhaps only 1% of the applied atrazine is lost by transport to rivers and lakes and another 1% to aerial transport, the large quantities applied together with efficient transport and slow transformation rates in water lead to accumulation in aquatic systems.

In contrast to most other POPs, the atmosphere plays little or no role in the removal of atrazine from large bodies of water. The reason for this is that atrazine is a polar solute with little tendency to volatilize because of its low vapor pressure and Henry's Law constant. Regional atmospheric transport and deposition in precipitation is an important pathway for atrazine's entry into the Great Lakes, but atmospheric pathways may not be as important for the global dispersal of atrazine as they are for semi-volatile organochlorines.

Case Study of Polychlorinated Biphenyls in the Baltic Sea

Description

The Baltic is a semi-enclosed sea of about 415,000 km^2 characterized by the interaction of fresh and saltwater sources (Table 7-2). It is the largest brackish body of water in the world and is distinguished by its division into a series of basins of varying depths, separated by shallow areas or sills. The connection to the North Sea is very narrow, the shallowest sill being only 18 m deep. Thus, inflows of salt water from the North Sea, which are forced by complex meteorological processes over northern Europe and the North Sea, must be extremely strong to penetrate and renew the deepest waters of the Baltic Sea. The brackish-water character of the Baltic Sea plays an important role in its ecology. Water circulation in the Baltic Sea is weak. Surface water movement is most affected by winds, a significant factor in water mixing and distribution of pollutants. In winter, the Baltic Sea is largely ice-covered, which renders it even more vulnerable to the effects of pollution. Finally, rivers bringing fresh water to the Baltic Sea also contribute contaminants including POPs. These rivers include Vistula, Odra, Daugava, Neman, and several other less-contaminated rivers.

Nine countries share the Baltic Sea coastline: Sweden and Finland to the north; Russia, Estonia, Latvia, and Lithuania to the east; Poland in the south; and Germany and Denmark in the west. About 16 million people live on the coast, and nearly 80 million live in the entire catchment area of the Baltic Sea. Major sources of contaminants to the Baltic include municipal and industrial wastewater treatment plants; industrial facilities (e.g., pulp and paper industrial plants and metal smelters); and runoff from agriculture, forestry, and livestock operations. Other sources of contaminants to the area include fish farming, mobile sources, combustion of fuel, and former military installations.

Polychlorinated biphenyls in the Baltic Sea

Throughout the 1960s and 1970s, the usage and thus the release of PCBs in northern Europe underwent large changes, while in the 1980s a slower development seemed to be established. Long-term trend studies on PCB levels in biota from the Baltic confirm relatively stable or only slightly decreasing concentrations of PCBs towards the late 1980s, when most of the measurements of the Baltic Sea Project were performed (Bignert et al. 1995).

Mass balance for polychlorinated biphenyls in the Baltic Sea

Contaminant fluxes have been used in a comprehensive non-steady-state fugacity model (Pacyna 1998) to assess the mass balances of POPs in various sub-basins of the Sea and their changes with time. Ideally the environmental behavior of each PCB congener should be modeled individually. However, this first and still-preliminary effort uses the sum of the 6 most-commonly reported PCB isomers 52, 101, 118, 138, 153, 180 (i.e., one tetra-, two penta-, two hexa- and one heptachlorobiphenyl congeners) for input and evaluation. The number of chlorine atoms in the PCB molecule strongly affects vapor pressure, aqueous solubility, and the octanol–water partition coefficient.

The selection of realistic degradation half-lives for POPs in the environment is problematic. Degradation rates established in the laboratory cannot be transferred with confidence to real environmental conditions. Two recent investigations of PCB profiles in sediment cores (Axelman et al. 1995; Kjeller and Rappe 1995) suggest that degradation of PCBs in Baltic sediments is either very slow or does not occur at all. Therefore, a long degradation half-life in surface sediment was used.

The model requires 3 types of source and emission information:

- average bulk air concentrations CA_i (vapor and aerosol-bound phase),
- average bulk river water concentrations CX_i (truly dissolved and particulate OC [POC]-bound phase), and
- an estimate of the direct discharge rate into water EW_i, which summarizes input from industrial and urban sources and sewage treatment plants.

The annually averaged air concentrations listed in Table 7-8 were estimated from recent measurements in numerous locations around the Baltic Sea (Brorström-Lundén 1995). Measurements in various rivers draining into the Baltic Sea showed highly variable PCB concentrations making a selection of an annually averaged value for all rivers draining into one sub-basin difficult. A typical value of 400 pg/L was

Table 7-8 Assumed average concentrations of PCBs in the atmosphere above and in river water draining into 7 sub-basins of the Baltic Sea. Also listed are the estimated direct discharge rates of PCBs and the concentration of aerosols in the atmosphere.

	Bothnian Bay	Bothnian Sea	Gulf of Finland	Gulf of Riga	Baltic Proper	Danish Straits	Kattegat
CA in pg/m^3	10	12	15	15	18	30	21
CX in pg/L	200	300	400	400	400	400	400
EW in kg/a	5	10	40	15	75	25	10
VAQ in m^3/m^3	1.0×10^{-11}	1.5×10^{-11}	2.5×10^{-11}	2.5×10^{-11}	2.0×10^{-11}	2.0×10^{-11}	2.0×10^{-11}

CA = average bulk air concentrations; CX = average bulk river-water concentrations (dissolved + particulate); EW = direct discharge of PCB by industrial and municipal discharges; VAQ = volume-to-volume ratio of aerosols in the atmosphere used for the entire Baltic Sea with the exception of the two northernmost basins. Information on direct releases of PCBs into the Baltic Sea is, to our knowledge, nonexistent.

The mass balance of PCBs (sum of 6 isomers) based on steady state calculations for the entire Baltic Sea (Figure 7-11), disregarding regional differences by giving sums of the fluxes in the various sub-basins, suggests that there is a total annual input to the Baltic Sea of 880 kg PCB, of which 47% is atmospheric deposition, 21% direct discharges, 18% riverine inflow, and 14% inflow from the North Sea. Part of the PCBs in riverine inflow may have originated from the atmosphere and, after deposition in the catchment area, eventually leached to the marine environment. The annual input is balanced by an equally high total output of 880 kg, of which 46% is buried in sediments, 34% is volatilized, 18% flows out to the North Sea, and 2% degrades in the water column.

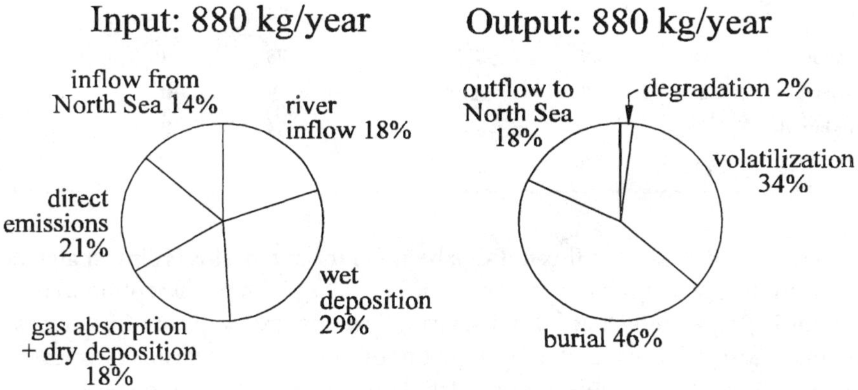

Figure 7-11 Input and output of PCBs (sum of 6 isomers) in the Baltic Sea estimated with the multi-compartment model POPCYCLING-BALTIC (Pacyna 1999).

A closer inspection shows a net annual air–water exchange of 117 kg into the water, i.e., the Baltic Sea is a sink for atmospheric PCBs. This does not exclude the possibility that PCB volatilization exceeds deposition during parts of the year. Major atmospheric deposition processes are wet and dry deposition of aerosol-associated PCBs. The net flux of 402 kg/y between water and surface sediment is also downward and major processes are the deposition and resuspension with POC. Therefore, PCBs in the Baltic Sea undergo a net downward flux from the atmosphere to the water to the surface sediment and finally to the deep sediment. This implies that the Baltic Sea is a sink for atmospheric PCBs, in contrast to recent investigations in the Great Lakes of North America that suggest that volatilization exceeds atmospheric deposition and that these lakes are presently a PCB source to the atmosphere (Hornbuckle et al. 1994, 1995; Jeremiason et al. 1994; Mackay et al. 1994).

Not all sub-basins show a net atmospheric input of PCBs (Table 7-9). The model calculations suggest that the Gulf of Finland acts as a source of PCBs to the atmosphere and that in the Bothnian Bay and the Gulf of Riga atmospheric deposition and volatilization are almost equal.

Table 7-9 Relative importance of various input and output pathways for PCBs in the sub-basins of the Baltic Sea (in percent). Overall input and output is summarized in Figure 7-11.

	Bothnian Bay	Bothnian Sea	Gulf of Finland	Gulf of Riga	Baltic Proper	Danish Straits	Kattegat
Input							
Direct emissions	11	8	32	31	17	12	4
River inflow	41	25	36	24	9	2	4
Atmospheric deposition	39	49	29	42	49	17	9
Interbasin flow	9	18	4	3	25	69	83
Output							
Volatilization	44	35	37	43	28	10	8
Degradation	2	2	2	1	3	0	0
Sediment burial	30	43	35	44	54	7	4
Inter-basin flow	24	20	26	12	15	83	88

Like the net water flux, the net flux of PCBs between the sub-basins is almost always directed from the upstream basins towards the North Sea with the exception that between Baltic Proper and Danish Straits, virtually no net exchange of PCBs occurs. Vertical fluxes within basins tend to be more important than interbasin transports. In the Baltic Proper, for example, the calculated gross flux of PCBs to the bottom water is five times as high as the outflow to all other basins together—a result of the very efficient scavenging of PCBs by settling POC. The highly dynamic marine system between Denmark and Southern Sweden (Basins 6 and 7) is an exception in that horizontal exchange dominates.

Riverine and atmospheric sources are of the same order of magnitude in most sub-basins, but a detailed analysis reveals that the sub-basins differ considerably in their source characteristics. In the large basin of the Bothnian Sea and Baltic Proper, PCB input is dominated by atmospheric deposition, while in the smaller and more densely populated Gulfs of Finland and Riga, direct emissions and riverine inflow become far more important (Table 7-9). As expected, the input of PCBs to the Danish Straits and the Kattegat is determined to a large extent by the inflow of water from the Baltic Proper and the North Sea. Finally, the Bothnian Bay has large contributions from both rivers and atmosphere. Differences are less pronounced with respect to PCB loss, as volatilization and sediment burial are the major loss processes for all basins except the Danish Straits and the Kattegat, where interbasin flow is by far the most important loss pathway. In the deep Baltic Proper, sediment burial becomes relatively more important.

The model suggests that the Gulf of Finland, and to a lesser degree the Gulf of Riga, should show elevated PCB concentrations in both water and sediment (Table 7-9) due to high riverine inflow and relatively high population and industry density. The model also suggests fairly high concentrations in the bottom water and deep sediments of the Baltic Proper. For the remaining basins, the model calculates rather uniform concentrations between 37 to 62 pg/L in seawater and 40 to 70 ng/g organic carbon in surface sediment.

Verification of model results

How do calculated concentrations and fluxes and their spatial patterns compare with measured concentration levels and transport rates? Two investigations of the Baltic Sea during the late 1980s revealed very heterogeneous concentrations of PCBs in seawater in both space and time. Schulz-Bull et al. (1994) reported concentrations (6 PCB isomers, dissolved phase and fraction adsorbed to suspended particles) between 8.5 pg/L and 1227 pg/L, with most determinations at the lower end of this range (10 to 100 pg/L). Modeled water concentrations ranged from 41 to 127 pg/L and are thus within the range of most measured values. In neither study was it possible to discern clear geographical or seasonal patterns, which would allow the evaluation of the calculated regional differences.

Polychlorinated biphenyl concentrations in surface sediments of the Baltic Sea are less variable than the water concentrations. Østfeldt et al. (1994) reported values (6 PCB isomers) between 67 and 238 ng/g loss on ignition (LOI), while Broman measured values between 50 and 600 ng/g C (D. Broman, Stockholm University, unpublished data). The modeled concentrations that ranged from 39 to 161 ng/g organic carbon are within the range of measured values but are on the low side. Measurements indicate slightly decreasing concentrations towards thenNorth, but large variations are found within rather small areas, especially within the Baltic Proper. Concentrations are less variable in northern basin sediments, and the measurements suggest higher concentrations in the Gulf of Finland compared to the Gulf of Bothnia, though there is not as clear a difference as that predicted in the model.

Polychlorinated biphenyl concentrations in precipitation deduced from the model calculations are between 0.6 ng/L in the Bothnian Bay and 1.6 ng/L in the Danish Straits region. Calculated PCB deposition fluxes (i.e., sum of wet deposition and dry particle deposition) are between 1.3 and 4.4 ng/m^2 d. These values agree well with measurements made for 16 stations around the Baltic (1.2 to 5.6 ng/m^2d; Agrell 1999) and in the Kattegat (1.6 to 2.6 ng/m^2d; Brorström-Lundén 1995).

Lessons learned: Persistent organic pollutants in the Baltic

It is increasingly recognized that a multi-compartmental approach is required in order to understand the environmental behavior of POPs in the marine ecosystems, particularly in terms of POP sea–air–land interactions. Concentrations in the atmosphere and in aquatic and terrestrial environments are not independent but, in fact, influence each other. In addition to riverine chemical transport from land to sea, chemicals evaporate from terrestrial regions and are deposited onto the marine environment and vice versa. The relative importance and magnitude of these pathways are poorly known and are subject to change with time. A multi-compartment model such as "POPCYCLING-BALTIC" (Pacyna 1999), which includes the entire catchment area rather than being restricted to the aquatic environment, enables the investigation of the influence of these sea–air–land interactions.

The contaminant behavior in the Baltic Sea is influenced by temporal changes that extend over long time periods. Most importantly, the use and/or production of most POPs have experienced substantial variations in northern countries and in the rest of Europe throughout the last few decades. It can be assumed that these variations have resulted in changing emissions and environmental concentrations as well. Steady-state models used to date are therefore unlikely to describe the long-term chemical behavior properly. An example would be the episodic intrusion of North Sea water replenishing the deep water in the Baltic Proper Basin, a non-steady-state process that might be of particular importance for the dynamical behavior of POPs in the Baltic.

Western Mediterranean Sea Case Studies

Description of Mediterranean marine system

The western Mediterranean, with a surface area of 8.5×10^5 km^2, is linked to the Atlantic Ocean and the eastern Mediterranean basin through the Gibraltar and Sicilian Straits, respectively (Table 7-2). A net inflow of 2500 km^3 y^{-1} of surface low-salinity water from the Atlantic Ocean enters through the Gibraltar Strait to compensate for high evaporation rates in the region, while a net outflow of 1800 km^3 y^{-1} of western Mediterranean water is exported eastward through the Sicilian Strait (Lipiatou et al. 1997). In brief, these processes induce a cyclonic circulation that results in an eastward current along the Morocco and the Algerian coasts and that bends north- and westward along the Italian, French, and Spanish coasts (Beckers et al. 1997). The lack of tides and slow circulation in the basin reduces advective transport of pollutants from river inputs and urban runoff to the open sea. Indeed, atmospheric transport may be the main route by which pollutants enter the open sea basin (UNEP/WHO 1992). The western Mediterranean is oligotrophic, with primary productivity ranging from 50 to 150 gC/ m^{2}y. However, higher productivity both in the Gulf of Lions and the Alboran Sea is supported by riverine nutrient inputs and shallow water in the former and upwelling in the latter.

The northwestern Mediterranean basin is highly industrialized, receiving multiple sources of contamination through continental runoff, sewage sludge disposal, and river inputs (Rhone and Ebro Rivers). The population in the western Mediterranean countries is 360 million and is expected to grow to 515 million by the year 2025 (Grenon and Batisse 1989). Most of the population increase will occur on the coast.

Case study of polycyclic aromatic hydrocarbon in the western Mediterranean

Description

Polycyclic aromatic hydrocarbons (PAHs) are 2- to 8-ring aromatic hydrocarbons present in fossil fuels, such as petroleum and coal, and are formed by the condensation of hydrocarbon fragments during the partial combustion of fuels (Seinfeld and Pandis 1998).

Polycyclic aromatic hydrocarbons exhibit a widespread distribution in the western Mediterranean, with multiple sources such as combustion, oil spills, and diagenetic processes. Here, budgets for 4 representative PAHs (phenanthrene, pyrene, benzofluoranthene and benzo[a]pyrene) are presented (Table 7-10). The selected PAHs, which have a wide range of physicochemical properties (Table 7-1) and varied sources including uncombusted and combusted fossil fuel, well represent the anthropogenic PAHs.

The PAH composition in aerosols from the western Mediterranean region is well-characterized (Marty et al. 1984; Sicre et al. 1987; Simo et al. 1991; Aceves and Grimalt 1993; Lipiatou et al. 1993; Lipiatou and Albaigés 1994). Polycyclic aromatic hydrocarbons profiles contain large amounts of phenanthrene, fluoranthene, pyrene, and benzofluoranthene and lesser amounts of other parent and methylated PAHs. Low concentrations of alkyl-PAHs in the aerosols from the western Mediterranean suggests that combustion is the prevalent source, in contrast to the eastern Mediterranean aerosols, which show a predominance of uncombusted sources (Gogou et al. 1994). There is a high variability of PAH concentrations depending on the origin of air masses and the location and season of sampling, with higher concentration in urban-influenced zones and in northern and northwestern air masses (Lipiatou et al. 1997). Gas-phase concentration data for PAHs are scarce; the few reported measurements (Masclet et al. 1988) show higher concentrations of the low-molecular-weight PAHs such as phenanthrene and pyrene.

Polycyclic aromatic hydrocarbon budget for the western Mediterranean

Atmospheric dry deposition to the western Mediterranean has been estimated from particle-size distributions as described elsewhere (Lipiatou and Albaigés 1994; Lipiatou et al. 1997). For wet deposition, the gas-phase concentrations have to be

Table 7-10 PAH and PCB budgets for the Western Mediterranean Sea. PAH is the sum of phenanthrene, pyrene, benzo(a)pyrene, and benzofluoranthenes

		PAH		ΣPCB	
		In (kg/y)	Out (kg/y)	In (kg/y)	Out (kg/y)
Rivers		3850	—	1000	—
Atmosphere	Precipitation	13,350	—	4470	—
	Dry deposition	5050	—	7930	—
Gas exchange	Absorption	4290	—	32,000	—
	Evaporation	—	198	—	22,000
Water exchange	Gibraltar	34,290	17,535	5100	960
	Sicily	10,620	27,650	1830	5360
Particle flux		76,600		14,000	
Sediment accumulation	Shelf	—	41,000	—	2950
	Basin	—	47,600	—	2390
Degradation			???		???
Total		71,450	133,983	52,330	33,660

	Inventories (kg)	Residence time (years)
PAH		
Surface water	57,7000	
Deep water	109,600	
Sediment	681,000	
PCB		
Surface water	21,850	0.6
Deep water	23,560	2.2
Sediment	38,650	

estimated from empirical relationships between gas or particle partitioning and subcooled liquid vapor pressure (Lipiatou et al. 1997). According to these estimations, dry and wet deposition account for an input of 5050 and 13,350 kg/y of PAHs to the western Mediterranean, respectively.

Dachs, Bayona, Raoux et al. (1997) provided an order-of-magnitude estimation of air–water exchange using the annual average PAH concentration in the dissolved phase, the calculated gas-phase PAH concentrations, and an assumed average wind speed of 5 m/s and temperature of 15 °C. The estimated values of absorption and volatilization were 4290 and 198 kg/y, respectively. Thus, the absorption input from air–water exchange is about the same magnitude as that from dry deposition. The relative importance of air–water exchange for PAHs may be higher in the open sea where the influence of land-based sources and advection through the Gibraltar and Sicilian Strait is less.

Riverine input for the Rhone River was estimated for four seasons using volumetric flow and corresponding compound concentrations. The input from Ebro river was simply estimated from the mean PAH concentrations and the annual water discharge (Lipiatou et al. 1997).

The concentration of PAHs varies widely in surface waters, ranging from 211 pg/L to 1100 pg/L (sum of 16 PAHs). Lower concentrations, where observed, tend to be found in the open sea. However, many of the open sea samples exhibit PAH concentrations similar to those found on the shelf and in localized zones such as offshore from Barcelona, where high PAH concentrations are found. The relatively high concentrations in the open sea, together with the PAH pattern, which is dominated by the low-molecular-weight PAHs in surface water, point to long-range atmospheric transport as the source. Indeed, the PAH profiles in the photic zone of the water column are analogous to those found in the gas phase implying that air–water exchange must play a major role. In contrast to the surface water PAH profiles, the sediments are dominated by combustion PAHs such as pyrene and benzofluoranthene (Tolosa et al. 1996). This apparent contradiction can be explained by examining the ratio of low-molecular-weight to high-molecular-weight PAHs, which shows higher ratios in surficial waters and a decrease with depth (Figure 7-12). The depletion of particulate phenanthrene with depth suggests that degradation and/or solubilization affects low-molecular-weight PAHs. Pyrolytic PAHs, on the other hand, are usually captured inside soot particles that protect them from degradation and/or partitioning to the dissolved phase. They therefore show little degradation from source to burial in sediments (Simo et al. 1997; Dachs, Bayona, Raoux et al. 1997). A selective degradation of alkyl PAH compared to parent PAH also has been described for sinking particles in the western Mediterranean (Lipiatou et al. 1993).

Polycyclic aromatic hydrocarbons show surface-enrichment, depth-depletion vertical profiles with a maximum at about 30 to 50 m concurrent with the phytoplankton maximum (Dachs, Bayona, Raoux et al. 1997). These vertical profiles are characteristic of compounds that enter the surface and are transported downwards by sinking particles (Chester 1993). A number of studies have illustrated that organic matter fluxes can be used as surrogates for the flux of low-molecular-weight PAHs such as phenanthrene, whereas high-molecular-weight PAHs are associated with soot particles (Lipiatou et al. 1993; Dachs et al. 1996). Vertical flux estimated from a large number of sediment trap studies in the northwestern Mediterranean and in the Alboran Sea (Lipiatou et al. 1993; Raoux et al. 1995; Dachs et al. 1996) accounts for a loss of 76,600 kg/y of PAHs out of the surficial waters. This flux approximately matches the observed burial in the sediment (88,600 kg/y). However,

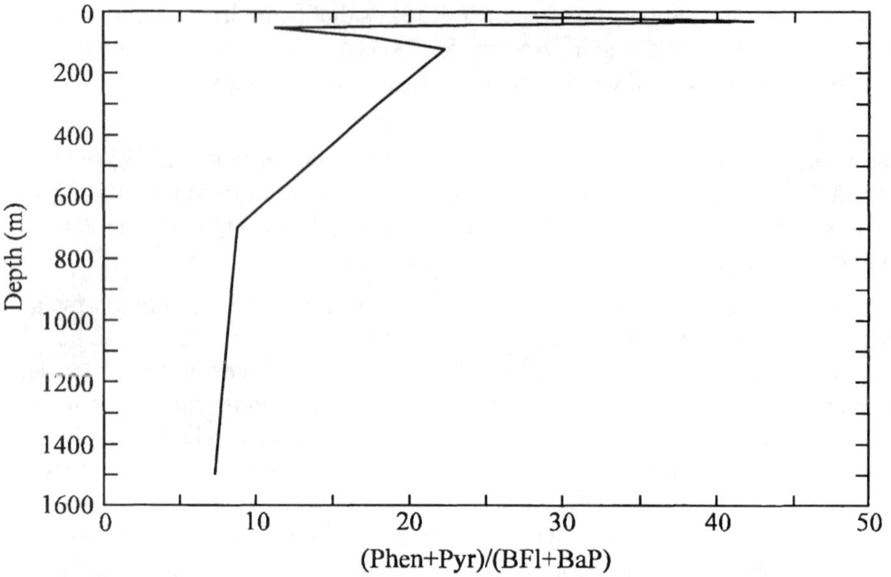

Figure 7-12 Vertical profile of the ratio of low-molecular-weight (phenanthrene + pyrene) to high-molecular-weight (benzofluoranthene + benzo(a)pyrene), particle-associated PAHs in the Western Mediterranean Sea. Ratios are higher in surficial waters and decrease with depth as a result of depletion of phenanthrene via degradation.

a detailed analyses of PAHs concentrations and accumulation rates in the northwestern Mediterranean basin pointed out the importance of coastal sediments as a sink of PAHs with about half of the accumulation occurring on the continental shelf (Tolosa et al. 1996; Lipiatou et al. 1997). This may be because of more effective removal in the higher-productivity coastal waters and because of the prevalent circulation pattern with currents parallel to the coast. The advective transport through the Gibraltar and Sicily Straits results in a net input and output of 16,755 and 1730 kg/y, respectively (Figure 7-13a; Dachs, Bayona, Raoux et al. 1997). This exchange is strongly influenced by the vertical distribution in the water column with net input or output being in the direction of the surficial waters.

Polycyclic aromatic hydrocarbons inventory in the water column has been estimated to be 57,700 kg (0 to 200 m) and 109,600 kg (deep water) (Dachs, Bayona, Raoux et al. 1997) a fact that underscores the importance of deep sea water as a sink and reservoir of PAHs in this marine environment. Even though deep sea waters contain an important fraction of total PAHs in the basin, most of the inventory for PAHs is in the sediment. The top 1 cm alone accounts for 680,000 kg, but the total sediment inventory, which would be much higher, has not yet been estimated (Tolosa et al. 1996).

The residence time of PAHs in surface waters (0.5 y) is much shorter than in deep waters (1.3 y) because of the importance of sinking particles as a removal process of

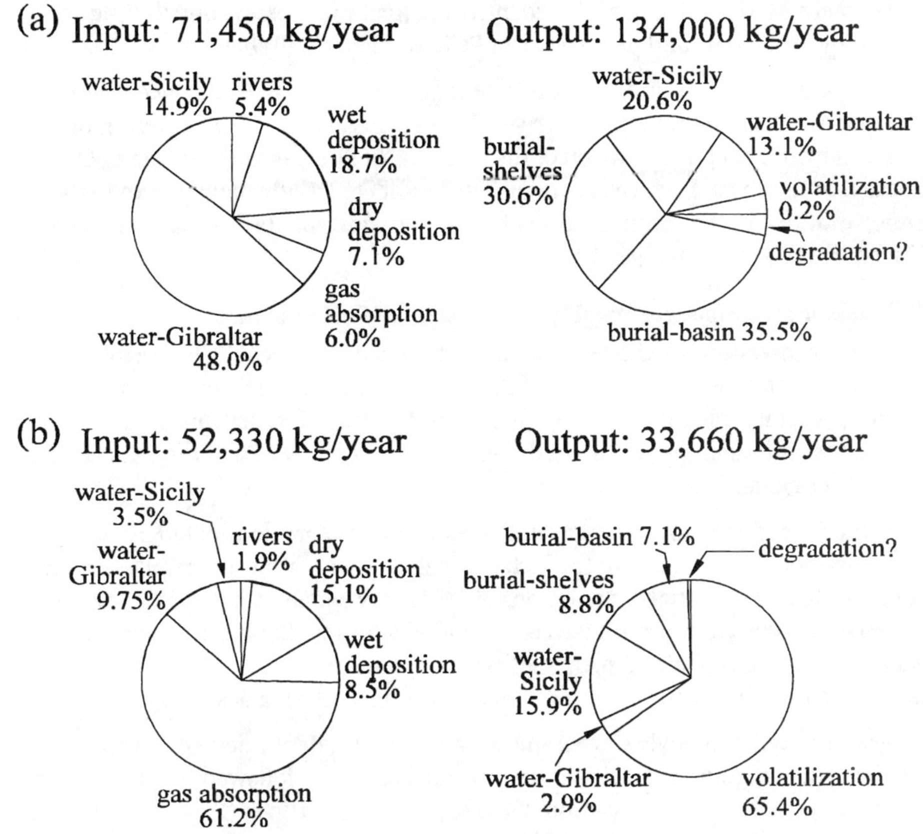

Figure 7-13 (a) Inputs and outputs of ΣPAHs in the Western Mediterranean (ΣPAHs = phenanthrene + pyrene + benzo(a)pyrene + benzofluoranthrenes) and (b) inputs and outputs of ΣPCBs in the Western Mediterranean Sea

PAHs from the photic zone. In fact, the actual residence times may be even shorter, since degradation was not taken into account because of a lack of data.

Case study of polychlorinated biphenyl in the western Mediterranean

Description

Polychlorinated biphenyls were produced intensively during the 1960s and 1970s until they were banned in the European Union in 1987. The estimated total production of PCBs in France, Italy, and Spain is about 300,000 t for the period 1954–1984, although the amounts used in the basin are not known and could be significantly lower. Sampling has been focused in certain areas (northwestern Mediterranean) and in certain compartments, with a consequent lack of data for other compartments such as the atmospheric gas phase or sub-basins such as the southern Mediterranean. Nevertheless, there are enough data to carry out a mass balance of the western Mediterranean and to identify the major processes contributing to long-range transport and persistence of PCB in the basin (Table 7-10).

Inputs from the Rhone and Ebro Rivers were estimated from PCB concentration in the river waters and the respective river flows (Tolosa et al. 1997). However, owing to the general circulation pattern of the western Mediterranean with currents parallel to the coast, both rivers and urban runoff contribute mainly to the PCB concentrations in the continental shelf waters and sediments (Tolosa et al. 1995; Dachs, Bayona, and Albaigés 1997).

Polychlorinated biphenyl budget for the western Mediterranean

For the northwestern Mediterranean, higher concentrations of PCBs are found in sediments from the continental shelf than in those in the deep basin. As a consequence, coastal sediments account for 55% of the total PCBs accumulated in sediments, although the continental shelf contributes only 15% of the total surface area (Tolosa et al. 1995; Tolosa et al. 1997).

Polychlorinated biphenyl patterns in the water column are dominated by the PCB congeners with 3 to 5 chlorines, which is consistent with atmospherically derived sources. However, spatial distributions of PCBs vary widely, ranging from 120 pg/L in urban influenced waters off Barcelona and Valencia to 20 pg/L in open sea waters. As in the case of sediments, PCB water concentrations are usually higher near the coast than in the open sea (Dachs, Bayona, and Albaigés 1997).

Polychlorinated biphenyls show a surface-enrichment, depth-depletion vertical profile, which is consistent with entry into surface waters followed by a rapid downward transport in association with large sinking particles (Fowler and Knauer 1986; Chester 1993; Dachs et al. 1996). Sediment-trap measurements in open sea and coastal waters (Tolosa et al. 1997) suggest that the removal flux of PCBs from the photic zone by sinking particles (14,000 kg/y) is much higher than the accumulation flux in the sediment (5350 kg/y), which implies recycling in deep waters and at the sediment-water interface, as described previously for lake Superior (Baker et al. 1991).

The distribution of PCBs, with surface-water enrichment, controls the direction of net PCB exchange in Gibraltar and Sicilian Straits. Even though concentrations of organic pollutants are generally lower in the Atlantic than in the Mediterranean, it is the enriched surface Atlantic water that enters the Mediterranean, whereas deep waters with low PCB concentrations exit through Gibraltar Strait to the Atlantic Ocean. A similar phenomenon explains the net outflow to the eastern Mediterranean through the Sicilian Strait (Dachs, Bayona, and Albaigés 1997).

There are very few estimates of wet and dry deposition of PCBs in the northwestern Mediterranean (Villeneuve and Cattini 1986). The extrapolation of these measurements to the rest of the basin provide estimates of 7930 kg/y (wet) and 4470 kg/y (dry). Absorption inputs and volatilization outputs have been estimated to be

22,000 and 32,000 kg/y, respectively. While the coastal zone seems to be a source of PCBs to the atmosphere, the open sea is a sink for atmospheric PCBs because of the low PCB concentration in the dissolved phase (Tolosa et al. 1997; Dachs, Bayona, and Albaigés 1997). However, there are major uncertainties (about 100%) in the estimations of atmospheric inputs and outputs, mainly as a result of the very few available measurements of gas-phase concentrations.

Finally, the inventory of PCBs in the water column of the western Mediterranean (Figure 7-13b) is 21,850 kg (0 to 200 m) and 23,560 kg (deep waters). This inventory was estimated from the vertical profiles of particulate PCBs, together with dissolved-phase measurement for the photic zone. Dissolved concentrations in deep sea waters were estimated from the concentration in the particulate phase (Dachs, Bayona, and Albaigés 1997). Despite higher concentrations of PCBs in surface waters, the larger volume of the deep water basin makes its inventory larger. The PCB inventory in the water column is of the same order of magnitude as the inventory in surface sediment (38,650 kg) (Tolosa et al. 1995) but probably is much smaller than the complete sediment inventory, which cannot presently be estimated.

The residence time of PCBs in surface water (0.6 y) is much shorter than that of deep sea waters (2.2 y), due mainly to the lack of removal processes in the deep waters. This implies that deep sea waters and sediments are both important sinks for PCBs, contributing to their persistence in the Mediterranean.

Lessons learned: Persistent organic pollutants in the Mediterranean

Atmospheric long-range transport and deposition into surface waters play a major role in the widespread distribution of PAHs in the western Mediterranean. Phase association explains the high persistence of soot-related PAHs. Biological processes, especially the sinking of organically rich particles, contribute to observed distributions of PAHs in the water column. An important fraction of the PAH inventory is in deep sea waters, where slower degradation may increase PAH persistence in the environment.

Sediments provide the dominant sink for combustion PAHs, whereas other PAHs such as alkylated phenanthrene seem to be efficiently degraded. For PCBs, both advective transport and particle flux in the water column are important to the assessment of long-range transport, occurrence, and fate in the western Mediterranean. These processes explain higher concentrations and higher accumulation rates in the continental shelf as well as the vertical profiles found in the basin. Advection together with vertical distribution accounts for the net the input of PCBs through the Gibraltar and Sicilian Straits.

Atmospheric long-range transport and deposition probably account for most of the PCBs in open sea water. Even though PCB concentrations are relatively low in deep sea waters, the large volume provides for an important fraction of the PCB inventory of the basin. This leads to greater persistence of these compounds owing to the longer residence times in deep waters. Furthermore, the capture of POPs by deep waters prevents their return to the atmosphere, where they are more likely to be degraded.

Case Studies in the Tropics

Case study of organochlorine pesticides in Lake Malawi (southern Africa)

Description

Lake Malawi is a large oligotrophic lake situated between Malawi and Tanzania in the rift valley of south central Africa between 3° and 10° S (Figure 7-1). The lake is noted for its large number of unique species of cichlidae fishes. With an area of 22,490 km^2 and volume of 8400 km^3, Lake Malawi is one of the world's largest lakes, similar in size to Lake Ontario or Lake Michigan. The lake is permanently anoxic below 200 m and has a water-residence time of 750 y because of very low outflow volumes (Bootsma and Hecky 1993).

Pesticide import surveys for Malawi indicate that some chlorinated pesticides (e.g., aldrin, lindane, and endosulfan) continue to be used (Malawi Government 1998). Other POPs are used inadvertently because they are impurities that are found in widely used pesticides (e.g., DDT has been found in dicofol and HCB is found in chlorthal-dimethyl and lindane). DDT also continues to be used in southern Africa for malaria vector control. Polychlorinated biphenyls continue to be used as flame retardant chemicals in older (pre-1980s) electrical equipment and hydraulic systems.

Fish and food-web samples from Lake Malawi have low levels of a range of POPs (Kidd et al. 1998), which indicates that Lake Malawi, like other large freshwater systems, is receiving these contaminants from long-range transport as well as from local use within the watershed. Recent studies have examined water, air, and precipitation samples for POPs to determine inputs to the lake from the atmosphere, the state of equilibrium between air and lake water, and possible sources of these contaminants (Muir et al. 1998).

Organochlorine pesticide concentrations and fluxes

The organochlorine pesticides were detected in Lake Malawi surface waters at low pg/L levels. Consistently higher levels of p,p'-DDT and HCH isomers were found in surface waters compared with 40 and 80 m depths. Lindane was present at similar or higher levels than was α-HCH in surface waters but was generally lower than the α- isomer at depth. This may imply recent input from the use of lindane. Hexachlorobenzene and dieldrin were present at similar concentrations in surface and deep waters.

The concentrations of organochlorine pesticides in Lake Malawi surface waters were generally lower than reported for the Laurentian Great Lakes. The largest differences between Malawi and the Great Lakes, and indeed for lakes in the North America generally, arc for α- and γ-HCH and dieldrin. Dieldrin levels are less than one percent those reported for Lakes Superior and Ontario.

Air measurements at a land-based site near the shore at the southern end of the lake (Senga Bay) and over the lake detected a wide range of chlorinated pesticides. The greatest variation in air concentrations of POPs occurred for lindane and dieldrin. Lindane concentrations ranged from < 0.1 pg/m^3 in mid-March to 176 pg/m^3 in mid-June. By contrast, α-HCH levels varied less and were not correlated well with those of lindane. Dieldrin levels in April, May, and June paralleled those of lindane, although a high concentration was also observed in early March. These observations suggest local use of lindane and dieldrin. Locally applied pesticides typically exhibit very large shifts in airborne concentrations related to the application period for the chemicals (Hoff et al. 1992a; Rawn et al. 1999). ΣDDT did not show a major shift in concentration during the February–June sampling period. A high proportion of o,p'-DDT and p,p'-DDT in ΣDDT was found, indicating a technical DDT source. Endosulfan and pentachloroanisole also were detected in all air samples but did not vary greatly in concentration between February and June 1997.

In general, concentrations of organochlorine pesticides observed in air samples from Malawi were lower than in other tropical regions. For comparison, ΣHCH concentrations in river and estuarine air in India in 1989 ranged from 13 to 11 × 10^6 pg/m^3 in 1989 (Iwata et al. 1994). The airborne concentrations of HCB and α-HCH in Malawi were about one-tenth of those observed in the Laurentian Great lakes region. For comparison, annual mean concentrations of HCB ranged from 80 to 130 pg/m^3 at 4 Great Lakes sites in 1994, and α-HCH concentrations ranged from 79 to 160 pg/ m^3 (Hoff et al. 1996). Dieldrin levels were generally higher in Malawi (3 to 287 pg/ m^3) than in the Great Lakes (14 to 34 pg/ m^3). ΣDDT in Malawi air site (30 to 300 pg/m^3) was higher than at Eagle Harbor (Lake Superior) but lower than at three other sites in the Great lakes region in 1994 (25 to 106 pg/m^3). Ngabe and Bidleman (1992) found the concentration of mean ΣDDT to be 2800 ng/m^3 and of ΣHCH to be 420 pg/m^3 in Brazzaville, Congo, in continuous sampling during August versus a range of 30 to 300 pg/m^3 for ΣDDT and 2 to 186 pg/m^3 in Malawi over a 1-y period. Concentrations of ΣDDT ranged from 22 to 12,000 pg/m^3 in rural and urban air in India in the study by Iwata et al (1994). DDT concentrations may have large temporal variations as a result of use patterns that are not captured in the short studies carried out to date in Malawi.

Air–water equilibrium in Lake Malawi was evaluated using fugacity ratios (FR_{AW}) and gas-exchange fluxes for selected organochlorine pesticides (Table 7-11). FR_{AW}

calculations suggested that lake waters were close to equilibrium with the atmosphere in February for α-HCH and p,p'-DDT, with FR_{AW} values ranging from 2.4 to 4.6. Dieldrin and lindane had much higher FR_{AW}, reflecting higher air concentrations relative to water concentrations than would be expected at equilibrium and implying net gas absorption into Lake Malawi.

Table 7-11 Fugacities and fluxes of selected POPs in Lake Malawi (February 1997)

Parameter		α-HCH	γ-HCH	Dieldrin	p,p'-DDT
Henry's Law constant[1]	Pa m^3/mol	0.87	0.15	5.90	2.36
Mass transfer coefficient (Kol)[2]	m/d	0.10	0.01	0.27	0.18
Fugacity ratio[3]		2.9	20	24	4.6
Absorption[4]	ng/m^2 d	3.9	3.45	8.55	2.64
Volatilization[5]	ng/m^2 d	1.7	0.21	0.38	0.80
Net flux[6]	ng/m^2 d	2.2	3.24	8.17	1.83

[1] Henry's Law constant at 25 °C (from Mackay et al. 1997 and Hoff et al. 1996) [2] Kol = air–water mass-transfer coefficient (m/day) (calculated according to Schwartzenbach et al. 1993) [3] Fugacity ratio = (Ca/Za)/(Cw/Zw) where Z = fugacity capacity in air and water, Cw = water concentration (dissolved), and Ca = air concentration. Both blank corrected. [4] Absorption = Kol × Ca × RT/H (where R = gas constant (8.32 Pa m^3/mol K), T = temperature Kelvin [5] Volatilization = Kol × Cw;
[6] Net flux = Absorption – Volatilization (negative value indicates net loss from the lake)

Hexachlorocyclohexane budget for Lake Malawi

Inventories of organochlorine pesticides in Lake Malawi are low (assuming all chemical is confined to the upper 200 m oxic layer) compared with net air–water exchange (Table 7-12). This mass balance does not include sedimentation, which could be important for DDT, especially in shelf areas. Concentrations in inflow waters and precipitation are also poorly determined. For the estimation, they were assumed to be 2 times higher than surface lake water concentrations. Gas absorption and volatilization are shown to be the major input and output pathways, respectively, for HCH isomers, dieldrin, and p,p'-DDT (Table 7-12, Figure 7-14). Lake Malawi is presently a sink for HCH and dieldrin. Degradation via hydrolysis and microbial transformation may be major removal processes in the lake along with volatilization (Figure 7-14).

Case Study: Hexachlorocyclohexane in the Vellar River Estuary (Southern India)

The Vellar River joins the Bay of Bengal at Parangipettai (11° 29' N 79° 46' E) on the east coast of India. The watershed encompasses agricultural fields, mainly rice paddies. In the estuary, river water can flow into mangrove-lined channels parallel to the coast and also directly through a narrow channel in the Bay of Bengal. The mangroves receive seawater from the Bay of Bengal through the Vellar and nearby

Table 7-12 Preliminary mass balance of POPs in Lake Malawi

	α-HCH	γ-HCH	Dieldrin	p,p'-DDT
Water column[1] (kg)	13.1	7.0	5.3	1.7
Outflow (kg/y)	0.1	0.1	0.0	0.0
Precipitation (kg/y)	3.9	3.9	3.9	0.4
Inflow (kg/y)	0.5	0.6	0.2	0.1
Gas absorption (kg/y)	41.3	36.3	89.9	27.7
Volatilization (kg/y)	17.9	2.2	4.0	8.4
Output (kg/y)	18.0	2.3	4.0	8.5
Input (kg/y)	45.7	40.8	94.0	28.2
Net (input – output)	27.7	38.5	90.0	19.8

[1] Calculated using lake-wide average concentrations for 1–20 m, 20–40 m, 40–200 m multiplied by volume of each layer. Amount in lower depths is assumed to be negligible because of permanent stratification.

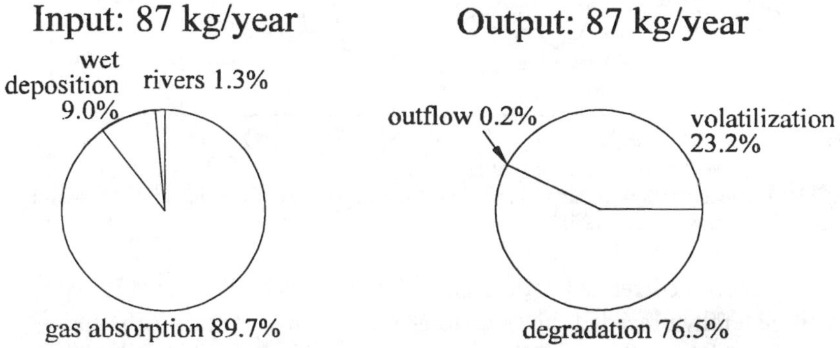

Figure 7-14 Inputs and outputs of ΣHCH estimated for Lake Malawi (southern Africa) using the data in Table 7-12 and balancing the budget assuming degradation equals (input – output)

Coleroon River mouths and freshwater from irrigation channels and small drains from nearby paddy fields. Hexachlorocyclohexane was intensively used in the watershed during the study period (1988). Annual consumption of HCH during the 1980s in India was estimated to be about 45,000 t (Ramesh et al. 1989). By determining HCH levels in air and surface waters in the Vellar estuarine area and by estimating river flow and water volume in the estuary, Takeoka et al. (1991) obtained a budget for HCH isomers in the Vellar River watershed.

Hexachlorocyclohexane budget

Air samples were collected near the Vellar estuary every 2 weeks for a 13-month period. Water samples were collected at the same schedule as air samples at one station and at 6 sites on the river in both the dry season (July) and the wet season (January). Large seasonal variations of airborne HCH were observed from September to December. Maximum HCH levels (134 ng/m^3) were observed in mid-December (Figure 7-15). Highest concentrations in Vellar river water (3300 ng/L) were observed in December as well (Figure 7-15). The high concentrations reflected large-scale use of HCH, especially during the period from November to

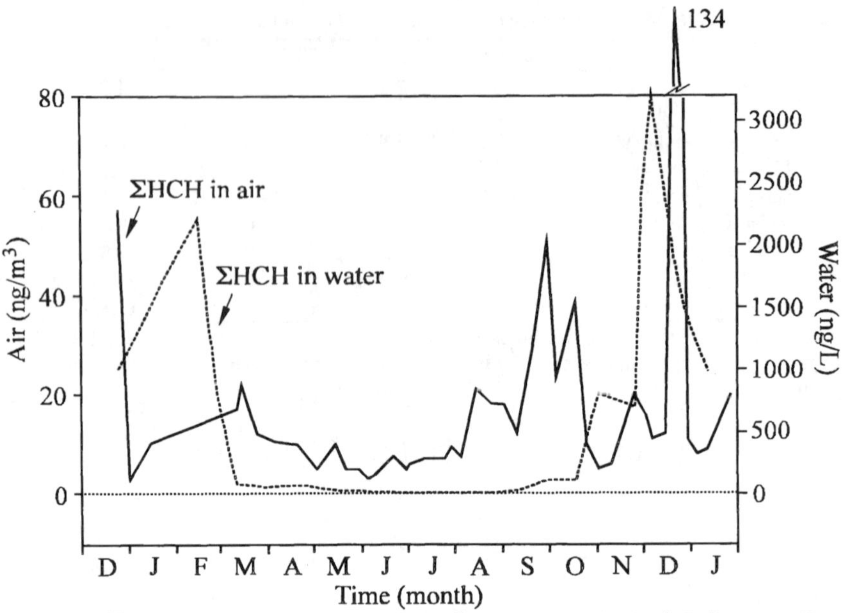

Figure 7-15 ΣHCH concentrations in air and in Vellar River water in South India over a 13-month period in 1988–1989 (Takeoka et al. 1991)

February, and were not correlated with river flow. High concentrations of HCH isomers in air generally preceded the elevations in river water (Figure 7-15) and estuary water. This delay reflects the slow movement of the pesticide residue in paddy field waters to reach the mouth of the estuary as opposed to the rapid travel time of airborne HCH (Takeoka et al. 1991).

River discharge to the estuary was negligible throughout the study and, therefore, it was assumed that the outflow of HCH to the sea was carried only by the tidal exchange at the mouth of the estuary. Air–water exchange fluxes of HCH in the estuary (surface area 6 km^2) were determined using the two-film model (Schwarzenbach et al. 1993). The total amount of HCH that did not reach the estuary was estimated by subtracting the total flux from the total amount of HCH applied to the catchment area. Rapid volatilization of HCH from paddy fields and low uptake by plants have been observed in other studies (Tanabe et al. 1991).

The total amount of HCH applied to the catchment area in 1988 was 42,000 kg. Takoeka et al. (1991) estimated that 99.6% of the applied HCH was volatilized and that only 0.4% drained into the estuary. Moreover, 75% of the flux to the estuary was volatilized. Thus, only about 0.1% of the applied HCH ultimately reached the Bay of Bengal by the aquatic pathway (Figure 7-16).

Shallow water depth and high temperatures favor volatilization of HCH from paddy fields. Removal of HCH from the paddy fields is assumed to be mainly by way of

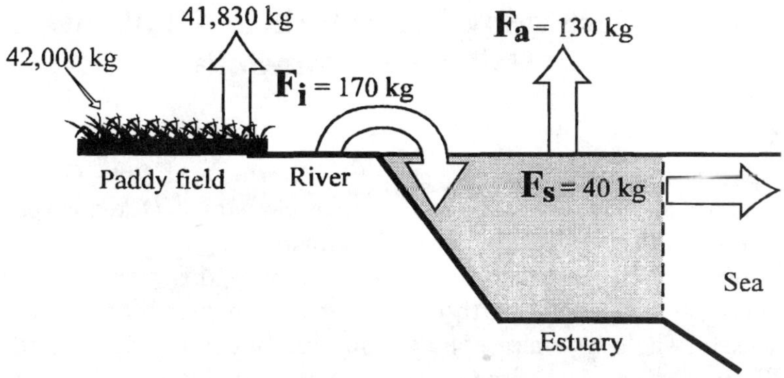

Figure 7-16 Schematic model of fluxes of HCH in the Vellar River watershed, South India, during the late 1980s (Takeoka et al. 1991). Forty-two t of HCH were applied regionally, of which 41.83 t were estimated to have volatilized or degraded in the paddies. River inputs (F_I = 170 kg) are balanced by volatilization (F_a) and exchange with, or outflow to, the sea (F_S).

volatilization and not degradation (hydrolysis, microbial transformation) because volatilization rates are more rapid than degradation rates.

Lessons learned: Persistent organic pollutants in the tropics

Volatilization and degradation may be much more important removal processes for some POPs and current-use pesticides in the tropics than in temperate and arctic regions. Tropical regions thus may act both as global sources and sinks because of rapid removal processes. In addition to chlorinated pesticides, Iwata et al. (1994) found elevated concentrations of PCBs in some urban areas in tropical southeast Asia, which they attributed to expansion of the use and disposal of electrical equipment and higher prevailing temperatures. The results for the Vellar River show that despite high use of HCH within the watershed, the river was not a major pathway for HCH to reach the Indian Ocean because most of HCH volatilized or degraded. This is likely to be true of other tropical agricultural watersheds, and for other pesticides of similar physicochemical properties to HCH that are applied to soils or paddy waters. Unlike many water bodies in the Northern Hemisphere, the direction of flux for current-use compounds such as lindane and dieldrin was into the water suggesting that Lake Malawi was a sink rather than a source. Both Lake Malawi and Vellar River water were undergoing net absorption of HCH isomers according to prevailing air and water concentrations. In general, concentrations of persistent organochlorine pesticides in air in cities and rural areas in southeast Asia were high during seasonal use and not in equilibrium with relatively low concentrations in nearby estuarine and seawaters (Iwata et al. 1994, 1993). These results imply a major transfer pathway of organochlorine pesticides during periods of use for agricultural and urban pest control is through air to nearby waters rather than through runoff.

Terrestrial Case Study: Polychlorinated Biphenyls in United Kingdom Soils

Description

Recently, Jones and co-workers (Alcock et al. 1995; Harner et al. 1995; Gevao et al. 1997; Coleman et al. 1997; Lee et al. 1998; Thomas, Sweetman, Ockenden et al. 1998; Thomas, Smith et al. 1998; Thomas, Sweetman, and Jones 1998; Sweetman and Jones 2000) have attempted to determine the sources, time trends, environmental distribution, and fluxes of PCBs through the environment in the UK. This work provides a useful whole-system case study, which can be developed into models. Figure 7-17 summarizes some of the data and concepts arising from this group's work.

Polychlorinated biphenyl budget in the terrestrial environment of Great Britain

The production and use of PCBs during the 1950s to 1980s has been estimated by Sweetman and Jones (2000) and broken down into different use types (i.e., from either "open" or "closed" uses) which essentially influence whether PCBs can then enter the environment and in what amounts, in what proportions, and over what time periods. Because PCB use was restricted in the UK (and elsewhere) in the 1970s, we can view these chemicals as a "pulse" that enters the environment over time and then transfers and partitions into different environmental compartments and slowly dissipates out of the UK, either by degradation or by advection elsewhere. Measurements on archived soils (e.g., Alcock et al. 1995), sediment cores (e.g., Gevao et al. 1997), and air monitoring (e.g., Coleman et al. 1997) provide evidence of the changing environmental burdens and fluxes of PCBs over time (see table in the lower left hand corner of Figure 7-17). These measurements can be compared to the concentrations predicted and modeled in these various media by Sweetman and Jones (2000): the top left hand corner of Figure 7-17 illustrates how the "pulse" of PCB transfers through different environmental compartments at different rates. Central to the issue of environmental cycling is the "buffering effect" from the storage of PCBs in soils following atmospheric transport and net deposition through the 1950s to 1970s (Harner et al. 1995). As primary emissions to atmosphere declined in the 1970s to 1990s, the soil has become a reservoir capable of resupplying the atmosphere by volatilization with the result that soil outgassing is believed to maintain present air concentrations of PCBs (Lee et al. 1998).

The right hand side of Figure 7-17 gives some estimates of burdens and fluxes in terrestrial compartments from a series of measurement programs. Taking 1 m^2 of land surface and the grass and air above it, it is clear that the bulk of the PCB burden still resides in the soil (e.g., ~ 5×10^5–10^6 ng in the top ~ 10 cm of soil). In grasslands ~ 100 to 500 ng ΣPCB is held in a standing crop of vegetation, a much smaller mass than that in the soil (Thomas, Sweetman, Ockenden et al. 1998; Thomas, Sweetman, and Jones 1998; Thomas, Smith et al. 1998).

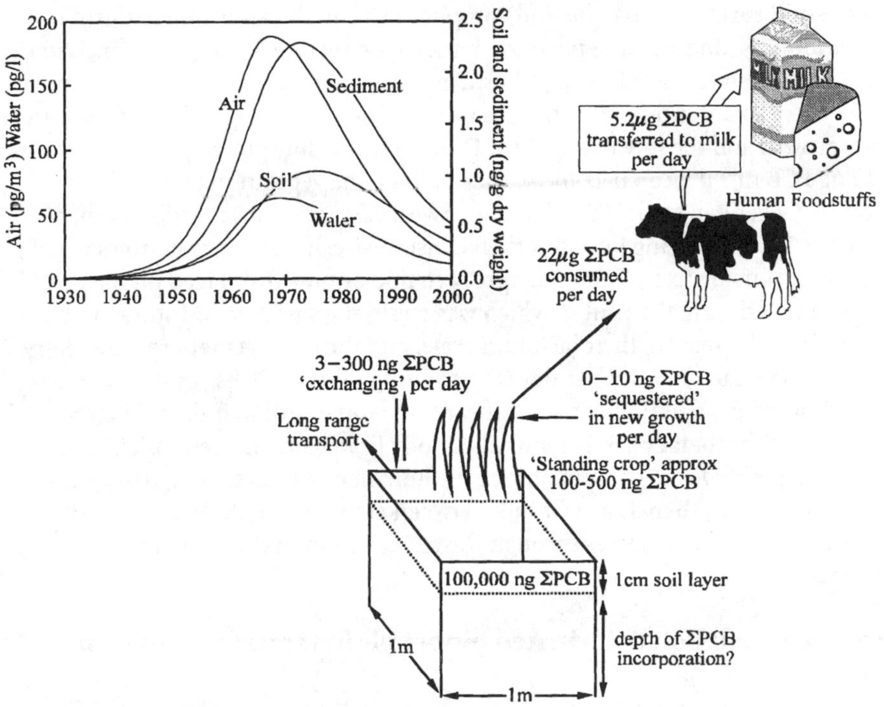

UK PCB production, use and historical environmental data
ΣPCB tonnes

Period	Production	Transformers	Large Capacitors	Other uses	Exported
1955-1980	66748	5742	8017	37250	15739

PCB-101 (dry weights where applicable)

	Soil (ng/g)[a]*	Sediment (ng/g)[b]	Air (pg/m^3)[c]	Grass (pg/g)[d]	Soil burden (t)
1942	0.16				5.4
1948		0.47			
1951	0.14				4.7
1960	0.93	2.24			31.4
1967	0.91				30.7
1972	1.55				52.3
1980	0.47				15.9
1981		2.24			
1984	0.22	2.01			7.4
1990		0.97			
1992	0.15		20		5.0
1993	0.15±0.11		10		50±3.7
1995			22		
1997				75±30	

[a] Alcock et al. 1995; [b] Gevao et al. 1997; [c] Coleman et al. 1997; [d] Thomas, Sweetman, Ockenden et al. 1998
*Adjusted

Figure 7-17 Model of the historical environmental fate of PCB-101 in the UK. The historical emission pulse of PCBs transfers through different environmental compartments at different rates. As a result of the buffering effect from the storage of PCBs in soils, this reservoir now resupplies the atmosphere by volatilization and soil outgassing. In this scheme, PCBs from soils enter the food chain through the grass to accumulate eventually in cow's milk.

The terrestrial surface (grass plus soil) supplies PCBs to the atmosphere during warm conditions, but PCBs re-partition back to the surface during cool conditions (Lee et al. 1998). As new grass grows in spring and summer, the increased biomass provides waxy surfaces onto which PCBs can partition, and, depending on the rate of grass growth and the rate of supply of PCBs to the freshly growing grass, as much as ~ 10 ng PCB m^{-2} d^{-1} can become sequestered by the vegetation (Thomas, Sweetman, Ockenden et al. 1998; Thomas, Sweetman, and Jones 1998; Thomas, Smith et al. 1998). Grazing livestock then consume fresh pasture as it grows when they are kept outside (e.g., from mid-spring to mid-autumn) and feed off harvested or silaged grass during the winter, which was cut in the summer and stored in barns. Depending on the season, their lactation cycle, and the management regime, dairy cows consume ~ 20 µg ΣPCB per day (the equivalent of the PCBs present in ~ 100,000 m^3 of air) and transfer a significant fraction of this to milk lipids (the remainder may be metabolized or stored in body lipid compartments). This, in turn, is incorporated into components of the human diet (meat, milk, dairy products). Consequently, there is a clear link between emissions of PCBs to air from primary and secondary sources through the terrestrial food chain to human exposure.

Lessons learned: Polychlorinated biphenyls in terrestrial environments

This study illustrates the importance of soil and plant reservoirs of PCBs for resupplying the atmosphere by way of volatilization. Similarly, soil and vegetation must provide an important reservoir for other POPs with similar physical properties such as CHBs, DDT-related compounds, chlordane and lower-molecular-weight PAHs. On a larger scale the soil and plant reservoir will be an important factor in determining the global trend in levels of POPs in terrestrial and aquatic environments as use of POPs is reduced or eliminated.

Summary

Each of our case studies show chemical persistence to be the outcome of interaction between a chemical's properties and regional or global biogeochemical cycles (Table 7-13). Traditionally, persistence has been viewed as the spread of a chemical into the environment, but a more compelling indication of persistence is given by the response of environmental chemical concentrations to source reduction. Chemicals like PCB, HCH, and toxaphene have been released as a transient pulse starting somewhere between 1930 and 1950, rising to a maximum in 1970 to 1980, and thereafter being reduced by complete or partial bans. For these chemicals we are presently in the environmental clearance, or redistribution, phase. During this phase, large environmental reservoirs including the oceans, lakes, soils, and

Table 7-13 The conspiracy between chemical properties and environmental characteristics to produce persistence

Location/ Chemical	Chemical properties	Environmental characteristics
Arctic Ocean/HCH	Low H, high volatility, low degradation in cold water, low particle reactivity	Cold water, ice cover stratification, low particle fluxes
Arctic Ocean/ Toxaphene	Low H, low degradation in cold water, low particle reactivity	Cold water, ice cover, stratification, low particle fluxes, biomagnification
Lake Superior/PCB	Moderate volatility, low degradation, high particle reactivity	Large terrestrial soil reservoir, high atmospheric exchange, high cycling with organic carbon, poor burial efficiency, biomagnified
Mediterranean Sea/PCB	Moderate volatility, low degradation, high particle reactivity	Large terrestrial soil reservoir, high atmospheric exchange, large vertical flux, accumulation in deep ocean, poor burial efficiency, biomagnified
Lake Michigan/ Atrazine	High use, high solubility in water, poor volatility, high degradation in soil, poor degradation in water	Small leakage from soil to water, long residence time in water, poor atmospheric exchange
Arctic Ocean/ PCB	Moderate volatility, low degradation, high particle reactivity	Cold water, ice cover, stratification, low particle fluxes, lag in delivery from temperate regions, biomagnification
Mediterranean/ PAH	Two phase associations: 1) gaseous - volatile, exchange important, degrade?, adsorb on carbon-rich particles; 2) soot particulate - involatile, refractory, behave as particles, enter biota but do not biomagnify.	Gaseous: air–sea exchange important, degradation probably important?, enter deep ocean during organic carbon recycling; Soot particulate: vertical flux important, sequestered in sediments, unavailable?
Lake Malawi/ HCH	HCH may biodegrade rapidly in 25 °C and higher water.	Long water-residence time, great depth, high water temperatures
Vellar estuary/ HCH	HCH undergoes biodegradation and volatilization in rice paddy waters	Shallow rice paddy waters and estuary enhance volatilization losses
UK soils/PCB	PCBs have strong affinity for to soil and plant wax; volatilize with temperature cycling	Soils and plants form large reservoir that continues to supply PCBs to the air. Grass provides direct link to terrestrial food web (cows, humans)

vegetation can resupply the atmosphere and thereby delay clearance for long periods of time measured in decades. For other chemicals, like atrazine and PAHs, emissions continue to grow or have levelled off, and environmental reservoirs continue to be loaded.

For PCB, the atmospheric pathway is a common and enduring control for persistence during both loading and clearance phases. For remote (Arctic Ocean), regional

(Baltic, Mediterranean), and local (Lake Superior) reservoirs, water–air fluxes dominate, and, ultimately, it is the slow turndown in atmospheric concentrations that biota record in their PCB trends (Jeremiason et al. 1994; Bignert et al. 1998). We are presently observing the continued global redistribution of PCB fed by locally and regionally contaminated soils, vegetation, and aquatic reservoirs. The removal of PCB from temperate environmental cycles is slow but steady. Clearance of PCB from surface waters also can be effected by particle transport. However, large vertical fluxes may not always be accompanied by large burial rates because OC is effectively mineralized and PCB may thus be released back to the water. In the case of Lake Superior, where vertical mixing is rapid relative to the vertical flux, permanent burial of PCB is small, and virtually all of the observed reservoir decreases are explained by PCB efflux to the atmosphere. In contrast, the burial flux of PCBs in Baltic sediments is significant and is supported by a matching atmospheric influx to Baltic Sea surface waters. In this case, sediment burial allows the Baltic to become a PCB sink. We conclude, therefore, that strong partitioning onto carbon-rich particles is not always a guarantee that PCB will be sequestered in sediments. For deeper reservoirs, like the Mediterranean Sea, particle fluxes are clearly effective at transferring PCB from the upper ocean to the deeper ocean, and these fluxes ultimately must contribute globally to atmospheric decreases. However, PCB then accumulates in the deep ocean and, to a lesser extent, in sediments. The deep ocean reservoir, therefore, is presently in the loading phase, and the relevant questions are what concentration could ultimately be produced by such a process and is the rate of degradation/burial of PCB faster than the deep ocean's turnover time (centuries to millennia)? For the coming century, PCB persistence will be supported by its recalcitrance to degradation and poor burial behavior resulting from recycling with OC. Aquatic reservoirs will reduce their PCB burdens at a rate dependent almost entirely on the atmospheric concentration, which is supported by soil and vegetation inventories. The observed reductions in PCB concentration in temperate regions are the net product of large fluxes between the atmosphere and soils, lakes, and oceans. It is not clear whether the decreasing trend of PCB in temperate ecosystems is, as yet, being followed in polar regions, which lag in the global redistribution (Muir et al. 1997; Macdonald et al. 2000). An important lesson here is that the Arctic Ocean is slow to load and slow to clear but has one of the greatest potentials to biomagnify. Thus, ultimately, the higher trophic levels in the Arctic Ocean, which includes humans, may prove to have the longest half-life for global PCB reduction.

Hexachlorocyclohexane and CHBs have, as their critical geochemical property, a low H. The consequence for both these chemicals is that while air may be the important transport medium initially, water inevitably will become an important mode of transport once reservoir concentrations build up. The initial loading of the upper ocean with HCH depends on 3 crucial factors: 1) low degradation rate in water, 2) low vertical mixing and 3) low vertical particle flux. The Arctic Ocean provides all three of these characteristics in addition to cold water temperatures, which exacerbate partitioning of HCH from the atmosphere. Sharp reductions in HCH emissions have been followed by a rapid decrease in atmospheric concentrations (Figure 7-5). As a consequence, large parts of the upper ocean have become supersaturated and are presently outgassing their HCH contents back to the atmosphere (Jantunen and Bidleman 1995). Ice cover in the permanent pack of the Arctic Ocean hinders this exchange with the result that the turndown of HCH concentration will be affected mostly by outflow of surface water into the Archipelago. Turndown, therefore, will follow surface-water residence times and will be slowed because replacement water continues to carry HCH, albeit at lower concentrations.

Data from the tropics provide an important contrast to the Arctic Ocean. Lake Malawi has very low, undersaturated HCH concentrations, which it can maintain only by having a loss term in the water. The most plausible loss term is degradation (hydrolysis and metabolism). Thus, warm lakes and, indeed, warm oceans (compare Iwata et al. 1993) likely provide important locations for mineralization of HCH. The story for CHBs in the Arctic Ocean is less clear because we do not have such an extensive dataset. Chlorobornane mirrors HCH in that water has become an important reservoir and medium of transport. However, CHB also parallels PCB in that contaminated soils provide a large reservoir that continues to supply the atmosphere. Therefore, we predict for CHBs that the turndown time of aquatic reservoirs will be substantially slower than that of HCH, especially for reservoirs directly downwind of contaminated soil sites (e.g., Lake Superior). Chlorobornane biomagnifies strongly (water to top predator -10^7 [Muir and Norstrom 1994]), making it much more of a concern than is HCH in aquatic environments.

Atrazine, initially, appears to have several environmentally attractive properties. It has a relatively short half-life in soil (30 to 90 d) and a low vapor pressure suggesting that it will not form a large soil reservoir (like PCB and CHB) and that it will not readily enter the atmosphere to be broadly distributed. Unfortunately, atrazine's high solubility and long half-life in water (10 to 27 y) provide an alternative route for long-range transport within drainage basins where it has been applied (e.g., see Isensee et al. [1988] and Kolpin et al. [1996]). Only about 1% of the applied chemical escapes from the soil into water, but in view of the large amounts being used and the long residence time, this rate of leakage has led to a significant build up in Lake Michigan within a decade. In contrast to the other POPs discussed here, the atmosphere is not required to transport atrazine to aquatic systems and it likely plays little or no role in clearance. The world's oceans are the final reservoir in the global drainage basin. We can safely predict that with a half-life of more than 10 y, atrazine will be delivered to the surface ocean where it will accumulate until steady state is achieved (i.e., flux to the ocean will equal the total rate of degradation). Because atrazine is not particle reactive, we expect it to behave much like ^{137}Cs in the ocean, e.g., providing a tracer of water from drainage basins where atrazine has been applied.

Polycyclic aromatic hydrocarbons provide an informative contrast to PCBs. From the Mediterranean case study and elsewhere (compare Macdonald et al. 2000), it is clear that the lighter PAH compounds are transported long distances in the atmosphere and that, as in the case of PCBs, air–water exchange is an important mechanism for delivery to aquatic reservoirs. However, there is also a particulate component of heavier PAHs carried by soot. Because the soot PAHs are delivered as particles, they behave differently than the lighter PAHs; they are resistant to degradation, participate strongly in vertical particle flux and become sequestered in sediments. The soot pathway is not unique to the Mediterranean (compare Gustafsson et al. 1997). Although soot PAH appears to be very persistent, the biological availability and consequences are less well-understood. Lighter PAHs enter deep water, probably following carbon cycling as observed for PCBs, but they also may be significantly degraded.

References

Aagaard K. 1989. A synthesis of the Arctic Ocean circulation. Rapports et Proces-verbaux des Reunion du Conseil International pour l'Exploration de la Mer. 188:11–22.

Aagaard K, Carmack EC. 1989. The role of sea ice and other fresh water in the Arctic circulation. *J Geophys Res* 94 (C14):485–498.

Aagaard K, Swift JH, Carmack EC. 1985. Thermohaline circulation in the Arctic Mediterranean seas. *J of Geophys Res* 90:4833–4846.

Aceves M, Grimalt JO. 1993. Seasonally dependent size distributions of aliphatic and polycyclic aromatic hydrocarbons in urban aerosols from densely populated areas. *Environ Sci Technol* 27:2896–2908.

Agrell C. 1999. Atmospheric transport of persistent organic pollutants to aquatic ecosystems [Ph.D. dissertation]. Lund, Sweden: Lund University, Department of Ecology and Ecotoxicology. 92 p.

Alcock RE, McGrath SP, Jones KC. 1995. The influence of multiple sewage sludge amendments on the PCB content of an agricultural soil over time. *Environ Toxicol Chem* 14:553–560.

Anderson PN, Hites RA. 1996. OH radical reactions: the major removal pathway for polychlorinated biphenyls from the atmosphere. *Environ Sci Technol* 30:1756–1763.

Anderson DJ, Bloem TB, Blankenbaker RK, Stanko TA. 1998. Concentrations of polychlorinated biphenyls in the water column of the Laurentian Great Lakes:Spring 1993. *J Gt Lakes Res.* 25:160–170.

Atkinson R, Arey J. 1994. Atmospheric chemistry of gas-phase polycyclic aromatic hydrocarbons: formation of atmospheric mutagens. *Environ Health Perspect* 102:(suppl 4)117–126.

Axelman J, Bandh C, Broman D, Carman R, Jonsson P, Larsson H, Linder H, Näf C, Pettersen H. 1995. Time-trend analysis of PAH and PCB sediment fluxes in the Northern Baltic proper using different dating methods. *Mar Freshw Res* 46:137–144.

Baker JE, Eisenreich SJ, Eadie BJ. 1991. Sediment trap fluxes and benthic recycling of OC, PAHs, and PCB congeners in Lake Superior. *Environ Sci Technol* 25:500–509.

Beckers J.-M, Brasseur P, Nihoul JCJ. 1997. Circulation of the western Mediterranean: from global to regional scales. *Deep Sea Res II* 44:531–549.

Beltrán FJ, González M, Rivas F, Alvarez P. 1996. Aqueous UV radiation and UV/H_2O oxidaton of atrazine first degradation products: deethylatrazine and deisopropylatrazine. *Environ Toxicol Chem* 15:868–782.

Bidleman TF, Alegria H, Ngabe B, Green C. 1998. Trends of chlordane and toxaphene in the ambient air of Columbia, South Carolina. *Atmos Environ* 32:1849–1856.

Bidleman TF, Falconer RL, Walla MD. 1995. Toxaphene and other organochlorine compounds in air and water at Resolute Bay, NWT, Canada. *Sci Total Environ* 160/161:55–63.

Bidleman TF, Jantunen LM, Falconer RL, Barrie LA, Fellin P. 1995. Decline of hexachlorocyclohexanes in the Arctic atmosphere and reversal of air-sea gas exchange. *Geophys Res Letters* 22:219–222.

Bidleman TF, Zaranski MT, Walla MD, 1988. Toxaphene: usage, aerial transport and deposition. In: Schmidtke NW, editor. Volume 1, Toxic contamination in large lakes. Chelsea MI: Lewis. p 257–284.

Bignert A., Litzén K, Odsjö T, Olsson M, Persson W, Reutergårdh, L. 1995. Time-related factors influence the concentrations of DDT, PCBs and shell parameters in eggs of Baltic Guillemot (Uria aalge), 1861–1989. *Environ Pollut* 89:27–36.

Bignert A, Olsson M, Alsberg T. 1998. Temporal trends of organochlorines in Northern Europe, 1967–1995. Relation to global fractionation, leakage from sediments and international measures. *Environ Pollut* 99/2:177–198.

Bishop CA, Weseloh DV, Burgess NM, Struger J, Norstrom RJ, Turle R, Logan KA. 1992. An atlas of contaminants in eggs of fish-eating colonial birds of the Great Lakes (1970–1988). Technical Report Series No. 152, Canadian Wildlife Service: Ontario Region.

Bodo BA. 1991. Trend analysis and mass-discharge estimation of atrazine in southwestern Ontario Great Lakes tributaries: 1981–1989. *Environ Toxicol Chem* 10:1105–1121.

Bootsma H, Hecky R. 1993. Conservation of the African Great Lakes: a limnological perspective. *Conserv Biol* 7:644–656.

Brorström-Lundén E. 1995. Measurements of semivolatile organic compounds in air and deposition [Ph.D. dissertation]. Göteborg, Sweden: Chalmers University of Technology.

Brubaker Jr WW, Hites RA. 1998. OH reaction kinetics of gas-phase: hexachlorocyclohexane and hexachlorobenzene. *Environ Sci Technol* 32:766–769.

Capel PD, Eisenreich SJ. 1985. PCBs in Lake Superior, 1978–1980. *J Gt Lakes Res* 11:447–461.

Carmack EC, Aagaard K, Swift JH, Macdonald RW, McLaughlin FA, Jones EP, Perkin RD, Smith J, Ellis K, Kilius L. 1997. Rapid changes of water properties and contaminants within the Arctic Ocean. *Deep-Sea Res* 44(8):1487–1502.

Carson R. 1962. Silent spring. New York: Houghton Mifflin.

Chernyak SM, McConnell LM, Rice CP. 1995. Fate of some chlorinated hydrocarbons in arctic and far eastern ecosystems in the Russian Federation. *Sci Total Environ* 160/161:75–85.

Chester R. 1993. Marine geochemistry. Cambridge: Chapman and Hall. 698 p.

Ciba-Geigy Corporation. 1994. Environmental fate reference data source book for atrazine. Greensboro NC: Ciba-Geigy.

Coleman PJ, Lee RGM, Alcock RE, Jones KC. 1997. Observations on PAH, PCB, and PCDD/F trends in UK urban air: 1991–1995. *Environ Sci Technol* 31:2120–2124.

Cotham W, Bidleman TF. 1991. Estimating the atmospheric deposition of organochlorine contaminants to the Arctic. *Chemosphere* 22:165–188.

Cromwell AE, Thurman EM. 1995. Atmospheric transport and deposition of triazine herbicides at Isle Royale National Park, Lake Superior. In: Meyer MT, Thurman EM, editors. Chemistry of herbicide metabolites in surface and ground water. Washington DC: American Chemical Society.

Dachs J, Bayona JM, Albaigés J. 1997. Spatial distribution, vertical profiles and budget of organochlorine compounds in western Mediterranean seawater. *Mar Chem* 57:313–324.

Dachs J, Bayona JM, Fowler SW, Miquel J-C, Albaigés J. 1996. Vertical fluxes of polycyclic aromatic hydrocarbons and organochlorine compounds in the western Alboran sea. *Mar Chem* 52:75–86.

Dachs J, Bayona JM, Raoux C, Albaigés J. 1997. Spatial, vertical distribution and budget of polycyclic aromatic hydrocarbons in the western Mediterranean seawater. *Environ Sci Technol* 31:682–688.

Donald DB, Stern GA, Muir DCG, Miskimmin BM, Fowler BR, Bailey R. 1997. Chlorobornanes in water, sediment and fish from toxaphene treated and untreated lakes in western Canada. *Environ Sci Technol* 32:1391–1397.

Eisenreich SJ. 1987. Chemical limnology of nonpolar organic contaminants in large lakes: PCBs in Lake Superior. In: Hites RA, Eisenreich SJ, editors. Series No. 216, Sources and fates of aquatic pollutants. Advances in Chemistry. Series No. 216. Washington DC: American Chemical Society. p 394–469.

Eisenreich SJ, Looney BB, Thornton JD. 1981. Airborne organic contaminants in the Great Lakes ecosystem. *Environ Sci Technol* 15:30–38.

Falconer RL, Bidleman TF. 1994. Vapor pressure and predicted particle-gas distribution of polychlorinated biphenyl congeners as a function of temperature and ortho-chlorine substitution. *Atmos Environ* 28:547–554.

Falconer RL, Bidleman TF, Gregor DJ. 1995. Air–water gas exchange and evidence for metabolism of hexachlorocyclohexanes in Resolute Bay, NWT. *Sci Total Environ* 160/161:65–74.

Fowler SW, Knauer GA. 1986. Role of large particles in the transport of elements and organic compounds through the oceanic water column. *Prog Oceanogra* 16:147–194.

Gevao B, Hamilton-Taylor J, Murdoch C, Jones KC, Kelly M, Tabner BJ. 1997. Depositional time trends and remobilisation of PCBs in lake sediments. *Environ Sci Technol* 31:3274–3280.

Giardina MC, Giardi MT, Filacchoini G. 1980. 4-amino-2-chloro-1,3,5-triazine: A new metabolite of atrazine by a soil bacterium. *Agric Bio Chem* 44:2067–2072.

Giardina MC, Giardi MT, Filacchoini G. 1982. Atrazine metabolism by nocardia: Elucidation of initial pathway and synthesis of porential metabolites. *Agric Bio Chem* 46:1439–1445.

Giardi MT, Giardina MC, Filacchioni G. 1985. Chemical and biological degradation of primary metabolites of atrazine by a nocardia strain. *Agric Bio Chem* 49:1551–1558

Glassmeyer ST, Brice KA, Hites RA. 1999. Atmospheric concentrations of toxaphene on the coast of Lake Superior. *J Gt Lakes Res* 25:492–499.

Glassmeyer ST, De Vault DS, Myers TR, Hites RA. 1997. Toxaphene in Great Lakes fish: a temporal, spatial and trophic study. *Environ Sci Technol* 31:84–88.

Glotfelty DE, Williams GH, Freeman HP, Leech MM. 1990. Regional atmospheric transport and deposition of pesticides in Maryland. In: Kurtz DA, editor. Long range transport of pesticides. Cheslea MI: Lewis. p 199–222.

Gogou A, Stephanou EG, Stratigakis N, Grimalt JO, Simo R, Aceves M, Albaigés J. 1994. Differences in lipid and organic salt constituents of aerosols from eastern and western Mediterranean coastal cities. *Atmos Environ* 28:1301–1310.

Goolsby DA, Thurman EM, Pomes ML, Battazlin WA. 1997. Herbicides and their metabolites in rainfall: transport and deposition patterns across the midwestern U.S. *Environ Sci Technol* 31:1325–1333

Grenon M, Batisse M. 1989. Future of the Mediterranean basin. In: Volume XIX, The Blue Plan. London: Oxford University. p 86–106.

Gustafsson O, Haghseta F, Chan C, Macfarlane J, Gschwend PM. 1997. Quantification of the dilute sedimentary soot phase: Implications for PAH speciation and bioavailability. *Environ Sci Technol* 31:203–209.

Hainzl D, Burhenne J, Parlar H. 1994. Theoretical consideration of the structural variety in the toxaphene mixture, taking into account recent experimental results. *Chemosphere* 28:245–251.

Hargrave BT, Barrie LA, Bidleman TF, Welch HE. 1997. Seasonality in exchange of organochlorines between arctic air and seawater. *Environ Sci Technol* 31:3258–3266.

Hargrave BT, Vass WP, Erickson PE, Fowler BR. 1988. Atmospheric transport of organochlorines to the Arctic Ocean. *Tellus* 40B:480–493.

Harner T, Bidleman TF. 1996. Measurements of octanol-air partition coefficients for polychlorinated biphenyls. *J Chem Engineering Data* 41:895–899.

Harner T, Bidleman TF. 1998. Octanol-air partition coefficient for describing particle/gas partitioning of aromatic compounds in urban Air. *Environ Sci Technol* 32:1494–1502

Harner T, Bidleman TF, Jantunen LMM, Mackay D. 2000. Soil-air exchange model of persistent pesticides in the U.S. cotton belt. *Environ Toxicol Chem* (submitted).

Harner T, Jantunen L MM, Bidleman TF, Barrie LA, Kylin H, Strachan WMJ, Macdonald RW. 2000. Microbial degradation is a key elimination pathway of hexachlorocyclohexanes from the Arctic Ocean. *Geophys Res Lett* 27:1155–1158.

Harner T, Kylin H, Bidleman TF, Strachan WMJ. 1999. Removal of α- and γ-hexachloro-cyclohexane and the enantiomers of α-hexachlorocyclohexane in the eastern Arctic Ocean. *Environ Sci Technol* 33:1157–1164.

Harner T, Mackay D, Jones KC. 1995. Model of the long-term exchange of PCBs between soil and the atmosphere in the southern UK. *Environ Sci Technol* 29:1200–1209.

Harner T, Wideman JL, Jantunen LMM, Bidleman TF, Parkhurst WJ. 1999. Residues of organochlorine pesticides in Alabama soils. *Environ Pollut* 106: 323–332.

Hillery BR, Basu I, Sweet CW, Hites RA. 1997a. Temporal and spatial trends in a long-term study of gas-phase PCB concentrations near the Great Lakes. *Environ Sci Technol* 31:1811–1816.

Hillery BR, Basu I, Sweet CW, Hites RA. 1997b. Temporal and spatial trends in a long-term study of gas-phase PCB concentrations near the Great Lakes. *Environ Sci Technol* 31:1811–1816.

Hillery BR, Hoff RM, Hites RA. 1997. Atmospheric contaminant deposition to the Great Lakes determined from the integrated atmospheric deposition networks. In: Baker JE, editor. Atmospheric deposition of contaminants to the Great Lakes and coastal waters. Pensacola FL: SETAC. p. 277-291.

Hinckley DA, Bidleman TF, Foreman WT, Tuschall JR. 1990. Determination of vapor pressures for nonpolar and semipolar organic compounds from gas chromatographic retention data. *J Chem Engineering Data* 35:232–237.

Hines N. 1996. Atrazine and metabolites in tributaries and open waters of Lake Michigan [M.S. Thesis]. Minneapolis MN: University of Minnesota.

Hoff RM, Bidleman TF, Eisenreich SJ. 1993b. Estimation of PCC loadings from the atmosphere to the Great Lakes. *Chemosphere* 27:2047–2055.

Hoff RM, Muir DCG, Grift NP. 1992a. Annual cycle of polychlorinated biphenyls and organohalogenated pesticides in air in southern Ontario. 1. Air concentration data. *Environ Sci Technol* 26:266–275.

Hoff RM, Muir DCG, Grift NP. 1992. Annual cycle of polychlorinated biphenyls and organohalogenated pesticides in air in southern Ontario. 2. Atmospheric transport and sources. *Environ Sci Technol* 26:276–283.

Hoff RM, Muir DCG, Grift NP, Brice KA. 1993. Measurement of PCCs in air in southern Ontario. *Chemosphere* 27:2057–2062.

Hoff RM, Strachan WJM, Sweet CW, Chan CH, Shackleton M, Bidleman TF, Brice KA, Burniston, DA, Cussion S, Gatz DF, Harlin K, Schroeder WH. 1996. Atmospheric deposition of toxic chemicals to the Great Lakes: a review of data through 1994. *Atmos Environ* 30:3505–3527.

Hornbuckle KC, Jeremiason JD, Sweet CW, Eisenreich SJ. 1994. Seasonal variations in air-water exchange of PCBs in Lake Superior. *Environ Sci Technol* 28:1491–1501.

Hornbuckle KC, Sweet CW, Pearson RF, Swackhamer DL, Eisenreich SJ. 1995. Assessing annual water-air fluxes of polychlorinated biphenyls in Lake Michigan. *Environ Sci Technol* 29:869–877.

Howdeshall MJ, Hites RA. 1996. Historical input and degradation of toxaphene in Lake Ontario sediment. *Environ Sci Technol* 30:220–224.

Isensee AR, Helling CS, Gish TJ, Kearney PC, Coffman BC, Zhuang W. 1988. Groundwater residues of atrazine, alachlor, and cyanazine under no-tillage practices. *Chemosphere* 17:165–174.

Isnard P, Lambert S. 1989. Aqueous solubility and n-octanol/water partition coefficient correlations. *Chemosphere* 18:1837–1853.

Iwata H, Tanabe S, Sakai N, Nishimura A, Tatsukawa R. 1994. Geographical distribution of persistent organochlorines in air, water and sediments from Asia and Oceania, and their implications for global redistribution from lower latitudes. *Environ Pollut* 85:15–33.

Iwata H, Tanabe S, Sakai N, Tatsukawa R. 1993. Distribution of persistent organochlorines in the oceanic air and surface seawater and the role of ocean on their global transport and fate. *Environ Sci Technol* 27:1080–1098.

Iwata H, Tanabe S, Ueda K, Tatsukawa R. 1995. Persistent organochorine residues in air, water, sediments and soils from the Lake Baikal region, Russia. *Environ Sci Technol* 29:792–801.

Jantunen LM. 1997. Air-water gas exchange of toxaphene in arctic regions [M.A.Sc. thesis]. Toronto Canada: University of Toronto, Department of Chemical Engineering and Applied Chemistry. 88 p.

Jantunen LM, Bidleman TF. 1995. Reversal of the air-water gas exchange direction of hexachlorocyclohexanes in the Bering and Chukchi Seas. *Environ Sci Technol* 29:1081–1089.

Jantunen LM, Bidleman TF. 1996. Air-water gas exchange of hexachlorocyclohexanes (HCHs) and the enantiomers of a-HCH in arctic regions. *J Geophys Res* 101:28837–28846, corrections Ibid 102:19279–19282 (1997).

Jantunen LM, Bidleman TF. 1998. Organochlorine pesticides and enantiomers of chiral pesticides in Arctic Ocean water. *Arch Environ Contam Toxicol* 35:218–228.

Jantunen LM, Bidleman TF. 2000a. Henry's law constant for technical toxaphene as a function of temperature. *Chemosphere-Global Change Science* 2:225–231.

Jantunen LMM, Bidleman TF. 2000b. Air-water gas exchange of toxaphene in Lake Superior. *Environ Sci Technol* (submitted).

Jantunen LMM, Bidleman TF, Harner T, Parkhurst WJ. 2000. Toxaphene and other organochlorine pesticides in Alabama air. *Environ Sci Technol* (submitted).

Jeremiason JJ, Eisenreich SJ, Baker JE, Eadie BJ. 1998. PCB decline in settling particles and benthic recycling of PCBs and PAHs in Lake Superior. *Environ Sci Technol* 32:3249–3256.

Jeremiason JD, Hornbuckle K, Eisenreich SJ. 1994. Polychlorinated biphenyls (PCBs) in Lake Superior, 1978–1992: decreases in water concentrations reflect loss by volatilization. *Environ Sci Technol* 28:903–914.

Kidd K, Lockhart WL, Wilkinson P, Muir DCG. 1998. Metals, pesticides and other persistent in water, sediments and biota from Lake Malawi food web. In: Report on Lake Malawi Biodiversity Conservation Project.

Kjeller L-O, Rappe C. 1995. Time trends in levels, patterns, and profiles for polychlorinated dibenzo-p-dioxins, dibenzofurans, and biphenyls in a sediment core from the Baltic Proper. *Environ Sci Technol* 29:346–355.

Kolpin DW, Thurman EM, Goolsby DA. 1996. Occurence of selected pesticides and their metabolites in near-surface aquifers of the midwestern U.S. *Environ Sci Technol* 30:335–340.

Kruger EL, Somasundaram L, Kanwar RS, Coats JR. 1993. Movement and degradation of [^{14}C]atrazine in undisturbed soil columns. *Environ Toxicol Chem* 12:1969-1975.

Kucklick JR, Baker JE. 1998. Organochlorines in Lake Superior's food web. *Environ Sci Technol* 32:1192–1198.

Kucklick JR, Hinckley DA, Bidleman TF. 1991. Determination of Henry's law constants for hexachlorocyclohexanes in distilled water and artificial seawater as a function of temperature. *Mar Chem* 34:197–209.

Lee RGM, Hung H, Mackay D, Jones KC. 1998. Measurement and modelling of the diurnal cycling of atmospheric PCBs and PAHs. *Environ Sci Technol* 32:2172–2179.

Lerch RN, Thurman EM, Kruger EL. 1997. Mixed-mode sorption of hydroxylated atrazine degradation products to soil: a mechanism. *Environ Sci Technol* 31:1539–1546.

Li YF. 1999. Global technical hexchlorocyclohexane usage and its contamination consequences in the environment from 1948 to 1997. *Sci Total Environ* 232:123–160.

Li YF, Bidleman TF, Barrie LA, McConnell LL. 1998. Global hexachlorocyclohexane use trends and their impact on the arctic atmospheric environment. *Geophys Res Letters* 25:39–41.

Lipiatou E, Albaigés J. 1994. Atmospheric deposition of hydrophobic organic chemicals in the northwestern Mediterranean sea: comparison with the Rhone river input. *Mar Chem* 46:153–164.

Lipiatou E, Marty J-C, Saliot A. 1993. Sediment trap fluxes of polycyclic aromatic hydrocarbons in the Mediterranean sea. *Mar Chem* 44:43–54.

Lipiatou E, Tolosa I, Simo R, Bouloubassi I, Dachs J, Marti S, Sicre M-A, Bayona JM, Grimalt JO, Saliot A, Albaigés J. 1997. Mass budget and dynamics of polycyclic aromatic hydrocarbons in the Mediterranean Sea. *Deep-Sea Res II* 44:881–905.

Lockhart WL, Muir DCG. 1996. Food chain accumulation, biological effects and sediment contamination in Lake Laberge and other Yukon lakes. Indian and Northern Affairs Canada Synopsis of Research Conducted Under the 1994–1995 Northern Contaminants Program; Environmental Studies No. 73. Ottawa: Indian and Northern Affairs Canada. p 199–207.

Macdonald RW. 1996. Arctic awakenings. *Nature* 380:286–287.

Macdonald RW, Bewers JM. 1996. Contaminants in the arctic marine environment: priorities for protection. *ICES J Mar Sci* 53:537–563.

Macdonald RW, Barrie LA, Bidleman TF, Diamond ML, Gregor DJ, Semkin RG, Strachan WJM, Li YF, Wania F, Alaee M, Alexeeva LB, Backus SM, Bailey R, Bewers JM, Gobeil C, Halsall CJ, Harner T, Hoff JT, Jantunen LMM, Lockhart WL, Mackay D, Muir DCG, Pudykiewicz J, Reimer KJ, Smith JN, Stern GA, Schroeder WH, Wagemann R, Yunker MB. 2000. Sources, occurrence and pathways of contaminants in the Canadian Arctic: a review. *Sci Total Environ* 254:93–234.

Macdonald RW, Carmack EC, McLaughlin FA, Falkner KK, Swift JH. 1999. Connections among ice, runoff and atmospheric forcing in the Beaufort Gyre. *Geophys Res Letters* 26:2223–2226.

Macdonald RW, McLaughlin FA, Adamson L. 1997. The Arctic Ocean - the last refuge of volatile organochlorines. *Can Chem News* Sept:28–29.

Macdonald RW, McLaughlin FA, Carmack EC. 1996. Long-range transport of contaminants to the Canadian Basin. In: Murray JL, Shearer RG, Han SL, editors. Indian and Northern Affairs

Canada Synopsis of Research Conducted Under the 1994–1995 Northern Contaminants Program; Environmental Studies No. 73. Ottawa: Indian and Northern Affairs Canada. p 45–50.

Mackay D, Bentzen E. 1997. The role of the atmosphere in Great Lakes contamination. *Atmos Environ* 4045–4047.

Mackay D, Sang S, Vlahos P, Diamond M, Gobas F, Dolan D. 1994. A rate constant model of chemical dynamics in a lake ecosystem: PCBs in Lake Ontario. *J Gt Lakes Res* 20:625–642.

Mackay D, Shiu WY, Ma KC. 1992. Illustrated handbook of physical chemical properties and environmental fate for organic chemicals, Voumes I-II. Boca Raton FL: CRC/Lewis.

Mackay D, Shiu WY, Ma KC. 1997. Illustrated handbook of physical chemical properties and environmental fate for organic chemicals, Volumes I–V. Boca Raton FL: CRC /Lewis.

Malawi Government. 1998. Pesticide imports to Malawi 1992. Internal report.

Marty JC, Tissier MJ, Saliot A. 1984. Gaseous and particulate polycyclic aromatic hydrocarbons (PAH) from the marine atmosphere. *Atmos Environ* 18:2183–2190.

Masclet P, Pistikopoulos P, Beyne S, Mouvier G. 1988. Long range transport and gas/particle distribution of polycyclic aromatic hydrocarbons at a remote site in the Mediterranean sea. *Atmos Environ* 22:639–650.

McConnell, LL, Bidleman, TF, Cotham, WE, Walla, MD. 1998. Air concentrations of organochlorine insecticides and polychlorinated biphenyls over Green Bay, Wisconsin and the four lower Great Lakes. *Environ Pollut* 103/3:391–399.

McLaughlin FA, Carmack EC, Macdonald RW, Bishop JKB. 1996. Physical and geochemical properties across the Atlantic/Pacific water mass boundary in the southern Canadian Basin. *J Geophys Res* 101(C1):1183–1197.

Melnikov S, Vlasov S. 1995. Report to AMAP on organochlorines in seawater, suspended particulate matter from the Laptev, Pechora, Kara, East Siberian and Chukchi Seas and on river water suspended matter 1994–1995. St. Petersburg Russia: Centre for Monitoring Arctic.

Meylan WM, Howard PH. 1993. Computer estimation of the atmospheric gas phase reaction rate of organic compounds with hydroxyl radicals and ozone. *Chemosphere* 12:2293–2299.

Mills MS, Thurman EM. 1994. Preferential dealkylation reactions of s-triazine herbicides in the unsatuated zone. *Environ Sci Technol* 28:600–605.

Miskimmin, BM, Muir, DCG, Schindler, DW, Stern, GA, Grift, NP. 1995. Chlorobornanes in sediments and fish 30 years after toxaphene treatment of lakes. *Environ Sci Technol* 29:2490–2495.

Muir DCG, Baker BE. 1978. The disappearance and movement of three triazine herbicides and several of their degradation products in soil under field conditions. *Weed Res* 18:111–120.

Muir DCG, Grift NP. 1995. Fate of herbicides and organochlorine insecticides in lake waters. In: Ragsdale NN, Kearney PC, Plimmer JR, editors. Chemistry. Options 2000, ACS Conference Proceedings Series. Washington DC: American Chemical Society. p 141–156.

Muir DCG, Braune B, DeMarch B, Norstrom R, Wagemann R, Gamberg M, Poole K, Addison R, Bright D, Dodd M, Duschenko W, Eamer J, Evans M, Elkin B, Grundy S, Hargrave B, Hebert C, Johnstone R, Kidd K, Koenig B, Lockhart L, Payne J, Peddle J, Reimer K. 1997. Ecosystem uptake and effects. In: Jensen J, Adhare K, Shearer R, editors. Canadian Arctic Contaminants Assessment Report; Indian and Northern Affairs Canada; Ottawa. p. 183–294.

Muir DCG, Kidd K, Grift B, Smith S, Mwita J, Savoie D, Rosenberg B, Burniston D, Strachan W. 1998. Persistent organochlorine chemicals in air and water from the Lake Malawi area: a preliminary report. Burlington Ontario: National Water Res. Institute. Unpublished Report.

Muir DCG, Norstrom RJ. 1994. Persistent organic contaminants in Arctic marine and freshwater ecosystems. *Arctic Research of the U.S.* 8:136–146.

Muir, DCG, Segstro MD, Welbourne PM, Toom D, Eisenreich SJ, Macdonald CR. 1993. Patterns of accumulation of airborne organochlorine contaminants in lichens from the upper Great Lakes region of Ontario. *Environ Sci Technol* 27:1201–1210.

Murphy TJ, Mullin MD, Meyer JA. 1987. Equilibration of polychlorinated biphenyls and toxaphene with water and air. *Environ Sci Technol* 21:155–162.

Ngabe B, Bidleman TF. 1992. Occurrence and vapor particle partitioning of heavy organic compounds in ambient air in Brazzaville, Congo. *Environ Pollut* 76:147–156.

Ngabe B, Bidleman TF, Falconer RL. 1993. Base hydrolysis of α- and γ-hexachlorocyclohexanes. *Environ Sci Technol* 27:1930–1933.

Oreskes N, Shrader Frechette K, Belitz K. 1994. Verification, validation, and confirmation of numerical models in the earth sciences. *Science* 263:641–646.

Østfeldt P, Gustavson K, Jansson B, Jonsson P, Miettinen V, Ringstad O, Wesén C. 1994. Halogenated organic compounds in the marine environment 1989–1990. København: Nordic Council of Ministers (TemaNord, 1994:591).

Pacyna JM. 1998. Environmental cycling of selected persistent organic pollutants (POPs) in the Baltic region. The Second Annual Report. Kjeller Norway: The European Union (EU) Environment & Climate Research Programme.

Pacyna JM. 1999. Environmental cycling of selected persistent organic pollutants (POPs) in the Baltic region. Executive Final Summary Report. Kjeller Norway: The European Union (EU) Environment & Climate Research Programme.

Pearson RF, Swackhamer DL, Eisenreich SJ, Long DT. 1997. Concentrations, accumulations and inventories of toxaphene in sediments of the Great Lakes. *Environ Sci Technol* 31:3523–3529.

Pekarik C, Weseloh DV, Barrett GC, Simon M, Bishop CA, Petit KE. 1999. An atlas of contaminants in the eggs of fish eating colonial birds of the Great Lakes. Technical Report Series No. 321., Canadian Wildlife Service.

Pelizzetti E, Maurino V, Minero C, Carlin V, Pramauro E, Zerbinati O, Tosata ML. 1990. Photocatalytic degradation of atrazine and other s-triazine herbicides. *Environ Sci Technol* 24:1559–1565.

Pereira WE, Hostettler FD. 1993. Nonpoint source contamination of the Mississippi River and its tributaries. *Environ Sci Technol* 27:1542–1552.

Pereira WE, Rostad CE. 1990. Occurance, distributions, and transport of herbicides and their degradation products in the lower Mississippi River and its tributaries. *Environ Sci Technol* 24:1400–1406.

Petit KE, Bishop CA, Weseloh DV, Norstrom RJ. 1994. An atlas of contaminants in eggs of fish-eating colonial birds of the Great Lakes (1989–1992). Technical Report Series No. 193., Canadian Wildlife Service.

Ramesh A, Tanabe S, Tatsukawa R, Subramanian AN, Palanichamy S, Mohan D, Venugopalan VK. 1989. Seasonal variation of organochlorine insecticide residues in air from Porto Novo, South India. *Environ Pollut* 62:213–222.

Raoux S, Bayona JM, Miquel JC, Fowler SW, Teyssie J-L, Albaigés J. 1995. Temporal variability of vertical fluxes of aliphatic and aromatic hydrocarbons in the North-western Mediterranean Sea. EROS 2000, 5th workshop North-western Mediterranean Sea. Water Pollution Research Reports 32. European Commission, Luxembourg, 181–186.

Rapaport RA, Eisenreich SJ. 1986. Atmospheric deposition of toxaphene to eastern North America derived from peat accumulation. *Atmos Environ* 20:2367–2379.

Rawn DFK, Haldorson THJ, Muir DCG. 1999. A multi-year study of four herbicides in air and precipitation from a small prairie watershed. *J Environ Qual* 28:898–905.

Rejto M, Saltzman S, Acher AJ, Muszkat L. 1983. Identification of sensitized photooxidation products of s-triazine herbicides in water. *J Agric Food Chem* 31:138–142.

Repetto R, Baliga SS. 1996. Pesticides and the immune system: the public health risks. World Resources Institute. 103 p.

Rice CP, Evans MS. 1984. Toxaphene in the Great Lakes. In: Nriagu J, Simmons MS. editors. Toxic contaminants in the Great Lakes. New York: J Wiley. p. 163–194.

Rice CP, Sampson PJ, Noguchi G. 1986. Atmospheric transport of toxaphene to Lake Michigan. *Environ Sci Technol* 20:1109–1116.

Richards RP, Baker DB. 1993. Pesticide concentration patterns in agricultural drainage networks in the Lake Erie basin. *Environ Toxicol Chem* 12:13–26.

Rudels B, Jones EP, Anderson LG. 1996. Formation and evolution of the surface mixed layer and halocline of the Arctic Ocean. *J Geophys Res* 101:8807–8821.

Saleh MA. 1991. Toxaphene: chemistry, biochemistry, toxicity and environmental fate. *Rev Environ Contam Toxicol* 118:1–85.

Schottler SP, Eisenreich SJ. 1994. Herbicides in the Great Lakes. *Environ Sci Technol* 28:2228–2232.

Schottler SP, Eisenreich SJ. 1997. Mass balance model to quantify atrazine sources, transformation rates and trends in the Great Lakes. *Environ Sci Technol* 31:2616–2625.

Schottler SP, Eisenreich SJ, Capel PD. 1994. Atrazine, alachlor, and cyanazine in a large agricultural river system. *Environ Sci Technol* 28:1079–1089.

Schulz-Bull DE, Petrick G, Kannan N, and Duinker JC. 1994. Distribution of individual chlorobiphenyls (PCB) in solution and suspension in the Baltic Sea. *Mar Chem* 48:245.

Schwarzenbach RP, Gschwend PM, Imboden DM. 1993. Environmental organic chemistry. New York: J Wiley. 681 p.

Seinfeld JH, and Pandis SN. 1998. Atmospheric chemistry and physics. New York: J Wiley. 1326 p.

Seybold CA, McSweeney K, Lowery B. 1994. Atrazine adsorption in sandy soils of Wisconsin. *J Environ Qual* 23:1291–1297.

Shoeib M, Brice KA, Hoff RM. 1999. Airborne concentration of toxaphene congeners at Point Petre, Ontario, using gas chromatography: electron capture negative ion mass spectrometry (GC-ECNIMS). *Chemosphere* 39:849–871.

Sicre MA, Marty JC, Saliot A, Aparicio X, Grimalt JO, Albaigés J. 1987. Aliphatic and aromatic hydrocarbons in different sized aerosols over the Mediterranean sea: occurrence and origin. *Atmos Environ* 21:2247–2259.

Simo R, Colom-Altes M, Grimalt JO, Albaigés J. 1991. Background levels of atmospheric hydrocarbons, sulphate and nitrate over the western Mediterranean. *Atmos Environ* 25A:1463–1471.

Simo R, Grimalt JO, Albaigés J. 1997. Loss of unburned-fuel hydrocarbons from combustion aerosols during atmospheric transport. *Environ Sci Technol* 31:2697–2700.

Solomon KR, Baker DL, Richards RP, Dixon KR, Klaine SJ, La Point TW, Kendall RJ, Weisskopf CP, Giddings JM, Giesy JP, Hall LW, Williams WM. 1996. Ecological risk assesment of atrazine in North American surface waters. *Environ Toxicol Chem* 15:31–76.

Stern GA, Tomy GT, Billeck B, Lockhart WL, Muir DCG, Wilkinson P. 1998. Chlorinated Bornanes (Toxaphene) in Yukon Lakes: atmospheric or non-atmospheric source. *Organohalogen Compounds* 35:299–302.

Strachan WMJ, Eisenreich SJ. 1988. Mass balancing toxic chemicals in the Great Lakes: the role of atmospheric deposition. Windsor Ontario: International Joint Commission Report. 115 p.

Strachan WMJ, Eisenreich SJ. 1990. Mass balance accounting of chemicals in the Great Lakes. In: Kurtz DA. Long range transport of pesticides. Chelsea MI: Lewis. p 291–303.

Strachan WMJ, Teixeria M, Harner T, Bidleman TF, Kylin H. 1998. Concentrations and trend of organochlorine compounds and polycyclic aromatic hydrocarbons in the Arctic Ocean. *Organohalogen Compounds* 39:407–410.

Suntio, LR, Shiu WY, Mackay D, Seiber JN, Glotfelty D. 1988. Critical review of Henry's Law constants for pesticides. *Rev Environ Contam Toxicol* 103:1–60.

Swackhamer DL, Schottler SP, Pearson RF. 1998. CHBs in the Great Lakes. *Chemosphere* 37:2545–2561.

Swackhamer DL, Schottler SP, Pearson RF. 1999. Air-water exchange and mass balance of toxaphene in the Great Lakes. *Environ Sci Technol* 33:3864–3872.

Sweetman AJ, Jones KC. 2000. Declining PCB concentrations in the UK atmosphere: evidence and possible causes. *Environ Sci Technol* 34:863–869.

Takeoka H, Ramesh A, Iwata H, Tanabe S, Subramanian AN, Mohan D, Magendran A, Tatsukawa R. 1991. *Mar Pollut Bull* 22:290–297.

Tanabe S. 1988. PCB problems in the future: foresight from current knowledge. *Environ Pollut* 50:5–28.

Tanabe S, Ramesh A, Sakashita D, Iwata H, Tatsukawa R, Mohan D, Subramanian AN. 1991. Fate of HCH (BHC) in tropical paddy field: Application test in South India. *Int J Environ Anal Chem* 45:45–53.

Tateya S, Tanabe S, Tatsukawa R. 1988. PCBs on the globe: possible trend of future levels in the open ocean environment. In: Schmidtke NW, editor. Toxic contamination in Large Lakes, 3. Chelsea MI: Lewis. p. 237–281.

Thomas G, Sweetman AJ, Ockenden WA, Mackay D, Jones KC. 1998. Air-pasture transfer of PCBs. *Environ Sci Technol* 32:936–942.

Thomas GO, Sweetman AJ, Jones KC. 1998. Input-output balance of PCBs in a long-term study of lactating dairy cows. *Environ Sci Technol* 33:104–112.

Thomas GO, Smith KEC, Sweetman AJ, Jones KC. 1998. Further studies of the air-pasture transfer of PCBs. *Environ Pollut* 102:119–128.

Thurman EM, Goolsby DA, Meyer MT, Kolpin DW. 1991. Herbicides in surface waters of the midwestern U.S.: the effect of spring flush. *Environ Sci Technol* 25:1794–1796.

Thurman EM, Goolsby DA, Meyer MT, Mills MS, Pomes ML, Kolpin DW. 1992. A reconnaissance study of herbicides and their metabolites in surface water of the midwestern U.S. using immunoassay and gas chromatography/mass spectrometry. *Environ Sci Technol* 26:2440–2447.

Thurman EM, Meyer MT, Mills MS, Zimmerman L, Perry CA, Goolsby DA. 1994. Formation and transport of deethylatrazine and deisopropylatrazine in surface water. *Environ Sci Technol* 28:2267–2277.

Tolosa I, Bayona JM, Albaigés J. 1995. Spatial and temporal distribution, fluxes, and budgets of organochlorinated compounds in northwestern Mediterranean sediments. *Environ Sci Technol* 29:2519–2527.

Tolosa I, Bayona JM, Albaigés J. 1996. Aliphatic and polycyclic aromatic hydrocarbons and sulfur/oxygen derivatives in northwestern Mediterranean sediments: spatial and temporal variability, fluxes, and budgets. *Environ Sci Technol* 30:2495–2503.

Tolosa I, Readman JW, Fowler SW, Villeneuve JP, Dachs J, Bayona JM, Albaigés J. 1997. PCBs in the western Mediterranean: temporal trends and mass balance assessment. *Deep-Sea Res II* 44:907–928.

[UNEP/WHO]. United Nation Enviornment Program/World Health Organization. 1992. Airborne pollution of the Mediterranean Sea. Report and Proceedings of the second WHO/UNEP Workshop. MAP Technical Report Series 64, UNEP, Athens, 246 p.

Villeneuve JP, Cattini C. 1986. Input of chlorinated hydrocarbons through dry and wet deposition to the western Mediterranean. *Chemosphere* 15:115–120.

Voldner EC, Schroeder WH. 1989. Modelling of atmospheric transport and deposition of toxaphene to the Great Lakes ecosystem. *Atmos Environ* 23:1949–1961.

VonRumker R, Lawless AW, Meiners AF, Lawrence KA, Kelsco GC, Horay F. 1974. Production, distribution, use, and environmental impact potential of selected pesticides. U.S. Environmental Protection Agency. EPA–540/1–74–001.

Walker K, Vallero DA, Lewis RG. 1999. Factors influencing the distribution of lindane and other hexachlorocyclohexanes in the environment. *Environ Sci Technol* 33:4373–4378.

Wania F, Mackay D. 1995. A global distribution model for persistent organic chemicals. *Sci Total Environ* 160/161:211–232.

Wania F, Mackay D. 1999. Global chemical fate of a-hexachlorocyclohexane: 2. use of a global distribution model for mass balancing, source apportionment and trend prediction. *Environ Toxicol Chem* 18:1400–1407.

Wania F, Mackay D, Li YF, Bidleman TF, Strand A. 1999 Global chemical fate of α-hexachlorocyclohexane: 1. evaluation of a global distribution model. *Environ Toxicol Chem* 18:1390–1399.

Welhouse G, Bleam WF. 1993. Atrazine hydrogen-bonding potentials. *Environ Sci Technol* 27:494–505.

Wheeler PA, Gosselin M, Sherr E, Thibault D, Kirchman DL, Benner R, Whitledge TE. 1996. Active cycling of organic carbon in the central Arctic Ocean. *Nature* 380:697–699.

Winkelmann DA, Klaine SJ. 1991. Degradation and bound residue formation of atrazine in a western Tennessee soil. *Environ Sci Technol* 10:335–345.

Chapter 8

Framework and Criteria for Evaluation of Persistence and Long-Range Transport

Bob Boethling, John Buccini, Christina Cowan-Ellsberry, Diana Graham, Bjorn Hansen, Michael Murray, Larry Rampy, Bruce Rodan, Bo Wahlström

The principal goal of this chapter is to present a framework for evaluating the environmental behavior of chemical substances with respect to their persistence and long-range transport. The purpose of such a framework is to provide an approach for determining whether a substance is persistent and can undergo long-range transport. The primary focus is on current regional and global regulatory initiatives, as discussed in Chapter 1. However, in addition, it is envisioned that this framework could be used broadly as guidance for the identification, comparison, and prioritization of chemical substances relative to their potential for persistence and long-range transport. Thus, this framework could be useful at the national level as part of a procedure for early identification of potentially problematic substances and in industry for evaluating alternative substances before they are marketed.

The first part of this chapter discusses the framework and presents guidance on how it may be used to determine if a specific substance has properties that meet the definitions of persistence and may be subject to long-range transport. This framework is limited to defining the persistence and long-range–transport potential. A more detailed screening assessment for identifying persistent, bioaccumulative, and toxic substances (PBTs) and persistent organic pollutants (POPs) that may be of national, regional, or international concern includes additional considerations of bioaccumulation and toxicity.

Whether a PBT is subject to any regulatory or other action depends on the policy objectives of the regulatory mandate. Policy objectives determine the aspects of the environment and/or of human health that need to be protected and the degree of protection desired. Once the level of concern is established, criteria can be selected

Chapter Preview

Evaluation of Persistence and Long-Range Transport of Organic Chemicals in the Environment. Gary Klečka et al., editors.
 ISBN 1-880611-22-8

against which chemicals are assessed, taking into account persistence, bioaccumulation, toxicity, and physicochemical properties associated with the potential for long-range transport.

Factors that must be considered in the development of appropriate criteria for persistence and long-range transport—and various approaches that can be used to develop these criteria—are discussed in the second part of the chapter. In the final sections of the chapter, considerations such as flexibility, the need for transparency and openness, and data availability and quality are discussed.

Framework for Substance Evaluation

This section describes the framework that can be used to separate those chemicals that are not of concern as PBTs/POPs and can be set aside from those that meet the definition within the regulatory mandate, i.e., those chemicals having sufficient potential for persistence and long-range transport to merit being evaluated for other characteristics such as bioaccumulation potential and toxicity. The basic principles in this evaluation are that the process will 1) be based on sound scientific principles, 2) allow for expert judgment and flexibility, and 3) be precautionary. The precautionary principle is applied so that, in the face of uncertainty or a lack of strong evidence to the contrary, substances will be assumed to meet the definitions for persistence and long-range transport at the first screening stage. This approach is considered precautionary because the chemical may be found, upon detailed evaluation, not to meet the definition of a PBT/POP, despite having been initially identified by the framework as having these characteristics. The overall process for determining whether substances are PBTs/POPs, as illustrated in Figure 8-1, includes this framework, in which the evaluation proceeds from screening to a more detailed assessment, with consequent increases in the amount of data and level of effort required. A more detailed description of the first two parts of the framework, i.e., screening and detailed evaluation, is given below and includes a discussion of the use of primary chemical data, models, and the role of monitoring data. The last two parts of the framework, i.e., risk assessment and risk management, were not within the scope of the workshop and are not described further. However, these elements are essential for a complete evaluation of a substance such as a POP.

Applying screening criteria

The objective of the screening portion of the framework is to identify, as simply and efficiently as possible, substances that are clearly nonpersistent. This focus on persistence recognizes that a substance must be sufficiently persistent in the environment to be available for long-range transport. The measured or estimated half-life of the substance in the major environmental media is compared to the half-life criteria for those media. If the half-life of the substance is less than the criteria values for all of the relevant environmental media, the substance is determined to be

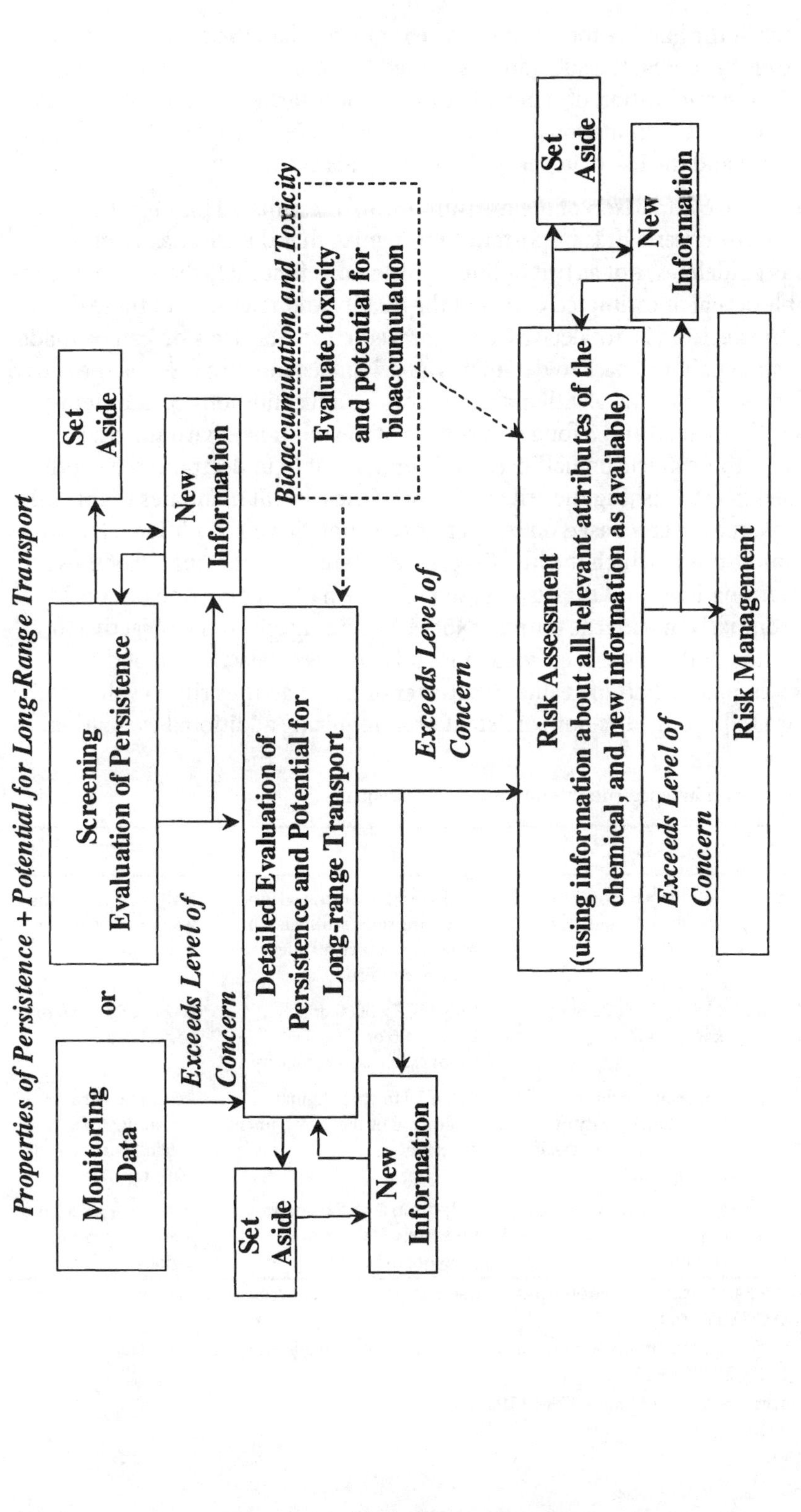

Figure 8-1 Framework for substance evaluation process

nonpersistent. If the half-life for the substance is greater than the criteria value for one or more of the media, the substance is passed to a more detailed evaluation. Conceptually, the application of criteria to evaluate substances for persistence has two components: the comparison of actual or estimated properties of the substance with the criteria and the use of expert judgment in this evaluation.

In conducting the comparison of the measured and/or estimated half-life values for a substance to the criteria, it is important to recognize that the criteria are meant to be applied as guidelines, not as bright lines. This means first that when the evidence (i.e., available actual or estimated data and the quality and relevance of these data) for making the decision is not decisive or is contradictory, the decision can be made that the substance should pass to detailed evaluation. Second, because the measured half-lives for a substance vary with many environmental conditions such as temperature (see Chapters 2 and 3 for a description of the relevant environmental conditions and their effects on half-lives), it is important to understand the conditions assumed in establishing the criteria. Most of the current mandates (Table 8-1) inherently assume criteria based on temperate (i.e., not Arctic) conditions. Furthermore, the conditions at which the half-lives were determined, the quality of these half-life determinations, and the distribution of these half-lives need to be considered in the comparison with the criteria. Notwithstanding, given the objective to apply precaution in the screening evaluation, it is not necessary that all of the measured or estimated half-life values be greater or less than the criteria values for the substance to be judged as nonpersistent or as requiring additional evaluation.

Table 8-1 Criteria used in representative international agreements

	NAAEC-SMOC[1]	UNECE-LRTAP[2]	Canadian TSMP[3]
Persistence	Half-life > 6 months water or soil; or 1 y sediment	Half-life > 2 months water or 6 months soils/sediment; or otherwise sufficiently persistent to be of concern	Half-life > 6 months water or soil; or 1 y sediment
Bioaccumulation	BAF/BCF> 5000 or Log K_{ow} > 5	BAF /BCF > 5000 or Log K_{ow} > 5 or factors such as high toxicity	BAF/BCF > 5000 or Log K_{ow} > 5
Toxicity	Acute and chronic (including toxicity of breakdown products, if appropriate)	Potential to affect human health and or the environment adversely	Acute and chronic (including toxicity of breakdown products, if appropriate)
Transport potential	Half-life in air > 2 days air or presence in remote regions	Half-life in air > 2 d and vapor pressure < 1000 Pa or presence in remote regions	Half-life in air > 2 d or presence in remote regions

[1] North American Agreement for Environmental Cooperation-Sound Management of Chemicals initiative [NAAEC-SMOC 1997].

[2] United Nations Economic Commission for Europe, Convention on Long-Range Transboundary Air Pollution [UNECE-LRTAP 1997].

[3] Toxic Substances Management Policy [TSMP 1995].

Rather, the complete set of data and evidence should be considered in making the decision.

Because of both the likely variability in the available half-life data for a substance in an environmental medium and the fact that the criteria are meant to be guidelines, expert judgment will play a critical role in the evaluation process, as well as throughout the application of the framework. Although expert judgment will not be required for evaluating all substances, it will be especially critical when the substance's half-lives are close to the criteria, the half-life values occur on both sides of the criteria, and/or data are missing.

Although the simplest application of screening would involve the comparison of the half-lives of the substance with the half-life criteria of all the environmental media, the accuracy and confidence in the evaluation can be greatly improved by including information on the likely distribution of the substance among the environmental media. Knowledge of the environmental distribution is important for determining which half-lives are most relevant for that substance. When partition coefficients indicate that a substance will reside predominantly (e.g., > 95%) in one compartment, comparison of the substance's half-life in that compartment to the corresponding half-life criterion will be sufficient to make a decision with a high degree of confidence regarding persistence. If the substance is a "multimedia chemical" (i.e., a chemical that exhibits a tendency to partition significantly into 2 or more environmental compartments), comparison of the half-lives with the relevant criteria for each medium will be required to make this decision. If all the substance's half-lives are less than the respective criteria for all the environmental media, the conclusion that the substance is nonpersistent can be made with a high level of confidence. This is true because the overall half-life of a multimedia substance cannot exceed the longest half-life in any environmental medium. If, on the other hand, one or more of the substance's half-lives exceeds the compartment-specific criteria, further evaluation (i.e., detailed evaluation) is required to judge whether the substance meets the overall requirements for designation as persistent. This concept is shown in simple form below. Criteria for classification in each of these environmental phases could be modified if required by differences in climatic conditions.

If T_X = the half-life of a chemical in a particular medium (A = air, W = water, S = soil and/or sediment), it can be compared to criteria half-lives as follows:

If $T_X < T_{X\text{-criteria}}$, the chemical is deemed nonpersistent;

For chemicals in the gas phase: $T_A < T_{A\text{-criteria}}$;

For chemicals in water: $T_W < T_{W\text{-criteria}}$;

For chemicals in soil and/or sediment : $T_S < T_{S\text{-criteria}}$; and

For chemicals in multiple media: $T_A < T_{A\text{-criteria}}$, $T_W < T_{W\text{-criteria}}$, and $T_S < T_{S\text{-criteria}}$.

To determine whether a substance resides primarily in one or more environmental media, the 3 major partition coefficients must be considered:

- air–water partition coefficient, $K_{AW,}$ or Henry's Law constant, H;
- octanol–water partition coefficient, K_{OW}; and
- octanol–air partition coefficient, K_{OA}, where $K_{OA} = K_{OW}/K_{AW} = K_{OW} * R * T/H$, where R is the gas constant, T is the temperature in degrees Kelvin, and H is the Henry's Law constant.

In addition to partitioning, the route of entry of a substance to the environment will play an important role in determining its transport and fate. As is discussed in Chapters 4 and 5, there are two different ways that this increased information on partitioning and route of entry to the environment can be incorporated into the screening process to provide greater confidence and discrimination. The first approach is to illustrate graphically the domain of the partition coefficients that leads to a substance residing primarily in any one of the environmental compartments. Because any 2 of these 3 partition coefficients determine the third partition coefficient, this can be shown as a two-dimensional plot as illustrated in Figure 5-4. Thus, depending on the values for the 2 partition coefficients and the region shown on the plot, the substance could be placed in 1 of the 4 categories above. Another approach is to present the relationship between partitioning into the environmental compartments and the half-life criteria that must be used in the determination of nonpersistence as a decision tree (See Figures 5-11 to 5-14). Both these approaches are discussed below and in Chapter 5.

Detailed evaluation process

In the detailed evaluation, substances that are found in the screening process to be potentially persistent undergo a more thorough review of the attributes of the substance that are necessary for determining persistence. In addition, the ability of the substance to undergo long-range transport is explicitly evaluated. As discussed in Chapter 5, the transport potential of a chemical is defined as the balance of the advective transport in air and water with the transformation losses in an open system. Each chemical has a scale or distance at which losses by transformation in the system equal advective transport out of the system. This is called the characteristic scale and is used to represent that chemical's potential for long-range transport.

The objectives of the detailed evaluation are

- to confirm, refute, and/or clarify the persistence results of the screening assessment;
- to determine, to the extent scientifically possible, whether the substance is persistent and has the potential for long-range transport, as defined by the provisions of the program or regulatory mandate; and
- to identify additional factors that warrant consideration, such as the impact of mode of entry into the environment, climatic and regional differences on persistence and long-range transport, the level of confidence in the available

scientific information, and the manner in which this information should be interpreted.

Such a detailed evaluation is essential to determining whether a substance meets the necessary persistence and long-range–transport criteria for a PBT/POP so that resources can be focused on those substances that have the highest probability of being of concern.

In detailed evaluation, the substance's overall persistence and long-range transport are compared to criteria for these two parameters. As with screening, substances that do not meet the criteria will be considered to be nonpersistent and/or not undergo long-range transport and are set aside (Figure 8-1). The detailed evaluation uses the same information that the screening evaluation uses on the substance's partition coefficients and half-lives in the various environmental media. However, it also includes information on the pathways of entry to the environment, which are not considered in screening. The detailed evaluation is best conducted using a multimedia model to determine the steady-state distribution of the substance among the major environmental compartments because the overall persistence is a weighted average of persistence in the individual environmental media, based on distribution of the substance's mass.

As discussed in Chapter 5, multimedia models have the ability to integrate properties of both the substance and the environment with information on use or release patterns to predict the steady-state environmental distribution of the substance and thus the overall persistence. Multimedia models can also provide insight into the relative contribution of degradation and advection to the overall persistence. Chapter 5 describes several approaches for using these multimedia models for determining the characteristic scale and thus the potential for long-range transport. These approaches include a regional, open model whose size is determined as that required for advective and transformation losses to be equal, as well as multiple multimedia models that are connected to represent a global model.

Role of monitoring data

Although predictive models are essential in the detailed evaluation, the emissions of widely used chemicals often are inadequately known, and these models cannot anticipate every environmental circumstance. Therefore, it is crucial to observe the environment for the presence and appearance of new chemicals and to attempt to understand the factors contributing to the presence of a substance in remote regions. Much can also be learned by observing and modeling the response when the emission of a chemical is curtailed or controlled. However, the mere presence of a substance in a remote region should not automatically lead to management of the substance. Analytical methods are becoming increasingly sensitive and able to detect minute concentrations for certain substances; therefore presence does not automatically indicate that an adverse effect will occur. The many ways that monitoring data could be used in the detailed evaluation are discussed in more detail in Chapter 6 and are illustrated using examples in Chapter 7.

As shown in Figure 8-1, monitoring data play an important role in the framework and in the process of evaluating whether a substance is persistent and undergoes long-range transport. Monitoring is shown with its own box in Figure 8-1 because such data have the potential on their own to indicate that a substance may be of concern with respect to persistence and long-range transport. It is emphasized that interpretation of monitoring data requires expert evaluation and may not be straightforward. Thus, it is anticipated that if appropriate monitoring indicates that the substance has been found in remote regions at environmentally significant concentrations, these substances will bypass screening and go directly to detailed evaluation. Substances that are initially determined to be nonpersistent, whether in the screening or detailed evaluation, may need to be reevaluated if field-monitoring data that demonstrate a substance's presence at a sufficient distance from the point of release become available during or after these evaluations.

Subsequent analyses

Screening and detailed evaluation are designed to determine if a substance is persistent and can undergo long-range transport. In order to determine if a substance is a PBT/POP that will require management, additional review is required that includes consideration of all of the attributes of the substance that determine its environmental fate, as well as its bioaccumulation, toxicity, and exposure concentrations. If the substance is determined to be a PBT/POP following this evaluation, risk management options need to be considered.

Methods for Selecting Criteria

Various approaches can be used to select single-medium half-life criteria for use in screening and overall persistence and long-range–transport criteria for use in detailed evaluation. Approaches described in this section emphasize empirical analysis and precedent, which have served as the basis for choosing criteria that are used in current national and international efforts to control PBTs/POPs. However, some new approaches and tools are also described that could further inform the selection of screening criteria and ensure that the criteria used are self-consistent. Further discussion of this topic may be found in Rodan et al. (1999).

The process of choosing single-medium half-life criteria and overall persistence and long-range–transport criteria must be distinguished clearly from the application of these criteria within the substance-evaluation process. The criteria-selection process focuses on determining specific criteria values for the relevant parameters that reflect the nature and magnitude of persistence and long-range transport that are defined to warrant concern within the pertinent mandate. In contrast, it is the substance-evaluation process that determines if a specific substance's properties exceed these criteria and, therefore, whether the substance should progress to additional analysis.

The approaches outlined can be used to select criteria for a broad spectrum of persistence and transport scenarios, ranging from local to regional and, ultimately, to global levels. In addition, the approaches can be applied to a wide range of political, regulatory, and programmatic goals. Taken together, these considerations suggest that no single set of criteria is likely to apply in all situations. This is especially true when the mandates apply to different climatic and environmental conditions. Selection of criteria for any particular application is, therefore, recognized to be a policy decision that must be informed to the maximum extent possible by sound science and an awareness of its limitations.

Precedent

One approach to choosing criteria for individual environmental media for use in screening is to adopt values that have been used in similar applications, such as those summarized in Table 8-1. These programs cover a range of geopolitical settings from individual countries (i.e., Canadian Toxic Substances Management Policy) to regional agreements (i.e., United Nations Economic Commission for Europe). The criteria selected for each of these mandates have benefited from considerable expert scientific input and analysis. Although adoption of such criteria is certainly one option, the goal of this workshop was to consider and recommend general approaches to criteria development that will provide a clearer technical basis for the criteria selection regardless of the intended nature of the program or its geographical scope.

Empirical analysis using existing data

A variety of empirical methods have been used to derive and inform the choice of criteria for half-lives in the individual environmental media. The most common empirical approach involves constructing tables and/or graphs that contain data for one or more of the environmental half-lives, bioconcentration factors (BCFs), and/or partition coefficients for a wide range of substances including currently identified PBTs/POPs. An example of this approach is shown in Figure 8-2 (Rodan et al. 1999). Criteria are then chosen for one or more of these parameters to ensure that the currently identified PBTs/POPs are correctly classified. The final criteria selection also includes consideration of data variability and the level of certainty desired or required in the classification of substances as PBTs/POPs.

Because of the high degree of variability in measured laboratory or field data for half-lives and partition coefficients, available data must be reviewed carefully to determine which values are relevant to the analysis of persistence in the environment. Many of the data points may fall along one portion of the range or, in some cases, there may be a partial convergence in data values. At a minimum, ranges

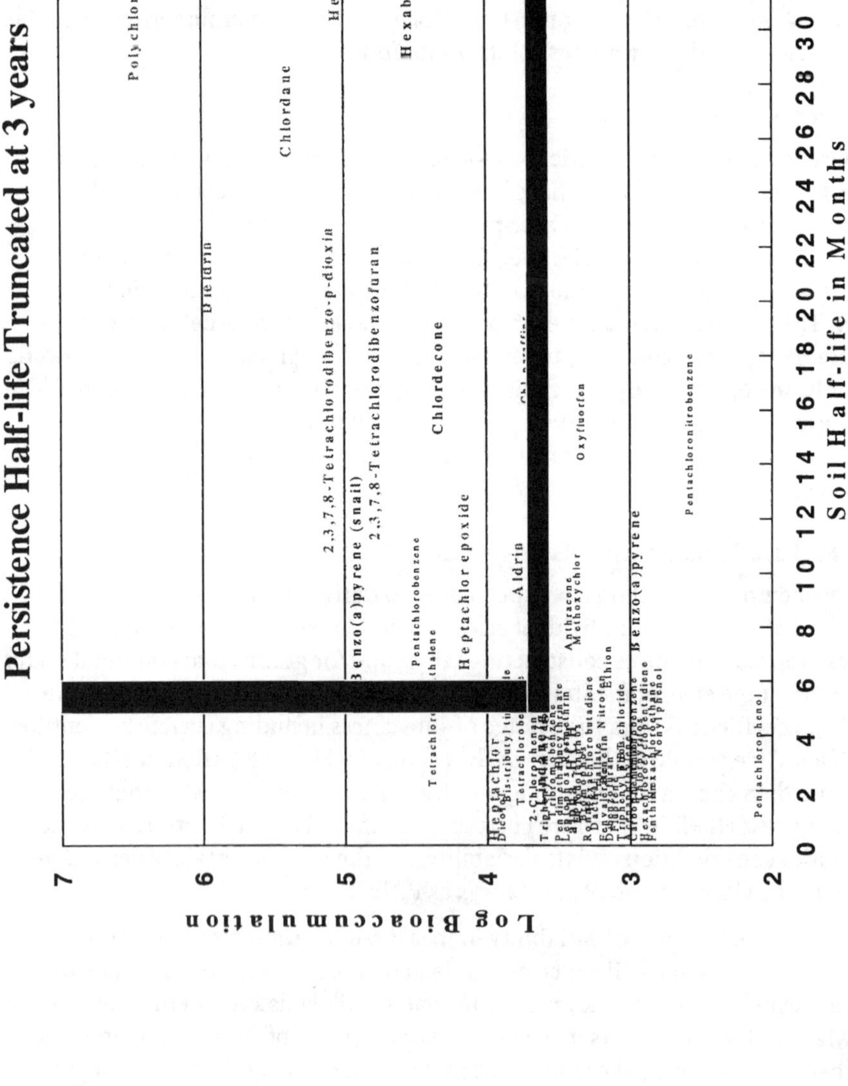

Figure 8-2 Log bioaccumulation factor versus persistence in soil for representative chemicals Reprinted with permission from Rodan et al. (1999) ©1999 American Chemical Society

rather than single point estimates are recommended for displaying the available data. Graphical and tabular displays of the empirical data are very useful in illustrating the inherent variability in half-lives, BCFs, and partition coefficients for a given substance as a result of different climatic and geographic conditions. One strength, and limitation, of these methods is the reliance on measured field and/or laboratory data for substances. As a result of this reliance, the analyses and the criteria selected using this approach unavoidably will be strongly influenced by the substances for which such data exist and by the quality of these data.

The most common graphical methods for presenting these analyses are bar charts and two- or three-dimensional plots. Two-dimensional plots of, for example, partition coefficient or BCF versus half-life for a specific environmental compartment can be used to illustrate the spread and interrelationships of these parameters and to assist in setting self-consistent criteria for more than one parameter.

Empirical analysis combined with modeling

Empirical data can be used in combination with results from models to advise the selection of criteria for half-lives in individual environmental compartments and/or for overall persistence and long-range transport. A simple illustration of this approach would be for a substance that is emitted into and primarily resides in one of the transport media (i.e., air or water). Data on the substance's persistence (either half-life or converted to residence time) in that medium could then be multiplied by the rate of movement of that medium to determine the distance that the substance would move before being reduced to a certain percentage of the initial level (i.e., 5%) of emitted mass as illustrated in Figure 5-8. Then a graph of the half-life versus the travel distance could be used with information on the geographic extent of the region of concern to choose a criterion value for the medium half-life.

Because many substances partition into more than one environmental medium, an overall persistence value and/or a long-range–transport distance calculated using multimedia models (see Chapters 4 and 5) can be graphed versus individual medium half-lives and/or partition coefficients. It would also be possible to graph the overall persistence and long-range–transport distances for several substances including the currently identified PBTs/POPs. As stated previously, these two-dimensional plots can be used to illustrate the spread and interrelationships of these parameters and to assist in setting self-consistent criteria or to identify constraints in the two criteria values.

Theoretical approach

A focus of the workshop was on identifying potential theoretical approaches that could be used to select criteria for screening and detailed evaluation. The approach described below has been applied to very simple examples, but the validity of this approach for more realistic and complex situations still needs to be demonstrated (see additional discussion in Chapter 5). This theoretical approach has 3 components:

1) Using the goal of the regulatory mandate, choose overall persistence and long-range–transport criteria;
2) Using a multimedia model calculate the overall persistence and long-range–transport distances for a hypothetical (yet realistic) range in chemical half-lives and partition coefficients and release patterns; and
3) Using the chosen criteria for persistence and long-range transport (Component 1), and the results of the multimedia model calculations, (Component 2), run computerized data-classification method (e.g., Classification and Regression Trees [CART]) to determine the individual compartment half-life criteria and partition coefficient decision tree.

The first component of the theoretical approach involves choosing overall persistence and long-range–transport criteria. The overall persistence criteria can be chosen directly, or an overall residence time criterion can be chosen instead. The overall residence time is a measure of how long the chemical remains in the environment after use or release is discontinued. Factors to consider in selecting an overall persistence or residence time criterion are the time frame within which it is desirable to have the chemical eliminated from the environment (e.g., one could select the generation time of humans, which is 20 to 30 y) and the level of removal that is considered to constitute "elimination." When this approach is applied on a global scale, the calculation of the overall persistence and residence time will need to consider the percent of the globe with different climatic conditions and/or be set low enough to effectively preclude transport of significant amounts (i.e., 1% to 5%) of a new substance into remote regions such as the Arctic. Long-range–transport criteria can be chosen by considering 1) geographic extent covered by the regulatory mandate and/or 2) concern for potential transport of substances to remote regions. For example, although persistence is a necessary condition for long-range transport, it is not a sufficient condition for long-range transport to occur. Persistence in various environmental media is a necessary condition in that persistence contributes both ensuring that there as a long-term source of the chemical and by providing sufficient time for transport to occur. In order for a chemical to be transported over significant distances, the residence time in one or more of the mobile media (i.e., air or water) needs to be relatively long.

In the second component of the theoretical approach, multimedia models are used to calculate the overall persistence and long-range–transport distances for a hypothetical range in chemical half-lives and partition coefficients and release patterns. It has been observed (Wania and Mackay 1996) that the 12 United Nations Environment Program POPs are so-called "multimedia" chemicals. For these multimedia chemicals, an understanding of environmental fate in all relevant environmental compartments is a prerequisite to understanding the substance's overall environmental persistence and potential for long-range transport. Therefore, multimedia models, such as a Mackay Level III model (Mackay 1991; Mackay, Paterson et al. 1992; Mackay et al. 1996) are useful, necessary tools for developing criteria that include implicit consideration of the environmental and substance characteristics that lead to persistence and long-range–transport potential. The multimedia model used in this evaluation must be relevant to the scale and characteristics of the area that is the focus of the program or regulatory mandate. For example, if the focus is either the local, regional, or global environment, it is necessary to consider the variations in temperature, humidity and other environmental characteristics that are consistent with the particular focus. The behavior of a large number of hypothetical chemicals with input parameters (i.e., half-lives and partition coefficients) that span the range of realistic values and a range of release patterns is then simulated using the model. Finally, for each of these hypothetical chemicals, both the overall persistence or residence time and the long-range–transport distance are calculated using the selected models.

In the third component of the theoretical approach, the chosen criteria for persistence and long-range transport (Component 1) and the results of the multimedia model calculations (Component 2) are combined to predict the individual compartment half-life and partition coefficient criteria. The focus should first be on persistence because the substance must be sufficiently persistent in order for long-range transport to be a concern. Several methods, including direct computer simulation and/or decision trees, can be used to select individual media half-lives and partition coefficients that are consistent with the overall persistence and long-range–transport criteria. For example the CART decision tree approach has been used in these type of applications (Bennett et al. 2000) and is described in Chapter 5. The final decision tree is dependent on the level of certainty required in the designation of persistence or long-range–transport potential (e.g., the need to be 95% certain of a decision). Another method uses the fact that the process for selecting the half-life criteria is simpler for those chemicals that partition almost exclusively (e.g., > 95 %) into only 1 of the 4 compartments. Based on these "single-medium" chemicals, criteria values for the individual environmental compartments could then be determined that are consistent with the "global" criteria for persistence or residence time and long-range–transport distance. However, it is important to recognize that these single-medium chemicals are not typical of the PBTs/POPs that are of greatest interest currently, all of which partition to multiple environmental media. Whichever method is used to select the individual media half-lives and relevant partition coefficients, the resulting decision process and criteria can be expressed in several ways, including

- a plot of two properties illustrating which region in these property domains leads to a designation of persistence, nonpersistence, or uncertainty;
- a decision tree with hierarchy of properties; and
- a set of statements that represent the decision tree and plot results.

Flexibility, Transparency, and Openness

Flexibility is an important characteristic of any conceptual model that is to be broadly applicable to a range of regulatory and non-regulatory goals and can change to incorporate advances in science. This holds true for the process of developing the criteria that are used in the framework as well as the application of this framework. For example, the framework must be able to accommodate different scales of environmental transport, different climatic and geographical conditions under which the persistence and transport are of concern, and different levels of certainty required in the outcome of the evaluation. The current Framework for Substance Evaluation is applicable to a broad spectrum of evaluation needs because it addresses the characteristics of flexibility listed in Table 8-2.

Table 8-2. Characteristics of a flexible framework for substance evaluation

Factor	Examples
Criteria	Vary with policy objective
Geographic scale	Local, national, regional, global
Input data	Nature and sources vary with level of review (screening versus detailed), geographic scale, etc.
Transport routes	Air, water and (possibly) migratory species
Environmental and climatic conditions	Temperature, humidity, soil and air movement pattern type, water type, etc

Geographic scale affects the range of environmental conditions that must be addressed in substance evaluation, but complexity does not necessarily increase as scale increases. This is because average parameter values can often be used at larger scales, e.g., in the evaluation of global persistence and transport potential. Conversely, local and regional models often are parameterized with more-complex data or data having a higher degree of spatial resolution provided such data are available. In addition, scale obviously determines the importance of long-range transport in substance evaluation. Persistence also needs to be evaluated within the context of the geographic area of concern, so that the range of environmental and climatic conditions within that geographic area is considered in the evaluation. The framework presented in this chapter allows consideration of environmental and climatic conditions in the evaluation of persistence and potential for long-range transport, e.g., through the influence of temperature on half-lives and partition coefficients, and thus ensures that the approach is applicable at global, regional, and local scales.

Transparency and openness are desirable characteristics in both the substance-evaluation and criteria-selection processes, regardless of the scope or policy objective. Transparency is defined here as the ability of educated laypeople to understand the process, whereas openness is defined as the opportunity of all interested parties to comment and provide input to the process on an ongoing basis. To further these goals, the proposed framework and criteria selection processes clearly describe the input parameters and the output and indicate at which points expert judgment and policy considerations are likely to be most important. The framework also promotes openness by indicating entry points for submission of data from interested parties.

In principle, the objective of transparency is met when the educated layperson is able to fully understand what has been done, why it was done, and how a particular outcome was achieved. Transparency is enhanced by the use of non-technical language whenever possible. A reasonable standard for transparency is whether an educated layperson could go through the process using the same data and arrive at a similar outcome. Related to transparency are the concepts of credibility and accountability. In any evaluation of chemical substances for persistence and long-range–transport potential, it should be clear to whomever is involved in the evaluation what the final outcome is and what the basis for this outcome is. If the process involves expert judgment, it should be clear who the experts were and how and by whom they were selected. With respect to data, there should be a clear description of which data were used and how they were obtained, what the reasons were for including or excluding specific values, and how the final report was prepared. Transparency is also enhanced by clearly stating the policy objective that drives the evaluation process and the criteria that must be met. Finally, credibility is enhanced by widely communicating and ensuring availability of the status and outcome of the evaluation. Openness is built into the substance-evaluation process by allowing for and encouraging input from all interested parties at several steps in the evaluation.

Data Availability and Quality Considerations

The effectiveness of any process for identifying PBTs/POPs and for selecting criteria could be severely limited by the availability and quality of the required data. The perceived and actual accuracy of data on chemical partitioning properties, half-lives, and release patterns will ultimately affect the level of confidence in any decisions that are made. Thus, determining the availability, accuracy, and confidence in the required data is an important element in the applicability of the framework.

Based on past experience, the screening of chemicals for persistence will be constrained by the availability and quality of partitioning and half-life data that are needed for comparison with screening criteria. Screening for persistence potential may require disappearance half-lives for air, water, soil, and sediment, as well as partition coefficients for air–water (i.e., H or K_{AW}) and n-octanol–water (i.e., K_{OW}).

In detailed evaluation, additional data may be required when multimedia models are used to estimate the steady-state distribution of a substance and its overall persistence and long-range–transport distance. Data on mode of entry are especially important, but such data may be the most difficult to obtain. Other data that may be required pertain to environmental characteristics and processes. These include, for example, rates of precipitation, aerosol and dry deposition, sediment deposition and resuspension, and soil water and solids runoff. Multimedia models often use default values for these parameters, which may be acceptable for the particular case being examined. However, these default values are chosen to represent a generic environment; and they may need to be changed to reflect specific geographic or climatic conditions relevant to the situation under evaluation (e.g., in modeling environmental fate and distribution of a chemical in the tropics).

Chemical-specific data (e.g., half-lives and partition coefficients: T_{air}, T_{water}, T_{soil}, $T_{sediment}$, K_{AW}, K_{OW}) are available from a wide variety of sources including on-line and personal computer-based databases, published compilations, and the research literature. Examples of widely available computer databases include IUCLID (Heidorn et al. 1996), the Hazardous Substances Data Bank (HSDB 1997), the Environmental Fate Data Base (Howard et al. 1982), and the U.S. Department of Agriculture's pesticide properties database (Wauchope et al. 1991). Published compilations include handbooks of environmental degradation rates (Howard et al. 1991) and physical properties of organic chemicals (Howard and Meylan 1997), the five-volume series of environmental data handbooks by (Mackay, Shiu et al. 1992), the Japanese chemical industry's handbook of biodegradation and bioaccumulation data (CITI 1992), and Kwok and Atkinson's compilation of atmospheric oxidation rate constants (Kwok and Atkinson 1995). In addition, manufacturers, processors and users of chemical substances often maintain their own files of data on physico-chemical properties and fate. Inclusion of these data in the substance-evaluation process should be not only allowed but also actively encouraged. However, none of these sources is exhaustive, and no endorsement of any specific source is implied. Further, there is no guarantee a priori that specific data in these or any other sources are of acceptable quality for a given application. This highlights the need for sound scientific judgment in the data collection and evaluation process.

In an ideal world, experimental data of acceptable quality would be available for all substances and parameters. The reality is that estimation methods or structure-activity relationships (SARs) are often needed to fill data gaps. This generalization holds for all of the chemical-specific parameters listed above with the possible exception of K_{OW}, for which measured values exist for many thousands of chemicals. Relatively reliable estimation methods exist for half-life in air (see Chapter 2) and the octanol–water partition coefficient, K_{OW}, and to a lesser extent for the air–water partition coefficient, K_{AW} or Henry's Law constant, H (see Chapter 4). Few reliable estimation methods exist for estimating half-lives in water, soil, and/or sediment. Biodegradation is often the dominant fate process in these compartments; unfortunately, existing estimation methods for biodegradability generally are not intended to predict half-lives and should be used with caution (see Chapter 3). In general, the appropriate role of estimation methods in the process of evaluating chemical substances for persistence and long-range–transport characteristics is a subject of current debate, and no government has yet taken an official position endorsing any specific approaches or methods. The interested reader should consult any of several recent reports for further information (Aronson and Howard 1999, Sijm et al. 1999, Tyle and Niemela 1999).

A general principle to use in selecting half-life data is that data obtained under environmentally realistic test conditions are preferred. This is especially important for biodegradation data, which collectively are derived from a wide variety of test protocols that vary in complexity and environmental relevance. Methods range from screening tests that use artificial media, to die-away tests and microcosms in which environmental (grab) samples are brought into the laboratory for study, to actual field studies. Based on this principle, half-life data from field studies are generally preferred over data from grab-sample studies, but only if removal processes other than transformation (i.e., partitioning or advection out of the study area) are either ruled out or accounted for in the study. A possible drawback to use of field data is that they may be too site-specific to be of general utility. Because of the difficulties inherent in designing and properly conducting field-scale experiments, high-quality field data exist for only a few substances. It is also possible to estimate half-lives for the environment based on available test data and basic principles (see Chapter 3).

Several other principles are important in data selection. First, measured parameter values are always preferred over estimates derived from SARs if the former are judged to be of sufficient quality and are representative of the conditions of interest. Second, it should be clearly communicated which input data are estimates and which are measured values. In both cases the sources and/or methods used to derive the data should also be given. Because the magnitudes of many chemical properties and fate parameters vary with temperature, special attention should be given to determining and reporting the temperature and other relevant environmental conditions at which a measurement was made or that were assumed in the estimation method (See Chapters 2, 3, and 4). Third, all parties with an interest in identification, selection, and subsequent action on an evaluated substance should be permitted and indeed encouraged to bring any and all relevant data into the process. The data that are considered should not be limited arbitrarily, e.g., restricting data considered for a given parameter to results from a particular experimental technique. Finally, every attempt should be made to accommodate new data, especially if the data increase confidence in the results.

It must be recognized that all evaluation systems will encounter difficulty when relevant data are unavailable or are considered to be of suspect accuracy. Quantitative structure-activity relationship (QSAR) techniques often fail for specific classes of chemicals that are underrepresented in the training sets. As the number of chemicals evaluated increases, this problem of data adequacy likely will become more critical.

A case can be made that the evaluation system should take this into account by seeking to use data that are most readily available, for which well established QSARs exist, or for which experimental determination is less expensive. If this argument is accepted, it is likely that early exploitation of partitioning data will prove advantageous.

In the detailed sequence of evaluation steps it seems appropriate to exploit the available data in this order of decreasing accuracy. If chemicals can be "set aside," as indicated in Figure 8-1, solely on the basis of partitioning data and a half-life in air, this will avoid excessive evaluation and facilitate the entire process. It is thus suggested that one approach is to obtain partitioning data, and a half-life in air, and to test whether the chemical is persistent, even if very long half-lives are assumed in other media. This could be done using a Level II fugacity calculation, a CART analysis, or even a graphical approach to identify the partitioning tendencies. The details and validity of such approaches are best determined by testing a variety of proposed methods; however, it is generally believed that partitioning data should be exploited as early in the process as possible.

Role of Experts and Expert Judgment

Expert judgment will play a critical role throughout the application of the framework and the development of criteria, in particular in deciding which data will be used, how the data are interpreted, and which approaches will be used to analyze the data. Although the involvement of experts in the process could enhance the scientific credibility of the end-results, involvement of experts can reduce the transparency of the evaluation process at all stages because the procedures used are not generally understandable to the lay public. Recognizing the technical nature of the process as well as the significant policy implications of selection decisions, it is important that in applying the framework and developing criteria, experts involved attempt to make substantial efforts to keep the process as transparent as possible. Because of the limitations in resources, expert involvement may need to be limited. It is therefore important at each step of the process to be able to identify where expert judgment can be used most effectively and efficiently.

References

Aronson D, Howard PH. 1999. Evaluating potential POP/PBT compounds for environmental persistence. Syracuse Research Corporation (SRC) prepared for the U.S. Chemical Manufacturers Association, on behalf of the Chlorine Chemistry Council. Syracuse NY: SRC. 41 p. Report No. SRC-TR-99-020.

Bennett DH, McKone TE, Kastenberg WE. 2000. Evaluating multimedia chemical persistence: classification and regression tree analysis. *Environ Toxicol Chem* 19:810–819.

[CITI] Chemicals Inspection and Testing Institute. 1992. Biodegradation and bioaccumulation data of existing chemicals based on the CSCL Japan. Chemical Industry Ecology-Toxicology and Information Center.

Heidorn CJA, Hansen BG, Norager O. 1996. IUCLID: a database on chemical substances information as a tool for the EU-Risk-Assessment program. *J Chem Inf Comput Sci* 36:949–954.

Howard PH, Boethling RS, Jarvis WF, Meylan WM, Michalenko EM. 1991. Handbook of environmental degradation rates. Chelsea MI: Lewis. 725 p.

Howard PH, Meylan WM. 1997. Handbook of physical properties of organic chemicals. Boca Raton FL: CRC/Lewis.

Howard PH, Sage GW, LaMacchia A, Colb A. 1982. The development of an environmental fate database. *J Chem Inf Comput Sci* 22:38–44.

[HSDB] Hazardous Substances Data Bank. 1997. Bethesda, MD: Toxicology Information Program Online Services, National Library of Medicine, Specialized Information Services, U.S. Department of Health and Human Services, Public Health Service, National Institutes of Health.

Kwok ESC, Atkinson R. 1995. Estimation of hydroxyl radical reaction rate constants for gas-phase organic compounds using a structure-reactivity relationship: an update. *Atmos Environ* 29:1685–1695.

Mackay D. 1991. Multimedia environmental models: the fugacity approach. Chelsea MI: Lewis. 257 p.

Mackay D, DiGuardo A, Paterson S, Cowan CE. 1996. Evaluating the environmental fate of a variety of types of chemicals using the EQC model. *Environ Toxicol Chem* 15:1627–1637.

Mackay D, Paterson S, Shiu WY. 1992. Generic models for evaluating the regional fate of chemicals. *Chemosphere* 24:695–717.

Mackay D, Shiu WY, Ma KC. 1992. Illustrated handbook of physical-chemical properties and environmental fate for organic chemicals (5 volumes). Chelsea, MI: Lewis.

[NAAEC-SMOC] North American Agreement for Environmental Cooperation-Sound Management of Chemicals. 1997. Draft process for identifying candidate substances for regional action under the SMOC initiative. Report to the North American working group on the sound management of chemicals by the task force on criteria, Commission for Environmental Cooperation (CEC). CEC-Montreal, Quebec, Canada, July 1997. Internet: http://www.cec.org.

Rodan BD, Pennington DW, Eckley N, Boethling RS. 1999. Screening for persistent organic pollutants: techniques to provide a scientific basis for POPs criteria in international negotiations. *Environ Sci Technol* 33:3482–3488.

Sijm D, Hulzebos E, Peijnenburg W. 1999. Estimating the PTB-profile. Rijksinstituut voor Volksgezondheid en milieu (RIVM) report 601503016, July 1999. Bilthoven, The Netherlands: RIVM. 112 p.

[TSMP] Toxic Substances Management Policy. 1995. Persistence and bioaccumulation criteria. Ottawa, Ontario, Canada: Environment Canada. No. En-40-499/2-1995E.

Tyle H, Niemela J. 1999. Use of QSARs for selection of POPs. Working paper: to the participants of CEG II meeting; 14–18 June 1999; Vienna. Danish EPA, June 10, 1999 draft.

[UNECE–LRTAP] United Nations Economic Commission for Europe-Convention on Long-Range Transboundary Air Pollution. 1997. Draft composite negotiating text for a protocol on persistent organic pollutants. EB.AIR/WG.5/R.72/Rev.

Wania F and Mackay D. 1996. Tracking the distribution of persistent organic pollutants. *Environ Sci Technol* 30:390A–396A.

Wauchope RD, Butler TM, Hornsby AG, Augustijn-Becker PW, Burt JP. 1991. The SCS/ARS/CES pesticide properties database for environmental decision-making. *Rev Environ Contam Toxicol* 123:1–35.

APPENDIX

Priority Persistent Organic Pollutants

Twelve persistent organic pollutants (POPs) have been identified by the United Nations Economic Commission for Europe for international regulation. In this appendix these substances are listed, and their sources, uses, and properties are discussed briefly. Properties are generally taken from the CD-ROM Handbook by Mackay et al. (1999).

The octanol–water partition coefficient (K_{OW}) characterizes the hydrophobicity of the substance, i.e., its tendency to partition from water to organic media such as lipids and natural organic matter. It also indicates the potential for bioconcentration or bioaccumulation in fish. A K_{OW} exceeding 10^5 is regarded as highly bioaccumulative. The air–water partition coefficient (K_{AW}) indicates the potential for evaporation from water. Substances with K_{AW} of 10^{-3} and higher are viewed as quite volatile, while those with a K_{AW} of less than 10^{-7} have little tendency to volatilize from water. The octanol–air partition coefficient (K_{OA}) characterizes partitioning from air to organic media such as soils and plant surfaces and to aerosols. When K_{OA} is less than about $10^{8.5}$, the gaseous chemical in air often approachs equilibrium with foliage. In the range $10^{8.5}$ to 10^{11}, there may be a kinetic limitation, i.e., equilibrium is approached but not reached. Beyond 10^{11} a substance becomes highly sorbed to aerosol particles in the atmosphere (McLachlan 1999). Note that K_{OA} is K_{OW}/K_{AW}.

Most of the priority POP substances are persistent, i.e., they have half-lives on the order of a week or more in air and months or years in water and soil. The half-lives given here should be regarded as only tentative because they vary with environmental conditions such as temperature.

It should be noted that several of these POPs are in reality mixtures of different substances with different individual properties. Some of these properties are difficult to measure, especially when values are extreme, e.g., high K_{OW} or K_{OA} or low K_{AW}, and reported data often cover a wide range. The values given should be treated as approximate rather than as accurate. If an accurate value is desired, current literature should be consulted.

References

Mackay D, Shiu WY, Ma KC. 1999. Physical-chemical properties and environmental fate and degradation handbook. (CD-ROM). Boca Raton FL: Chapman & Hall CRCnetBASE, CRC.

McLachlan MS. 1999. Framework for the interpretation of measurements of SOCs in plants. *Environ Sci Technol* 33:1799–1804.

There are eight pesticides.

Hexachlorobenzene (HCB) has been used as a fungicide but has also been produced as a byproduct in the synthesis of other organochlorine compounds including polyvinyl chloride. It is quite hydrophobic and fairly volatile; thus, it is susceptible to long range transport and is detected throughout the global atmosphere.

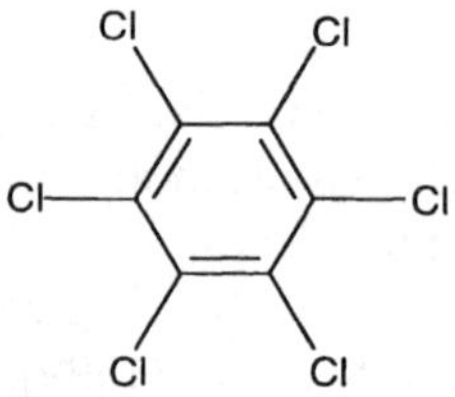

C_6Cl_6

Molar mass: 284.8

Properties: $K_{OW} \sim 10^{5.5}$ $K_{AW} \sim 10^{-1.28}$ $K_{OA} \sim 10^{6.78}$

Half-lives (h): Air (10,000–30,000), Water (> 30,000), Soil (> 30,000).

Mirex is an insecticide that was used mainly to control fire ants, but it also has been used as a fire retardant. It is a very high-molar-mass substance and is highly hydrophobic because of its low solubility in water. It undergoes fairly rapid photolysis to "photo-mirex," a similarly structured molecule.

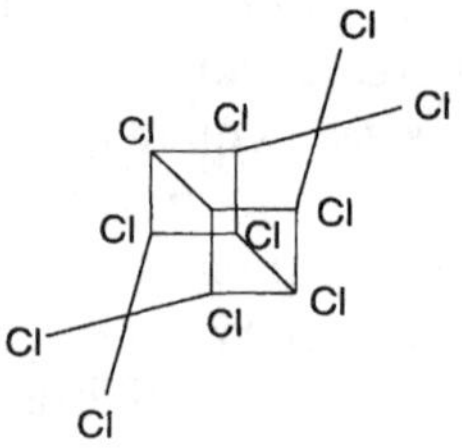

$C_{10}Cl_{12}$

Molar mass: 545.6

Properties: $K_{OW} \sim 10^{6.9}$ $K_{AW} \sim 10^{-0.47}$ $K_{OA} \sim 10^{7.37}$

Half-lives (h): Air (100–300), Water (3000–10,000), Soil (> 30,000).

Chlordane is an insecticide that has been widely used for agricultural and domestic purposes. It is a mixture, cis- and trans-chlordane and cis- and trans-nonachlor being the usual substances monitored.

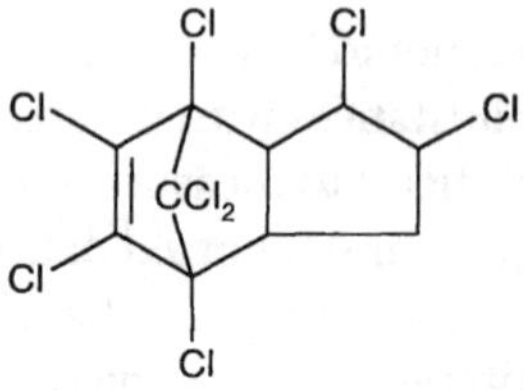

$C_{10}H_6Cl_8$

Molar mass: 409.8

Properties: $K_{OW} \sim 10^{6}$ $K_{AW} \sim 10^{-2.93}$ $K_{OA} \sim 10^{8.93}$

Half-lives (h): Air (30–100), Water (10,000–30,000), Soil (10,000–30,000).

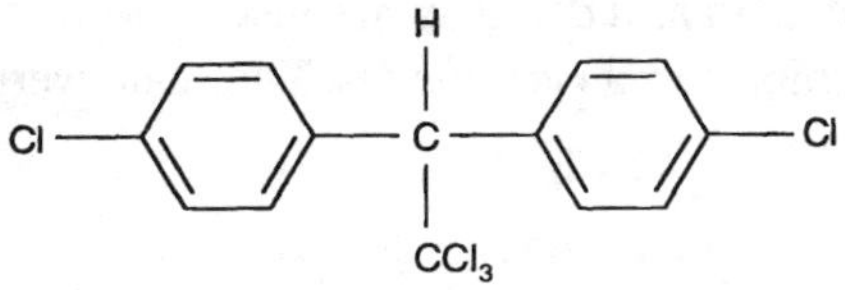

Dichlorodiphenyltrichloroethane (DDT) is an insecticide that was widely used on agricultural crops such as cotton and fruits but is now mainly used in tropical climates for malaria control by killing mosquitos. It degrades fairly quickly to a similar compound, dichlorodiphenyldichloroethylene (DDE), and then more slowly to a series of related substances.

$C_{14}H_9Cl_5$
Molar mass: 354.5
Properties: $K_{OW} \sim 10^{6.19}$ $K_{AW} \sim 10^{-3.28}$ $K_{OA} \sim 10^{9.47}$
Half-lives (h): Air (100–300), Water (3000–10,000), Soil (10,000–30,000).

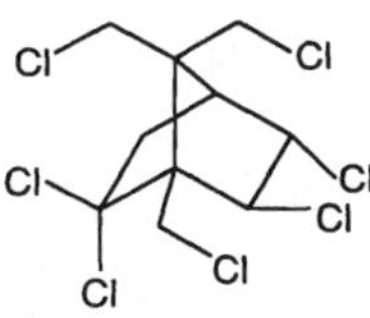

Toxaphene is a very complex mixture of several hundred substances produced by chlorination of camphene. It has been widely used in the U.S. as an insecticide on cotton crops and it has been used to kill unwanted fish in rivers and lakes. It is fairly volatile. Chemical analysis is very difficult because of the large number and variable proportions of the congeners.

$C_{10}H_{10}Cl_8$
Molar mass: 413.8
Properties: $K_{OW} \sim 10^{5.5}$ $K_{AW} \sim 10^{-3.52}$ $K_{OA} \sim 10^{9.02}$
Half-lives (h): Air (100–300), Water (> 30,000), Soil (> 30,000).

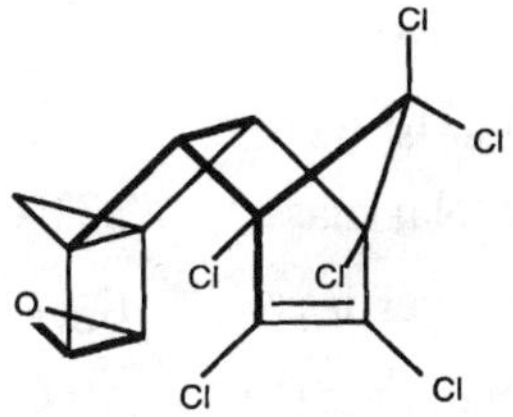

Endrin is a cyclopentadiene-derived insecticide that was used mainly on crops such as grains and cotton and for rodent control.

$C_{12}H_8Cl_6O$
Molar mass: 380.9
Properties: $K_{OW} \sim 10^{5.2}$ $K_{AW} \sim 10^{-4.87}$ $K_{OA} \sim 10^{10.07}$
Half-lives are uncertain.

Aldrin and **dieldrin** are insecticides that are, like endrin, in the general class of chlorinated cyclodienes. They also were used for termite control.

$C_{12}H_8Cl_6$

Molar mass: 364.9

Properties: $K_{OW} \sim 10^{6.5}$ $K_{AW} \sim 10^{-1.43}$ $K_{OA} \sim 10^{7.93}$

Half-lives (h): Air (< 10), Water (10,000–30,000), Soil (10,000–30,000).

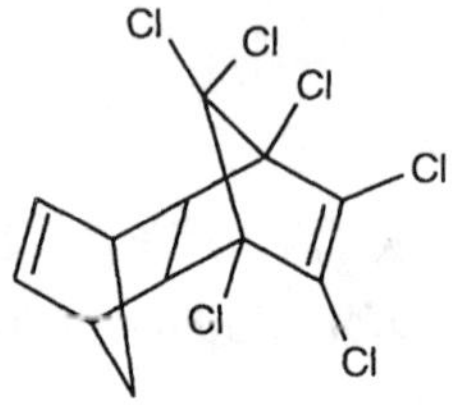

$C_{12}H_8Cl_6O$

Molar mass: 380.9

Properties: $K_{OW} \sim 10^{5.2}$ $K_{AW} \sim 10^{-3.34}$ $K_{OA} \sim 10^{8.54}$

Half-lives (h): Air (30–100), Water (10,000–30,000), Soil (10,000–30,000).

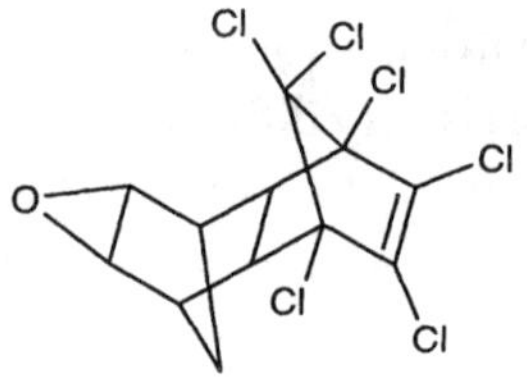

Heptachlor is an insecticide used in agriculture and for malaria control.

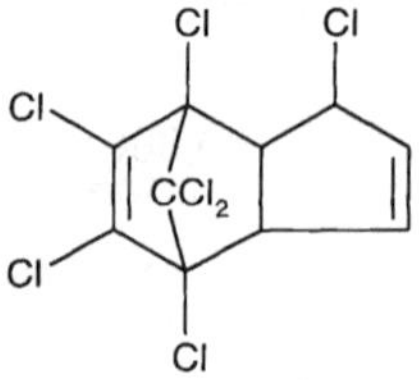

$C_{10}H_5Cl_7$

Molar mass: 373.4

Properties: $K_{OW} \sim 10^{5.27}$ $K_{AW} \sim 10^{-0.85}$ $K_{OA} \sim 10^{6.12}$

Half-lives (h): Air (30–100), Water (300–1000), Soil (1000–3000).

Two industrial chemicals are listed: HCB (discussed above) and polychlorinated biphenyls (PCBs).

Polychlorinated biphenyls consist of potentially 209 congeners obtained by chlorinating biphenyl with from 1 to 10 chlorines. Commercial mixtures, sold as Aroclors in North America, are widely used as heat transfer and dielectric fluids under closed or controlled conditions, but they have also been used as plasticizers and in capacitors and copy paper. Their properties vary greatly depending on the number and position of the chlorine substituents. The lower chlorine-number PCBs are fairly volatile and are widely distributed in the global atmosphere. They are very hydrophobic. This group of compounds is probably the most studied and monitored POP.

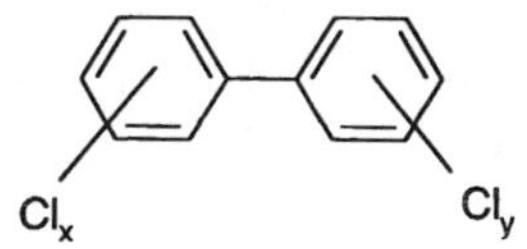

Molar mass: 188.7–498.7

Properties: $K_{OW} \sim 10^{6.5}$ $K_{AW} \sim 10^{-2.3}$ $K_{OA} \sim 10^{8.8}$

Half-lives (h): Air (100), Water (50,000), Soil (10,000).

Two byproducts of natural origin and human activity are the chlorinated **dioxins** and **furans**. They are formed in small quantities when organic matter burns with chlorine present. Dioxins and furans consist of 135 potential chlorinated dibenzofurans and 75 chlorinated dibenzo-*p*-dioxins with from 1 to 8 chlorine substituents. As with PCBs, they vary in hydrophobicity and volatility. The tetrachlorinated (2,3,7,8) congeners are regarded as particularly toxic. They are produced during combustion when there is limited oxygen availability and chlorine is present. The principal source is probably incineration, but they also have been produced during pulp bleaching, high temperature industrial processes, and a variety of combustion processes ranging from auto emissions to forest fires. The quantities generated are small, being generally only few kilograms for a large country with a population of some 100 million. Therefore, they are present at levels in the environment at typically one thousandth of that of PCBs. These low levels are offset by their higher toxicity.

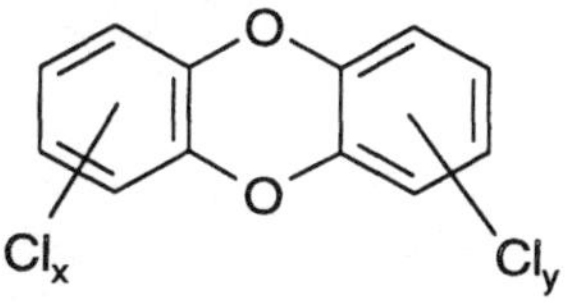

Molar mass: 218.5–460 (mono- to octa-Cl "dioxin)

Properties: $K_{OW} \sim 10^{6.9}$ $K_{AW} \sim$ $10^{-2.9}$ $K_{OA} \sim 10^{9.7}$ (a tetrachlorinated dioxin)

Half-lives (h): Air (100–300), Water (1000–3000), Soil (3000–10,000).

Molar mass: 202–443.8

Properties: $K_{OW} \sim 10^{6.1}$ $K_{AW} \sim$ $10^{-3.2}$ $K_{OA} \sim 10^{9.3}$ (a tetrachlorinated furan)

Half-lives (h): Air (100 to 300), Water (1000–3000), Soil (10,000–30,000).

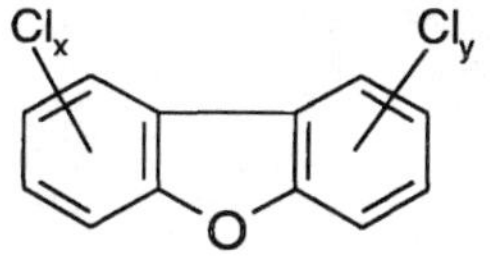

Abbreviations

ABL	atmospheric boundary layer
ARET	accelerated Reduction and Elimination of Toxics
BAF	bioaccumulation factor
BCF	bioconcentration factor
CART	Classification and Regression Tree
CHB	chlorobornane
CITI	Chemicals Inspection and Testing Institute (Japan)
COD	chemical oxygen demand
DDD	dichlorodiphenyldichloroethane
DDE	dichlorodiphenyldichloroethylene
DDT	dichlorodiphenyltrichloroethane
DEA	deethylatrazine
DIA	deisopropylatrazine
EAO	Eurasian Arctic Ocean
ECETOC	European Centre for Ecotoxicology and Toxicology of Chemicals
ELISA	enzyme-linked immunosorbent assay
EQC Model	Equilibrium Criterion Model
H	Henry's Law constant
HCB	hexachlorobenzene
HCH	hexachlorocyclohexane
HRGC-HRMS	high resolution gas chromatography coupled to high resolution mass spectrometry
IUCLID	International Uniform Chemical Information Database

K_{AW}	air–water partition coefficient
K_d	soil partition coefficient
K_{OA}	octanol–air partition coefficient
K_{OC}	organic carbon partition coefficient
K_{OW}	octanol–water partition coefficient
K_p	partition coefficient
K_{PA}	plant–air partition coefficient
K_{SA}	soil–air partition coefficient
K_{SW}	soil–water partition coefficient
LRTAP	long-range transboundary air pollution
MITI	Ministry of International Trade and Industry (Japan)
MW	molecular weight
NAAEC-CEC	North American Agreement for Environmental Cooperation - Commission for Environmental Cooperation
NMR	nuclear magnetic resonance
NAAO	North American Arctic Ocean
NPE	nonylphenol ethoxylate
OAQPS	Office of Air Quality Planning and Standards (USEPA)
OC	organic carbon
OECD	Organization for Economic Co-operation and Development
OPPT TRI	Office of Pollution Prevention and Toxics – Toxics Release Inventory (USEPA)
OSPAR	Oslo-Paris Convention
p^S	Saturation vapor pressure
PAH	polycyclic aromatic hydrocarbon
PBL	planetary boundary layer

PBT	persistent, bioaccumulative, and toxic substance
PCB	polychlorinated biphenyl
PCDD	polychlorinated dibenzo-p-dioxin
PCDF	polychlorinated dibenzofuran
POC	particulate organic carbon
POP	persistent organic pollutant
QSAR	quantitative structure-activity relationship
QSRR	quantitative structure-reactivity relationship
S_{aq}	aqueous solubility
SAR	structure-activity relationship
SVOC	semi-volatile organic compound
SPMDS	semipermeable membrane devices
SPM	suspended particulate matter
T_A	degradation half-life in air
$T_{A\text{-criteria}}$	degradation half-life in air, criteria value
$T_{A\text{-overall}}$	overall degradation half-life in air
T_W	degradation half-life in water
$T_{W\text{-criteria}}$	degradation half-life in water, criteria value
T_S	degradation half-life in soil/sediment
$T_{S\text{-criteria}}$	degradation half-life in soil/sediment, criteria value
TCDD	tetrachlorodibenzo-p-dioxin
TSMP	Toxic Substances Management Policy (Canadian)
UNECE	United Nations Economic Commission for Europe
UNEP	United Nations Environment Program
USEPA	U.S. Environmental Protection Agency

INDEX

A

Evaluation of Persistence and Long-Range Transport of Organic Chemicals in the Environment. *Gary Klečka et al., editors.*

C

D

G

H

I

J

K

L

M

N

O

P

Q

R

S

SETAC

A Professional Society for Environmental Scientists and Engineers and Related Disciplines Concerned with Environmental Quality

The Society of Environmental Toxicology and Chemistry (SETAC), with offices in North America and Europe, is a nonprofit, professional society that provides a forum for individuals and institutions engaged in the study of environmental problems, management and regulation of natural resources, education, research and development, and manufacturing and distribution.

Specific goals of the society are:

- Promote research, education, and training in the environmental sciences.
- Promote systematic application of all relevant scientific disciplines to the evaluation of chemical hazards.
- Participate in scientific interpretation of issues concerned with hazard assessment and risk analysis.
- Support development of ecologically acceptable practices and principles.
- Provide a forum for communication among professionals in government, business, academia, and other segments of society involved in the use, protection, and management of our environment.

These goals are pursued through the conduct of numerous activities which include:

- Hold annual meetings with study and workshop sessions, platform and poster papers, and achievement and merit awards.
- Sponsor a monthly scientific journal, Environmental Toxicology and Chemistry, SETAC newsletter, and special technical publications.
- Provide funds for education and training through the SETAC Scholarship/Fellowship Program.
- Organize and sponsor chapter forums for the presentation of scientific data and for the interchange and study of information about local concerns.
- Provide advice and counsel to technical and nontechnical persons through a number of standing and ad hoc committees.

SETAC membership is currently composed of nearly 6,000 individuals from government, academia, business, and public-interest groups with technical backgrounds in chemistry, toxicology, biology, ecology, atmospheric sciences, health sciences, earth sciences, and engineering.

If you have training in these or related disciplines and are engaged in the study, use, or management of environmental resources, SETAC can fulfill your professional affiliation needs.

All member receive a newsletter highlighting environmental topics and SETAC activities, and reduced fees for the Annual Meeting and SETAC special publications.

All members except Students and Senior Active Members receive monthly issue of Environmental Toxicology and Chemistry (ET&C), a peer-reviewed journal of the Society. Student and Senior active Members may subscribe to the journal. Members may hold office and, with the Emeritus Members, constitute the voting membership.

If you desire more information, contact the appropriate SETAC Office.

1010 North 12th Avenue
Pensacola, Florida, USA 32501-3367
T 850 469 1500 F 850 469 9778,
E setac@setac.org.

Avenue de la Toison d'Or 67
B-1060 Brussels Belgium
T 32 2 772 72 81 F 32 2 770 53 83
E setac@setaceu.org

http://www.setac.org

Environmental Quality Through Science®